Lecture Notes on Data Engineering and Communications Technologies

Volume 104

Series Editor

Fatos Xhafa, Technical University of Catalonia, Barcelona, Spain

The aim of the book series is to present cutting edge engineering approaches to data technologies and communications. It will publish latest advances on the engineering task of building and deploying distributed, scalable and reliable data infrastructures and communication systems.

The series will have a prominent applied focus on data technologies and communications with aim to promote the bridging from fundamental research on data science and networking to data engineering and communications that lead to industry products, business knowledge and standardisation.

Indexed by SCOPUS, INSPEC, EI Compendex.

All books published in the series are submitted for consideration in Web of Science.

More information about this series at https://link.springer.com/bookseries/15362

Eric C. K. Cheng · Rekha B. Koul ·
Tianchong Wang · Xinguo Yu
Editors

Artificial Intelligence in Education: Emerging Technologies, Models and Applications

Proceedings of 2021 2nd International
Conference on Artificial Intelligence
in Education Technology

 Springer

Editors
Eric C. K. Cheng 🄳
Department of Curriculum and Instruction
Education University of Hong Kong
Hong Kong, China

Rekha B. Koul 🄳
School of Education
Curtin University
Perth, WA, Australia

Tianchong Wang 🄳
Department of Curriculum and Instruction
Education University of Hong Kong
Hong Kong, China

Xinguo Yu 🄳
CCNU-UOW Joint Institute
Central China Normal University
Wuhan, Hubei, China

ISSN 2367-4512   ISSN 2367-4520   (electronic)
Lecture Notes on Data Engineering and Communications Technologies
ISBN 978-981-16-7529-4   ISBN 978-981-16-7527-0   (eBook)
https://doi.org/10.1007/978-981-16-7527-0

This Springer imprint is published by the registered company Springer Nature Singapore Pte Ltd.
The registered company address is: 152 Beach Road, #21-01/04 Gateway East, Singapore 189721,
Singapore

Conference Committees

Honorary Chair

Qingtang Liu, Dean of School of Educational Information Technology, Central China Normal University, China

Conference Chair

Xinguo Yu, Central China Normal University, China & University of Wollongong, Australia

Program Chairs

Eric C. K. Cheng, The Education University of Hong Kong, China
Yanjiao Chen, Zhejiang University, China

Program Co-chairs

Flora Amato, University of Naples Federico II, Italy
Rekha Koul, Curtin University, Australia
Cheng Siong Chin, Newcastle University, Singapore

Local Chairs

Lei Niu, Central China Normal University, China
Zhizhong Zeng, Central China Normal University, China

Local Organising Committees

Chao Sun, Central China Normal University, China
Bin He, Central China Normal University, China
Ting Zhang, Central China Normal University, China

Technical Program Committees

Jining Yan, China University of Geoscience, China

Technical Committees

Lawrence Gundersen, Jackson State Community College, USA
Michael Smalenberger, The University of North Carolina at Charlotte, USA
Norma Binti Alias, Technology University of Malaysia, Malaysia
Mohd Faisal Hushim, Universiti Tun Hussein Onn, Malaysia
Yansha Guo, Tianjin University of Technology and Education, China
Yueen Li, Shandong Jianzhu University, China
Sun zhong, Capital Normal University, China
Tuan Tongkeo, Suan Dusit University, Thailand
Bin Xue, National University of Defense Technology, China
Mah Boon Yih, Universiti Teknologi MARA, Malaysia
Zahra Naimie, University of Malaya, Malaysia
Tim Schlippe, IU International University of Applied Sciences, Germany
Ying-Hao Yu, National Chung Cheng University, Taiwan, China
Dariusz Jacek Jakóbczak, Technical University of Koszalin, Poland
Beligiannnis Grigorios, University of Patras, Greece
Loc Nguyen, International Engineering and Technology Institute (IETI), China
Mao Feng, Shanghai University of International Business and Economics, China
Wang Aiwen, Guangzhou Cicy Construction College, China
Liu Zheng-yan, Fuyang Normal University, China
Mah Boon Yih, Universiti Teknologi MARA, Malaysia
Sun zhong, Capital Normal University, China

Tuan Tongkeo, Suan Dusit University, Thailand
Zahra Naimie, University of Malaya, Malaysia
Nigar Zeynalova, Azerbaijan University of Languages, Azerbaijan
Marien Alet Graham, University of Pretoria, South Africa

Preface

Background of the Book

Rapid developments in artificial intelligence (AI) and AI's appalling potential in educational use have drawn significant attention from the education community in recent years. Many theoretical and practical questions concerning AI in education (AIED) are raised for educators entering this uncharted territory, and issues on AI's technical, pedagogical, administrative and socio-cultural implications are being debated.

This book is a compendium of selected research papers presented at the 2nd International Conference on Artificial Intelligence in Education Technology (AIET 2021)—held in Wuhan, China, on July 2–4, 2021. AIET establishes a platform for education researchers in AI to present research, exchange innovative ideas, propose new models and demonstrate advanced methodologies and novel systems.

This timely publication is well-aligned with UNESCO's *Beijing Consensus on Artificial Intelligence and Education*. It is committed to exploring how best to prepare our students and harness emerging technologies for achieving the *Education 2030* Agenda as we move towards an era in which AI is transforming many aspects of our lives. Providing broad coverage of recent technology-driven advances and addressing several learner-centric themes, the book is an informative and valuable resource for researchers, practitioners, education leaders and policy-makers who are involved or interested in AI and education.

The 26 papers in this book are divided into five main parts—(1) An Overview of AI in Education, (2) AI Technologies and Innovations, (3) Teaching and Assessment across Curricula in the Age of AI, and (4) Ethical, Socio-cultural and Administrative Issues in Education of an AI Era. The book provides a comprehensive picture of the current status, emerging trends, innovations, technologies, applications, challenges and opportunities of current AIED research through these sections.

An Overview of AI in Education

In chapter "Towards a Tripartite Research Agenda: A Scoping Review of Artificial Intelligence in Education Research", Tianchong Wang and Eric C. K. Cheng conduct a scoping review of research studies on AIED published over the last two decades (2001–2021). A wide range of empirical studies are collected in the selected databases related to the field, and they are analysed with content analysis and categorical meta-trends analysis. Three distinctive AIED research directions, namely *Learning from AI*, *Learning about AI*, and *Learning with AI,* are identified. By depicting the AIED research directions, this chapter serves as a blueprint for AIED researchers to position their up-and-coming AIED studies for the next decade.

In chapter "Analysis of Research Progress in the Field of Educational Technology in China-Research on Knowledge Graph Based on CSSCI (2015–2020)", Zhao Yaru, Li Chaoqian, Tian Xinyu, Chen Yunhong and Li Shuming present their analysis of research progress in the field of educational technology in China. Using the CNKI database as the primary source, visual and statistical analysis was conducted with 6,740 CSSCI journal articles published from 2015 to 2020. As a result, key research trends, gaps and possible future directions for China's educational technology research are revealed.

AI Technologies and Innovations

In chapter "A Two-Stage NER Method for Outstanding Papers in MCM", Shuting Li and Lu Han introduce a two-stage Named Entity Recognition (NER) method, a significant subtask for Natural Language Processing (NLP). By exploiting the NER approaches to recognise the entities about papers occurring in Mathematical Contest in Modelling (MCM)—the most influential international mathematical modelling competition—the authors provide some reference for the study of modelling methods in the future.

In chapter "Comprehensibility Analysis and Assessment of Academic Texts", José Medardo Tapia-Téllez, Aurelio López-López and Samuel González-López present a computational-linguistic analysis, evaluation and feedback for academic documents based on three comprehensibility measures: Connectivity, Dispersion, and Comprehensibility Burden. Such NLP measures are explored and validated. Visualisation feedback during the assessment process is proposed and illustrated, which may be applied to the comprehensibility assessment as it understands and resolves detected language deficiencies.

In chapter "Deep Learning Techniques for Automatic Short Answer Grading: Predicting Scores for English and German Answers", Jörg Sawatzki, Tim Schlippe and Marian Benner-Wickner investigate and compare state-of-the-art deep learning techniques for automatic short answer grading. The BERT models were proposed, and their evaluations show a significant performance improvement compared to the baseline system and related work. Such models could be applied to the automatic short answer grading correction process for its improved efficiency.

In chapter "Solving Chemistry Problems Involving Some Isomers of Benzene Ring", Xinguo Yu, Lina Yan, Hao Meng and Rao Peng present an auto-solver algorithm for understanding and solving the chemistry problems involving some isomers. The experimental results show that their proposed algorithm has a good performance.

In chapter "Automatic Question Answering System for Semantic Similarity Calculation", Minchuan Huang, Ke Chen, XingTong Zhu and Guoquan Wang present an automatic question answering system for semantic similarity calculation. Its system analysis, theoretical conception, system module design and core algorithm are introduced.

In chapter "Solving Shaded Area Problems by Constructing Equations", Zihan Feng, Xinguo Yu, Qilin Li and Huihui Sun present an AI-powered auto-solver that can generate the readable solution to shaded area problems, a kind of geometry reasoning problem that requires finding shaded area from a diagram with some constraints. The experimental results show that their proposed method is more accurate than the baseline one.

In chapter "Cross-Lingual Automatic Short Answer Grading", Tim Schlippe and Jörg Sawatzki analyse the AI-driven cross-lingual automatic short answer grading by leveraging the benefits of a multilingual NLP model. Their analysis on 26 languages demonstrated the potential of cross-lingual automatic short answer grading. It allows students to write answers in exams in their native language, and graders can rely on the system to score the responses.

In chapter "Application of Improved ISM in the Analysis of Undergraduate Textbooks", Meiling Zhao, Qi Chen, Qi Xu, Xiaoya Yang and Min Pan examine the use of an improved Interpretive Structural Model (ISM) in textbook analysis. Authors demonstrate the improved ISM can decompose the complex and messy elements in the textbook structure into a clear multi-layer hierarchical form. Such an analysis can aid teachers to establish their more explicit teaching goals and internal connections between different knowledge elements to optimise teaching.

In chapter "Design of Student Management System Based on Smart Campus and Wearable Devices", Jun Li, Yanan Shen, Wenrui Dai and Bin Fan taking a technological perspective introduce the system design of combining wearable devices and cadet Management in military academies. The key technologies behind them are also illustrated.

In chapter "Proving Geometric Problem by Adding Auxiliary Lines-Based on Hypothetical Test", Mingrui Zhou and Xinguo Yu propose an auxiliary line adding algorithm for plane geometry problems in automatic reasoning systems. Their experiment suggests that the method is more efficient in solving complex elementary mathematical and geometric problems and can avoid the unlimited growth issue of traditional algorithms.

In chapter "A Virtual Grasping Method of Dexterous Virtual Hand Based on Leap-motion", Xizhong Yang, Xiaoxia Han and Huagen Wan develop a new, inverse kinematics virtual grasping method, enabling users to use a dexterous virtual hand to operate the avatar in the virtual environment. The preliminary verification results show that their proposed method has some advantages over the traditional forward kinematics virtual grasping method.

In chapter "Performance Evaluation of Azure Kinect and Kinect 2.0 and Their Applications in 3D Key-Points Detection of Students in Classroom Environment", Wenkai Huang, Jia Chen, Xiaoxiong Zhao and Qingtang Liu make the performance evaluation of Azure Kinect and Kinect 2.0 and their applications in 3D key-points detection of students in a classroom environment. A multi-person experiment in the classroom environment is conducted. The results show that Azure Kinect outperform Kinect 2.0 in various aspects.

In chapter "Research on Remote Sensing Object Parallel Detection Technology Based on Deep Learning", Cheng Guang Zhang, Xue Bo Zhang and Min Jiang introduce a remote sensing object parallel detection technology based on deep learning. A parallel remote sensing target detection algorithm based on the YOLT detection model is proposed. The experimental results show that their parallel algorithm has high efficiency in remote sensing object detection.

In chapter "Research on Expression Processing Methods of Children with Autism in Different Facial Feature Types", Yishuang Yuan, Kun Zhang, Jingying Chen, Lili Liu, Qian Chen and Meijuan Luo explore expression processing methods by using AI-powered eye-tracking technology to study the facial processing methods of autism spectrum disorder (ASD) children. Differences in expression processing between ASD children and typically developing children are identified.

Teaching and Assessment Across Curricula in the Age of AI

In chapter "Improving Java Learning Outcome with Interactive Visual Tools in Higher Education", Yongbin Zhang, Ronghua Liang, Ye Li and Guowei Zhao present how to improve non-Computer Science undergraduate students' learning outcomes of java programming by adopting the integrated development tool *Eclipse* and interactive visual educational tool *BlueJ*. A semi-experiment is conducted with 47 junior students from the mechanical engineering major. The experiment shows that students in the experimental group achieved better performance than students in the control group. Furthermore, the post-course survey reveals that students find interactive learning tools helpful for mastering object-oriented programming concepts.

In chapter "Interdisciplinarity of Foreign Languages Education Design and Management in COVID-19", Rusudan Makhachashvili, Ivan Semenist and Dmytro Moskalov explore the interdisciplinary foreign languages education design and management during the COVID-19. A computational framework of foreign languages education interdisciplinarity is introduced in the study, and a survey is administered to evaluate the dimensions of interdisciplinarity, universality and trans-disciplinarity. With the survey results, recommendations for interdisciplinary foreign languages education are given.

In chapter "A Survey of Postgraduates' MOOC Learning Satisfaction Based on the Perspective of User Experience", Xiaoxue Li, Xinhua Xu, Shengyang Tao and Chen Sheng investigate postgraduates' MOOC learning satisfaction based on user experience. A questionnaire administered with 312 graduate students found that students' overall satisfaction with MOOC learning is relatively low, but there are differences between genders, academic levels and disciplines. With such findings, they put forward suggestions to improve MOOCs.

In chapter "A Goal Analysis and Implementation Method for Flipped Classroom Instructional Design", Bo Su, Wei Zhang, Yang Ren and Le Qi draw on the goal analysis technique to decompose the desired teaching goals into operable steps and quantifiable indicators in flipped classrooms to optimise instructional design.

In chapter "Instruction Models of Located Cognition and Their Effectiveness Using Flipped Learning in Initial Training Students", Fabiola Mary Talavera Mendoza, Fabián Hugo Rucano Paucar, Rolando Linares Delgado and Ysabel Milagros Rodríguez Choque analyse the relationship between the instructional cognitive model and its implications in using flipped learning to achieve meaningful learning. This study was carried out with 60 university students. The survey results reveal that using flipped classrooms may reduce the cognitive load and subsequently promote learning.

Ethical, Socio-cultural and Administrative Issues in Education of an AI Era

In chapter "Controversial Issues Faced by Intelligent Tutoring System in Developing Constructivist-Learning-Based Curriculum", Yue Wang discusses benefits and challenges when integrating AIED into contemporary Constructive-Learning-Based Curriculum design. It is noted that the Intelligent Tutoring System (ITS) could help to enhance learners' formative learning experience and promote learners' 21st-century skills. However, the author warns that biased data in the ITS may also give rise to learning recommendation bias and inequality. Also, it is argued that, as the algorithms are incapable of capturing unquantifiable learning information, ITS often emphasises quantifiable educational recommendations, which could deteriorate the construction of the constructivist-learning-based curriculum.

In chapter "Analysis of the Current Situation of Urban and Rural Teachers' Sense of Fairness in Online Education Equity", Qi Xu, Xinghong Liu, Xue Chen, Han Zhang and Min Pan present the urban and rural teachers' sense of fairness in online education equity. The study reveals urban and rural teachers present significant differences in their understanding of starting fairness, process fairness and outcome fairness. While some possible reasons are identified, the authors urge the improvement of teacher fairness senses in their online teaching.

In chapter "Construction of Multi-Tasks Academic Procrastination Model and Analysis of Procrastination Group Characteristics", Chao Zhou, Jianhua Qu and Yuting Ling construct a model that analyses student procrastination tendency in online courses. In their proposed model, online learning log data, static procrastination indicators and fluctuating procrastination indicators are extracted to analyse students' procrastination behaviour. By categorising behaviour characteristics of different procrastination groups, the authors offer suggestions that may reduce academic procrastination in online learning and subsequently improve learning.

In chapter "Alternative Digital Credentials—UAE's First Adopters' Quality Assurance Model and Case Study", Samar A. El-Farra, Jihad M. Mohaidat, Saud H. Aldajah and Abdullatif M. Alshamsi propose the VRSVAS, an outcome-focused quality assurance model, for Alternative Digital Credentials (ADCs). Their subsequent case study of applying VRSVAS in the United Arab Emirates constitutes a blueprint for the future of ADCs' provisions.

In chapter "Quality Analysis of Graduate Dissertations in Natural Science", Lin Yuan, Shuaiyi Liu, Shenling Liu, Yuheng Wang, Aihong Li and Yanwu Li study the quality of dissertations in China. They find that the total pages, the word counts, citation amount, English reference percentage and author's achievements have positive correlations with the quality of dissertations in different levels, while the rate of overlapping words negatively correlates with the quality. Their results can provide a reference for education management departments to evaluate dissertations qualities.

We hope this compilation of chapters written by authors worldwide can contribute to the advancement of knowledge, practice and technology in the emerging AIED research.

Hong Kong, China
Perth, Australia
Hong Kong, China
Wuhan, China

Eric C. K. Cheng
Rekha B. Koul
Tianchong Wang
Xinguo Yu

Contents

About the Editors

Eric C. K. Cheng is a specialist in Knowledge Management and Lesson Study. He is currently an associate professor of the Department of Curriculum and Instruction of the Education University of Hong Kong. He earned his Doctor of Education in education management from the University of Leicester. His research focuses on exploring organizational factors and management strategies that enable school leaders and teachers to leverage pedagogical knowledge assets. Such leveraging aims at improving student learning and capitalizing on the knowledge assets as school intellectual capital for sustainable development. His publication covers the areas of knowledge management, school management, Lesson and Learning Study.

Rekha B. Koul is Dean International, Faculty of Humanities and Associate Professor at STEM Research Group, School of Education, Curtin University, Australia. She has nearly three decades of teaching and research experience. Her expertise lies in the development, refinement and validation of questionnaires; investigations of the effects of classroom environments on student outcomes; evaluation of educational programs; teacher-action research aimed at improving their environments and evaluation of curriculum. She has successfully secured many local, national and international research grants to the value of over one million dollars. Her publication record includes two books authored, seven books edited, eight book chapters and many journal articles published in peer-reviewed journals. She has delivered invited keynote addresses and also conducts teacher professional learning workshops. Rekha convenes International Conference on Science, Mathematics and Technology Education (icSTEM), a biannual conference in different parts of the world and has established a Learning Environment Research Centre in Indonesia. She was the elected chair of Jumki Basu Scholarship (NARST 2016–18) and Programme Chair for Learning Environment SIG (AERA).

Tianchong Wang is currently a Post-doctoral Fellow in the Department of Curriculum and Instruction at The Education University of Hong Kong. He has obtained an Ed.D. degree from the Chinese University of Hong Kong, an M.Sc. (Information Technology in Education) degree from the University of Hong Kong, and a B.A. Hons (Teaching English as a Second Language) degree from Hong Kong Baptist University. Dr. Wang has a solid technical background in Information and Communication Technologies (ICTs); he holds several professional qualifications such as MCP, MCSA, MCSE, MCDBA, MCTS, MCITP, ACSP, ACTC, and THXCP. He is also a member of IEEE. As an advocate of using digital technologies to enhance education quality and inclusion, Dr Wang's current research interests include AI in education (AIEd), blended learning, mobile learning, flipped teaching, ICT-enabled learning at scale, and digital learning for development. He has participated in research and development projects commissioned or supported by international organisations (including UNESCO, IDRC, UKID and The HEAD Foundation). His research has been published in SSCI/EI-indexed journals, books chapters, and conference proceedings. Dr. Wang is also a passionate teacher educator. He is a PT Lecturer at the University of Hong Kong. He often serves as a resource person who conducts teacher capacity-building activities for NGOs and government agencies.

Xinguo Yu is the dean and Professor of CCNU Wollongong Joint Institute and Professor of National Engineering Research Center for E-Learning at Central China Normal University, Wuhan, China, senior member of both IEEE and ACM, and an adjunct professor of University of Wollongong, Australia. He is a member of steering board of PSIVT conference and a member of steering board of Smart Educational Technology Branch Society under Automation Society, China. He received Ph.D. degree in Computer Science from National University of Singapore. His current research mainly focuses on intelligent educational technology, educational robotics, multimedia analysis, computer vision, artificial intelligence, and virtual reality. He has published over 100 research papers. He is an Associate Editor of International Journal of Digital Crime and Forensics, was Guest Editor of Multimedia Systems and International Journal of Pattern Recognition and Artificial Intelligence. He is general chair of International Conference on Internet Multimedia Computing and Service 2012 and Pacific-Rim Symposium on Image and Video Technology 2017, and program chair of International Conference of Educational Innovation through Technology 2015, Pacific-Rim Symposium on Image and Video Technology 2015.

An Overview of AI in Education

Towards a Tripartite Research Agenda: A Scoping Review of Artificial Intelligence in Education Research

Tianchong Wang⊙ and Eric C. K. Cheng⊙

Abstract This paper reports on a scoping review of research studies on artificial intelligence in education (AIED) published over the last two decades (2001–2021). A wide range of manuscripts were yielded from the education and educational research category of the Social Sciences Citation Index (SSCI) database, and papers from an AIED-specialised journal were also included. 135 of those meeting the selection criteria were analysed with content analysis and categorical meta-trends analysis. Three distinctive and superordinate AIED research agenda were identified: *Learning from AI*, *Learning about AI*, and *Learning with AI*. By portraying the current status of AIED research and depicting its tripartite research agenda, gaps and possible directions were discussed. This paper serves as a blueprint for AIED researchers to position their up-and-coming AIED studies for the next decade.

Keywords Artificial intelligence in education · AIED · Scoping review

1 Introduction

Artificial Intelligence (AI), the genre of technologies that allow tasks to be done in a way similar to how a human would do them, has been developing rapidly and making significant impacts in recent years. Through a wide array of applications, the AI market size is at USD 62.35 billion in 2020 and is expected to expand at a compound annual growth rate (CAGR) of 40.2% from 2021 to 2028 [14]. With such an expansion rate and scale, AI is transforming the labour market [4] and overall economic structure [13]. At the same time, AI has also affected education at an unprecedented pace. In the 2021 Horizon Report [29], AI is prominently featured as one of the six key emerging technologies and practices in education. AI in Education (AIED)—the intersection of AI and education domains—has become a topic of keen interest among researchers. A growing number of research studies on AIED have

T. Wang (✉) · E. C. K. Cheng
Department of Curriculum and Instruction, The Education University of Hong Kong, Tai Po, N.T., Hong Kong S.A.R., China
e-mail: twang@eduhk.hk

© The Author(s), under exclusive license to Springer Nature Singapore Pte Ltd. 2022
E. C. K. Cheng et al. (eds.), *Artificial Intelligence in Education: Emerging Technologies, Models and Applications*, Lecture Notes on Data Engineering and Communications Technologies 104, https://doi.org/10.1007/978-981-16-7527-0_1

been conducted with different research focuses and directions. Given this dynamic development, as well as growing interest in the field among educators, an up-to-date scoping review of AIED literature is warranted.

Therefore, the study seeks to provide a current synthesis of AIED research. More specifically, we attempt to streamline key focuses and research directions. By doing so, we will be able to depict the widely covered research agenda of AIED. The findings of this study can serve as a springboard for AIED researchers to position their work when they conceptualise their AIED research. They also serve as a spur to initiate dialogue regarding up-and-coming AIED research themes in the next decade.

The rest of this paper is organised as follows. First, a background section draws on AI definitions and shows how AI is impacting education. The methodology section follows with descriptions of the search strategies, eligibility criteria, coding methods and analysis approaches. Our review findings, discussions and limitations come afterwards, strengthening the contribution of the study.

2   Background

2.1   What is AI, Anyway?

With the term being coined by John McCarthy during a summer conference at Dartmouth College in 1956 [23], roots of AI started. However, at present there is no universally accepted scholarly definition of AI [24], with its *status quo* being described as the parable of "the blind men and an elephant" [26].

While it is clear that AI involves human-made technologies, only a fair degree of consensus has been achieved as to what extent, or whether it is possible, that AI may display those properties that we equate with human intelligence [34]. Some experts use the Turing Test [37] to determine the intelligence of AI. Still, the practical value of this measure has been challenged due to its subjective nature [10]—not to mention the practical issues of the test such as how it determines cognitive capacity and the autonomy of agent behaviour [3]. Nevertheless, the necessity to determine fully human-like intelligence is being debated by the AI research community, and some simply distinguished weak/strong AIs (e.g., [31]) and narrow/general/super AIs (e.g., [12]) based on the capacity and the scale for usage.

Perhaps the most significant challenge to establish a well-accepted definition is that AI relates to different branches of the knowledge domain due to its historical development [39]. Because of the varied perspectives of each AI research field, answers to the questions of what constitutes AI and what is appropriate terminology have been inconsistent [24]. The relentless pace of technological advancement further complicates the definition of AI [16]. Many technologies under the AI banner—even for those powering highly sophisticated technologies such as natural language processing, computer vision, and recommender systems—are sometimes not labelled as AI anymore but become "just algorithms" when they are general enough and adopted by mainstream developers [26]. The relatively recent recognition of AI as a particular genre of pervasive technology extracted from ICT [32] has also blurred boundaries between associated concepts.

Therefore, before these boundaries are formed and a concept of "correct use" is established, it might be useful for AI to be tentatively conceived as a collective notion that covers all the current and future working definitions, while maintaining its status as being *"nothing but what the AI researchers have been doing"* ([39], p. 3). Keeping the definition open accurately reflects that AI domain is constantly evolving.

2.2 How is AI Impacting Education?

As AI continues to mature, how such kind of technology intersects with education is being explored. Coupled with current education reforms such as digital transformations of educational institutions and adoptions of education technologies, AIED is bringing many new opportunities. At present, AI is impacting education on the following perspectives.

First, AI can drive school efficiency and streamline teacher tasks through automation. For example, auto-grading can tremendously reduce workload for marking homework and tests. Although current AI systems can only grade certain assessment types such as multiple-choice questions and short-answer questions [42], teachers are freed from highly repetitive tasks and can spend more time engaging with students.

Second, AI may also address existing challenges in education, especially in terms of access and inclusion. For example, with the power of AI engines such as Automatic Captioning and Machine Translation, classrooms can be accessible to all, including those who speak different languages or have hearing impairments [30]. This also opens up possibilities for students who lack exposure to educational resources and who could not access learning without additional support.

Third, AI could fill needs gaps, extending learning opportunities outside the classroom. For example, AI tutoring systems and learning apps offer students instructions and support if they struggle with homework and when they are preparing for tests [15]. These programmes are becoming more advanced thanks to AI. New AI-powered features such as image and voice recognition further expand the applicability of the programmes in disciplines and learning scenarios where teacher presence traditionally plays a critical role (e.g., language learning and music education) [27].

Fourth, AI can empower both teachers and students with enhanced learning experiences [11], and can promote differentiated and personalised learning. Adjusting learning and teaching based on individual students' pace and needs is a perennial goal in the education community [35]. Unfortunately, school reality, such as classroom size and teacher-student ratio has often hindered the personalisation of learning. As AI becomes more sophisticated with machine learning, adaptation has become viable; students can be provided with learning content, tests, and feedback appropriate to their level.

Fifth, by leveraging the attributes of AI, teachers can further improve the quality of learning and teaching. the analytical feedback offered by AI allows teachers to identify knowledge gaps among students and offer support when appropriate. In other words, Assessment for Learning (AFL)—the ongoing assessment activities that allow teachers to monitor student learning on a day-to-day basis and adjust their teaching based on what the students need to be successful [5]—can be more effectively facilitated.

Sixth, the proliferation of AI in today's society is changing what knowledge, skills and values students need to acquire. Since the students of today will need to work in a future where AI might be ubiquitous, their exposure to the technologies and their ability to apply AI to solve real-world problems creatively may become crucial [40].

3   Methodology

This study employed the scoping review method for developing a current synthesis of AIED research. Proposed by Arksey and O'Malley [2] and further advanced by Levac et al. [20] and others, scoping review determines coverage of a body of literature on a given topic, clearly indicates the volume of literature available, and gives an overview of its focus. Scoping reviews are useful for examining emerging topics when it is still unclear what other, more specific questions can be addressed by a more detailed systematic literature review [25]. As indicated in the previous section, AIED research appears to be one such topic.

The methodology for this scoping review was based on the framework outlined by Arksey and O'Malley [2]. The review included the following five key phases: (1) identifying the research question, (2) identifying relevant studies, (3) study selection, (4) charting the data, and (5) collating, summarising, and reporting the results.

3.1   Research Questions

"What are the key themes of AIED research in the last two decades?" is asked as a research question. In addition, "What are the journal and year of publication of the articles, education settings, country context, learning domain, supported technologies, and AIED role/purposes" is also investigated.

3.2   Data Source

The electronic databases searched in this review included those identified as reputable sources that index research relevant to AIED. More particularly, as AIED may be considered as the interplay between science and social sciences domains, the Web

Table 1 Initial search strings

Topic	Search terms
Artificial intelligence	"artificial intelligence" OR "machine intelligence" OR "intelligent support" OR "intelligent virtual reality" OR "chatbot*" OR "machine learning" OR "automated tutor" OR "personal tutor*" OR "intelligent agent*" OR "expert system" OR "neural network" OR "natural language processing"
Education	"higher education" OR college* OR undergrad* OR graduate OR postgrad* OR "K-12" OR kindergarten* OR "corporate training*" OR "professional training*" OR "primary school*" OR "middle school*" OR "high school*" OR "elementary school*" OR "vocational education" OR "adult education"

of Science database and the Social Science Citation Index (SSCI) journals were selected to conduct the literature search. This is also because articles published in the SSCI database are generally considered as high-quality publications among education researchers. Additionally, as some newer AIED-focused journals may not be included in Web of Science database, additional search efforts were made to locate more targeted publications. A specialised journal, namely *Computers & Education: Artificial Intelligence*, was thus selected as an additional source.

3.3 Search Terms and Searching Strategies

The initial search was implemented in July 2021. The search terms and phrases for the title, abstract and keywords were searched and identified as related to AIED. Synonyms (e.g., 'smart', 'intelligent'), abbreviations (e.g., 'Artificial Intelligence', 'AI'), singular/plural/verbal/adjective forms (e.g., 'ITS', 'ITSs'), and broader/narrower terms (e.g., 'machine learning', 'classification') were also checked. The title and abstracts of the search results were assessed for relevance; another member of the research team verified this assessment.

The following Table 1 gives a summary of all initial search strings.

3.4 Eligibility Criteria

Our selections were restricted to high-ranking, peer-reviewed SSCI journal papers published in English language journals between the years 2001–2021 that were written on themes related to AI and its intersection with education. The duplicated records between databases were removed. The title and abstracts of the search results were assessed for relevance. Those citations deemed relevant after title and abstract screening were procured for subsequent review of the full-text article. Studies were excluded if they were found to have little connection to AIED. Studies that

merely focused on the development processes of AI without any educational implications and discussions were also excluded. The screening process was verified by
another member of the research team. Cohen's kappa [9] was not calculated, but the
researchers met to discuss and resolve any disagreement.

3.5 Coding and Analysis

Microsoft Excel spreadsheets were used to extract the data and manage the information obtained in an organised manner. A coding system was developed. Codes
included article information (year of publication, journal name, countries of author-
ship), the kind of AI technologies, how AI elements were involved, and subject
areas (if any). Descriptive statistics were calculated to summarise the data at meta-
levels. Inductive content analysis [19] was conducted to extract the key information,
particularly regarding the purpose and nature of AI involvement in the selected
papers.

4 Results

Within the 651 publications yielded by the initial search on Web of Science, 550
were SSCI journal articles. After screening the potentially relevant full-text papers,
327 were excluded for having a slight connection to AIED, 82 were excluded
for merely focusing on the AI-related technology development processes without
any educational implications and discussions, and six were excluded for not being
written in English. Meanwhile, 31 were selected from the AIED-specialised journal
Computers & Education: Artificial Intelligence. As a result, a total of 135 research
articles were included in this scoping review.

4.1 Journals Distribution and Article Count

The 135 AIED research articles reviewed were distributed in a total of 20 journals,
as depicted in Table 2. As the specialised journal of AIED, *Computers & Education:
Artificial Intelligence* published 31 AIED articles, having the highest result count.
Computers & Education, Educational Technology & Society, and the *British Journal
of Education Technology* were followed, with 18, 16 and 13 AIED articles published,
respectively. These journals are highly reputable and well-cited in the fields of Education and Education Technologies, given their high impact factors (8.538, 3.522 and
4.929 in 2020, respectively). It is worth noting that the cross-disciplinary, Open
Access journal *Sustainability* also contributed 12 AIED researches.

Table 2 Journal distribution

Journal	Result count
Computer & Education: Artificial Intelligence	31
Computers & Education	18
Educational Technology & Society	16
British Journal of Education Technology	13
Sustainability	12
IEEE Access	6
Education and Information Technologies	5
Learning, Media and Technology	5
Educational Technology Research and Development	4
Interactive Learning Environments	4
International Journal of Educational Technology in Higher Education	4
IEEE Transactions on Learning Technologies	3
Applied Artificial Intelligence	2
Computer Assisted Language Learning	2
Frontiers in Psychology	2
Journal of Educational Computing Research	2
Teaching in Higher Education	2
Educational Philosophy and Theory	1
Educational Studies	1
IEEE Transactions on Education	1
Science Education	1

As Fig. 1 shows, in the first decade of the New Millennium (2001–2010), there were minimal articles on AIED (n = 9). In the second decade (2011–2020), after a relatively slow development between 2011–2018 (n = 7), AIED research publications were flourishing in 2019–2020 (n = 58). In 2021, a total of 61 papers were published, marking an explosion of AIED research in the new decade.

4.2 Geographic Distributions

Based on author affiliations, a total of 35 countries and regions are represented in the 45 articles we reviewed, as listed in Table 3 and illustrated in Fig. 2. Most of the publications are affiliated with high- and upper-middle-income economies. More specifically, Taiwan, China was most represented, with 33 articles, followed by United States with 20, United Kingdom with 15, Hong Kong, China with 14, and Mainland China with 13. Lower-middle- and low-income economies only account for five articles included (3.7%).

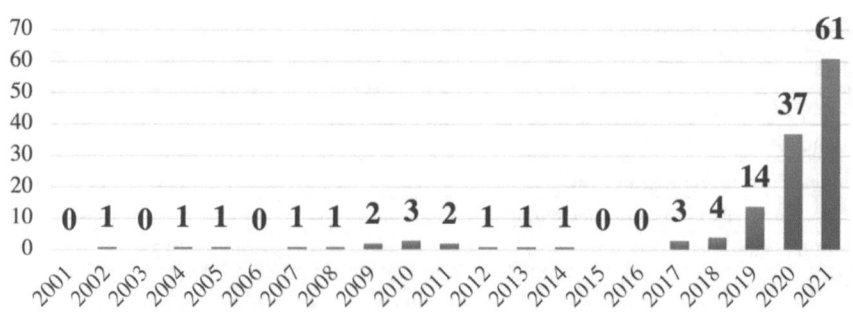

 Number of Publications

Fig. 1 Number of publications over time

Table 3 Geographic
distributions

Countries and regions	Numbers
Taiwan, China	33
United States	20
United Kingdom	15
Hong Kong, China	14
Mainland China	13
Australia	10
South Korea	6
Spain	5
Germany	4
Japan	4
Canada	3
Netherlands	3
Belgium	2
India	2
Mexico	2
Saudi Arabia	2
Singapore	2
Argentina	2
Brazil	1
Czech Republic	1
Ecuador	1
Greece	1
Iran	1
Ireland	1

(continued)

Table 3 (continued)

Countries and regions	Numbers
Malaysia	1
Norway	1
Pakistan	1
Portugal	1
Romania	1
Serbia	1
South Africa	1
Sweden	1
Switzerland	1
Turkey	1
Vietnam	1

Geographic Distribution

Numbers of AIED research
published

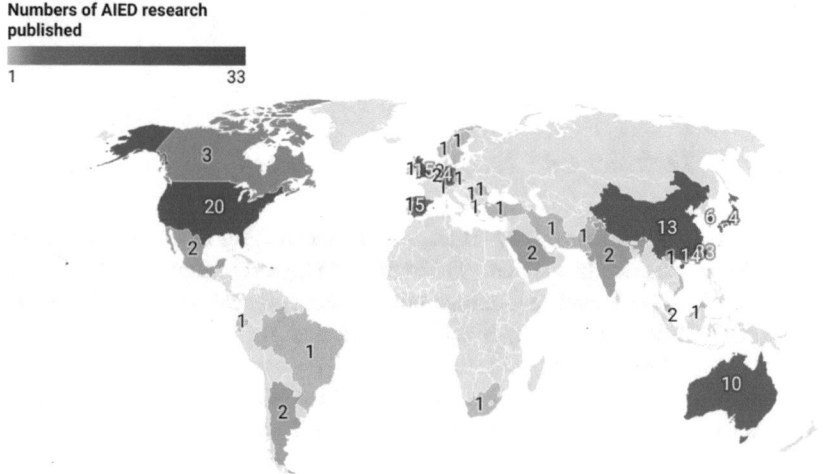

Fig. 2 Geographic distributions of AIED research

4.3 Subject Areas

AI was implemented and examined in a wide variety of subject areas. While many articles did not specify the subject areas addressed in the study, the presence of AI in computer science, language education, music, sports, mental health, medical education and the interdisciplinary field of General Education and STEM are clear.

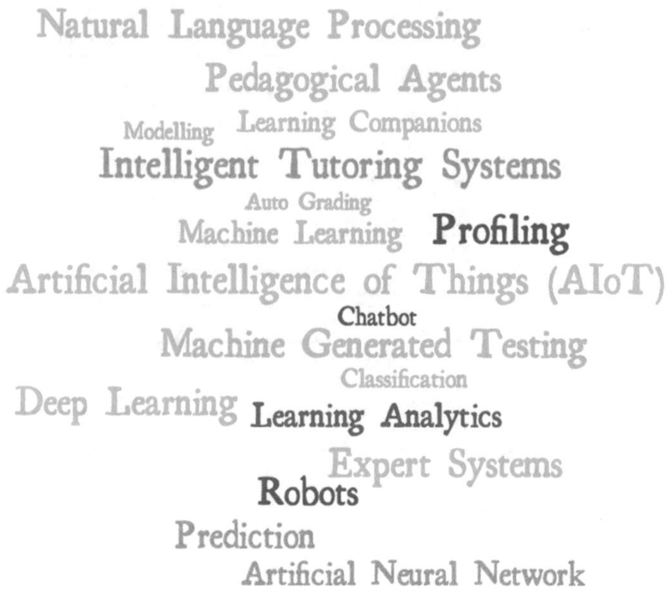

Natural Language Processing

Pedagogical Agents

Modelling   Learning Companions

Intelligent Tutoring Systems

Auto Grading

Machine Learning   **Profiling**

Artificial Intelligence of Things (AIoT)

Chatbot

Machine Generated Testing

Classification

Deep Learning Learning Analytics

Expert Systems

Robots

Prediction

Artificial Neural Network

Fig. 3 Word cloud of key AI technologies involved

4.4 Key AI Technics and Technologies Involved

We identified multiple kinds of key AI technics and technologies from the literature reviewed. More specifically, they are (1) Chatbot, (2) Expert Systems, (3) Intelligent Tutoring Systems, (4) Learning Companions, (5) Pedagogical Agents, (6) Robots, (7) Learning Analytics (LA), (8) Machine Generated Testing, (9) Profiling, (10) Modelling, (11) Classification, (12) Prediction, (13) Natural Language Processing, (14) Machine Learning, (15) Deep Learning, (16) Auto Grading, (17) Artificial Neural Network, and (18) Artificial Intelligence of Things (AIoT), as illustrated in Fig. 3.

4.5 Research Focus of AIED

Our review acknowledges the fact that AIED research involves diverse foci. Amongst the many themes identified from the papers that we reviewed, we identified three superordinate research agenda identified (see Fig. 4) besides the general discussion ($n = 16$) and AI educational philosophy ones ($n = 9$): (1) *Learning from AI*, (2) *Learning about AI* and (3) *Learning with AI*.

Category 1: Learning from AI. In this category of AIED research, AI serves as the principal means by which students learn, i.e., *Learning from AI*. In our review, a total

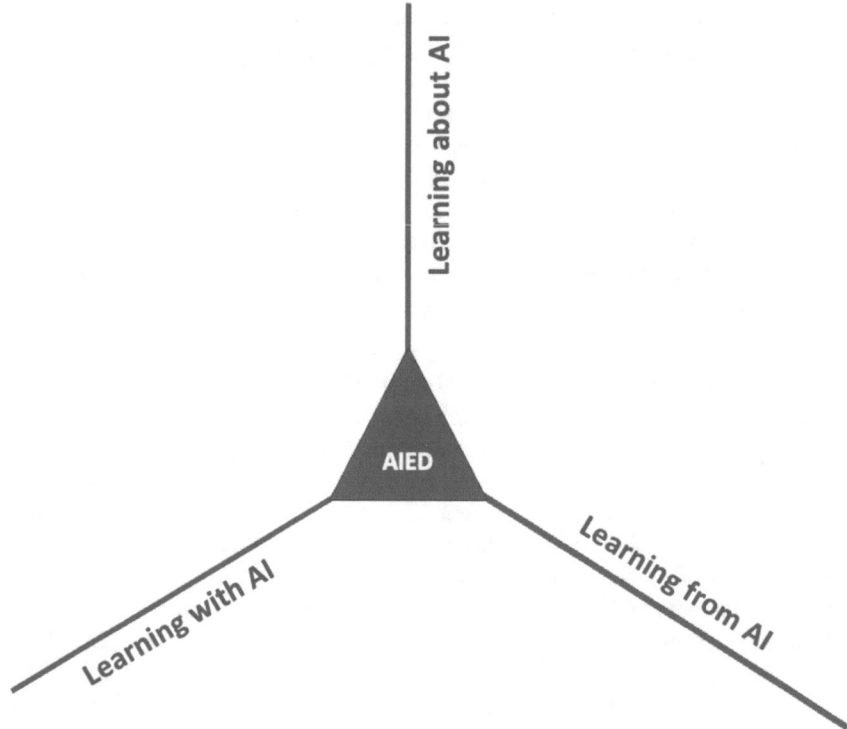

Fig. 4 A tripartite research agenda of AIED

of 48 articles falls into this category. Powered by rule-based or machine learning algorithms, these instructional platforms, notably Expert Systems and Intelligent Tutoring Systems (ITS), have the adaptive capacity to deliver customised content and learning paths according to student interest, aptitude, and behaviour profiling. A number of studies have found that these systems were effective sources of instruction and support for student learning, although the results could be challenged as the studies primarily focus on system development and only include pilot study results. In our review, the applicability of Expert Systems and ITS appeared to be limited, because the target subject area, knowledge taught, and the skills developed must be amenable to a rule-based or machine learning AI architecture. This could prevent the actualisation of these systems in a broader scale, wider contexts, or more varied learning scenarios.

Category 2: Learning about AI. This category of AIED research encompasses efforts toward equipping learners to thrive in an AI-saturated future, i.e., *Learning about AI* (interchangeably, AI education). Only 12 of our articles fall into this category, and they are mostly in higher education settings. The K-12 AI education is still very few and far between, and the education models are still being explored [8, 41]. Technical complexity was a common reason why AI has seldom been part of K-12 teaching and has been included only in higher education subjects [33]. However, the emergence of drag-and-drop, block-based platforms have made it possible for users to programme applications that involve AI elements without mastering the complex syntax of coding languages, as reported in Chiu and Chai [8]. A broader focus of AI education also calls for more comprehensive knowledge coverage. Long and Magerko [21] suggested a need to develop student knowledge for more general, non-technical AI Literacy. Touretzky et al. [36] envisioned what they think are the "big ideas" in AI that K-12 students should know: perception, representation and reasoning, learning, natural interaction, and societal impact. These recent proposals may be valid but are thus far under-investigated by researchers and educators.

Category 3: Learning with AI. The third category encompasses research on using AI tools to improve learning and teaching practices, i.e., *Learning with AI*. This category is the most-represented in the sample papers (n = 50). Similar to the previous category, most research in this direction was conducted in higher education settings, and some steps in this direction are being explored by K-12 schools, albeit slowly. A prominent example is teachers using AI-powered analytical systems to improve learning processes. Research in this direction in recent years has focused mainly on LA. This emerging type of AI application uses algorithms to analyse data about learners and their environments for helping users improving learning experiences and outcomes. For example, some education institutions piloted LA dashboards that display learning behaviour patterns so that teachers could provide just-in-time support to students [6]. Others have developed predictive LA to help identify students who may be considered "at-risk" of failing and which of those students may need additional support [22]. There are uses of LA that learn and validate students' language use patterns from their daily work so that academic integrity can be ensured [1]. These LA attempts have empowered teachers to make informed decisions, deliver responsive pedagogies, offer differentiated instruction, and provide timely feedback. While these LA may be highly effective in some contexts, several potential risks and issues are associated with using machine-learning approaches. For example, the risk of student misclassification may have serious negative consequences [28]. In addition, issues such as access to appropriate data to train AI models, biases in the models introduced through training data sets, and a lack of transparency into how the models work have also been documented in the literature [17].

5   Discussions and Implications

This paper conducted a scoping review of research studies on AIED published over the last two decades. Three distinctive, superordinate AIED research agendas were identified: *Learning from AI*, *Learning about AI*, and *Learning with AI*. From the results of our review, the following discussions and implications emerged.

First, AIED research has yet to catch up with the rapid advancement of AI technologies in a way that can guide and support evidence-based practices. During our screening process, we found many publications reporting on new advances in AI technologies, and yet a large percentage of those papers had to be excluded due to the scarcity of educational perspectives. This situation is likely to be improved, as our results highlight a significant jump in the number of AIED researches since 2019.

Second, our review reveals a general disparity of AIED research in terms of education settings. The majority of the current AIED studies were conducted in higher education contexts. K-12 AIED research is still in its nascent stage. This situation was noted by Knox [18].

Third, our review suggests AIED research in Global South contexts is under-represented. The predominance of AIED research from Taiwan, China, the United States, the United Kingdom, Hong Kong, China and the Mainland China is not surprising given the fact that these economies are possibly the biggest players in published literature worldwide. And it is possible that the language limitations of our study contributed to this disparity as we only included English language articles. Still, the issue of digital divide, as UNESCO's *Beijing Consensus* [38] heavily emphasised, merits attention.

Fourth, while research on *Learning from AI* reported its technical development, possible use and pedagogical impact in certain subject areas, there appeared to be an imbalanced development between disciplines. Therefore, more comprehensive coverage of different subject areas could lead to a better understanding of the overall landscape of *Learning from AI* research, as well as learning possibly varied opportunities and challenges across the disciplines. In addition, although Expert Systems and ITS facilitate self-directed learning with personalised learning experiences, the inconsistencies between AI tutors and classroom learning may also arouse conflicts. In view of this, discovering how these systems can be effectively used alongside classroom education in blended learning form, i.e., harnessing the best of both Learning about AI and Learning with AI together, could be an interesting future research direction.

Fifth, amongst the range of AIED research, the studies on *Learning about AI* (n = 12) are significantly less than those in the categories of *Learning from AI* (n = 48) and Learning with AI (n = 50). For example, a lot of research focused on intelligent tutors or personalized learning systems/environments and LA. In contrast, only a very limited number of research publications explored what and how to incorporate AI elements in the curriculum. There are pressing needs for in-depth investigations in this arena; without sufficient research, actionable guidelines for educators cannot be made, as pointed out by Wang and Cheng [40]. Therefore, we encourage researchers to conduct more studies on *Learning about AI*.

Sixth, while current research on *Learning about AI* investigated the appropriate content, educational models, and pedagogies for AI education, how they relate to existing educational approaches appeared to be under-investigated. Such studies would be highly valuable so that future AIED researchers do not "reinventing the wheel".

Seventh, ethical, bias, privacy, and security issues are critical for all *Learning from AI*, *Learning about AI* and *Learning with AI* research. All the AI technologies in our selected studies, such as profiling, performance prediction, modelling, adaptation, machine learning, involves data. How data are handled and used could arouse concerns, and in some extreme cases, cause serious ramifications (e.g., racism, as [7] reported). AIED Researchers need to be mindful of these issues, as they pose research opportunities and challenges. Furthermore, AIED practitioners should also be aware of these issues when engaging in *Learning about AI* implementations.

6 Limitations and Future Review

The main limitation of this study is methodological. The limited set of literature sources that we explored as relevant, and the specific strings used in the search efforts may not be considered as comprehensive, and this could be inherent to any scoping review. In addition, the research relied mainly on those published in the English language, which could also limit the scope of this review.

Future reviews may extend the search scope to include other reputable databases, specialised journals, or peer-reviewed conference proceedings beyond those in English and located in the SSCI-index. And additional keywords can be added as new AI technology emerges.

References

1. Amigud A, Arnedo-Moreno J, Daradoumis T, Guerrero-Roldan A-E (2017) Using learning analytics for preserving academic integrity. Int Rev Res Open Distance Learn 18(5):192–210
2. Arksey H, O'Malley L (2005) Scoping studies: towards a methodological framework. Int J Soc Res Methodol 8(1):19–32
3. Baker T, Smith L, Anissa N (2019) Educ-AI-tion rebooted? Exploring the future of artificial intelligence in schools and colleges. Retrieved 21 July, 2020, from https://media.nesta.org.uk/documents/Future_of_AI_and_education_v5_WEB.pdf
4. Bessen J (2019) Automation and jobs: when technology boosts employment. Econ Policy 34(100):589–626
5. Brown S (2005) Assessment for learning. Learn Teach High Educ 1(1):81–89
6. Chen CM, Wang JY, Hsu LC (2021) An interactive test dashboard with diagnosis and feedback mechanisms to facilitate learning performance. Comput Educ: Artif Intell 2:1–11
7. Cheuk T (2021) Can AI be racist? Color-evasiveness in the application of machine learning to science assessments. Sci Educ
8. Chiu TK, Chai CS (2020) Sustainable curriculum planning for artificial intelligence education: a self-determination theory perspective. Sustainability 12(14):1–18
9. Cohen J (1960) A coefficient of agreement for nominal scales. Educ Psychol Meas 20(1):37–46
10. Cohen PR (2005) If not Turing's test, then what? AI Mag 26(4):61
11. Cope B, Kalantzis M, Searsmith D (2020) Artificial intelligence for education: knowledge and its assessment in AI-enabled learning ecologies. Educ Philos Theory 1–17
12. Fjelland R (2020) Why general artificial intelligence will not be realised. Hum Soc Sci Commun 7(1):1–9
13. Goolsbee A, Hubbard G, Ganz A (2019) A policy agenda to develop human capital for the modern economy. The Aspen Institute, Washington, DC

14. Grand View Research (2021) Artificial intelligence market size, share & trends analysis report by solution, by technology (deep learning, machine learning), by end use, by region, and segment forecasts, 2021–2028. Grand View Research, San Francisco
15. Hao K (2019) China has started a grand experiment in AI education. It could reshape how the world learns. MIT Technol Rev 123(1)
16. Hoeschl MB, Bueno TC, Hoeschl HC (2017) Fourth industrial revolution and the future of engineering: could robots replace human jobs? How ethical recommendations can help engineers rule on artificial intelligence. In: 2017 7th world engineering education forum (WEEF), November. IEEE, pp 21–26
17. Kitto K, Knight S (2019) Practical ethics for building learning analytics. Br J Educ Technol 50(6):2855–2870
18. Knox J (2020) Artificial intelligence and education in China. Learn Media Technol 45(3):298–311
19. Krippendorff K (2019) Content analysis: an introduction to its methodology, 4th edn. Sage publications, Thousand Oaks
20. Levac D, Colquhoun H, O'Brien KK (2010) Scoping studies: advancing the methodology. Implement Sci 5(1):1–9
21. Long D, Magerko B (2020) What is AI literacy? Competencies and design considerations. In: Proceedings of the 2020 CHI conference on human factors in computing systems. ACM, New York, pp 1–16
22. Lykourentzou I, Giannoukos I, Nikolopoulos V, Mpardis G, Loumos V (2009) Dropout prediction in e-learning courses through the combination of machine learning techniques. Comput Educ 53(3):950–965
23. McCarthy J, Minsky ML, Rochester N, Shannon CE (2006) A proposal for the Dartmouth summer research project on artificial intelligence, August 31, 1955. AI Mag 27(4):12–12
24. Monett D, Lewis CWP (2018) Getting clarity by defining artificial intelligence—A survey. In: Müller VC (ed) Philosophy and theory of artificial intelligence 2017. Springer, Cham, pp 212–214
25. Munn Z, Peters MD, Stern C, Tufanaru C, McArthur A, Aromataris E (2018) Systematic review or scoping review? Guidance for authors when choosing between a systematic or scoping review approach. BMC Med Res Methodol 18(1):1–7
26. Nilsson NJ (2009) The quest for artificial intelligence: a history of ideas and achievements. Cambridge University Press, Cambridge
27. Noviyanti SD (2020) Artificial intelligence (AI)-based pronunciation checker: an alternative for independent learning in pandemic situation. ELT Echo: J Engl Lang Teach Foreign Lang Context 5(2):162–169
28. Okoye K, Nganji J, Hosseini S (2020) Learning analytics for educational innovation: a systematic mapping study of early indicators and success factors. Int J Comput Inf Syst Ind Manag Appl 12:138–154
29. Pelletier K, Brown M, Brooks DC, McCormack M, Reeves J, Arbino N, Bozkurt A, Crawford S, Czerniewicz L, Gibson R, Linder K, Mason J, Mondelli, V (2021) 2021 EDUCAUSE Horizon report teaching and learning edition. EDUCAUSE, Boulder
30. Phillips C, Colton JS (2021) A new normal in inclusive, usable online learning experiences. In: Thurston TN, Lundstrom K, González C (eds) Resilient pedagogy: practical teaching strategies to overcome distance, disruption, and distraction. Utah State University, pp 169–186
31. Russel S, Norvig P (2010) Artificial intelligence - a modern approach. Pearson Education, New Jersey
32. Schmidt A (2016) Cloud-based AI for pervasive applications. IEEE Comput Arch Lett 15(01):14–18
33. Steinbauer G, Kandlhofer M, Chklovski T, Heintz F, Koenig S (2021) A differentiated discussion about AI education K-12. KI-Künstliche Intell 1–7
34. Tegmark M (2018) Life 3.0: being human in the age of artificial intelligence. Penguin Books, London

35. Tomlinson CA (2014) The differentiated classroom: responding to the needs of all learners, 2nd edn. ASCD, Alexandria
36. Touretzky D, Gardner-McCune C, Martin F, Seehorn D (2019) Envisioning AI for K-12: what should every child know about AI? In: Proceedings of the AAAI conference on artificial intelligence, vol 33, no 01, pp 9795–9799
37. Turing AM (2009) Computing machinery and intelligence. In: Epstein R, Roberts G, Beber G (eds) Parsing the Turing test. Springer, Dordrecht, pp 23–65
38. UNESCO (2019) Beijing consensus on artificial intelligence and education. UNESCO, Paris
39. Wang P (2019) On defining artificial intelligence. J Artif Gen Intell 10(2):1–37
40. Wang T, Cheng ECK (2021) An investigation of barriers to Hong Kong K-12 schools incorporating artificial intelligence in education. Comput Educ: Artif Intell
41. Wong GK, Ma X, Dillenbourg P, Huan J (2020) Broadening artificial intelligence education in K-12: where to start? ACM Inroads 11(1):20–29
42. Zhang L, Huang Y, Yang X, Yu S, Zhuang F (2019) An automatic short-answer grading model for semi-open-ended questions. Interact Learn Environ 1–14

List of the Articles Reviewed

Abbas M, Hwang GJ, Ajayi S, Mustafa G, Bilal M (2021) Modelling and exploiting taxonomic knowledge for developing mobile learning systems to enhance children's structural and functional categorization. Comput Educ: Artif Intell 2

Auerbach JE, Concordel A, Kornatowski PM, Floreano D (2018) Inquiry-based learning with robogen: an open-source software and hardware platform for robotics and artificial intelligence. IEEE Trans Learn Technol 12(3):356–369

Baker RS, Gašević D, Karumbaiah S (2021) Four paradigms in learning analytics: why paradigm convergence matters. Comput Educ: Artif Intell 2

Berendt B, Littlejohn A, Blakemore M (2020) AI in education: learner choice and fundamental rights. Learn Media Technol 45(3):312–324

Breines MR, Gallagher M (2020) A return to Teacherbot: rethinking the development of educational technology at the University of Edinburgh. Teach High Educ 1–15

Cavalcanti AP, Diego A, Carvalho R, Freitas F, Tsai YS, Gašević D, Mello RF (2021) Automatic feedback in online learning environments: a systematic literature review. Comput Educ: Artif Intell

Chai CS, Lin PY, Jong MSY, Dai Y, Chiu TK, Qin J (2021) Perceptions of and behavioral intentions towards learning artificial intelligence in primary school students. Educ Technol Soc 24(3):89–101

Chatterjee S, Bhattacharjee KK (2020) Adoption of artificial intelligence in higher education: a quantitative analysis using structural equation modelling. Educ Inf Technol 25(5):3443–3463

Chen B, Hwang GH, Wang SH (2021) Gender differences in cognitive load when applying game-based learning with intelligent robots. Educ Technol Soc 24(3):102–115

Chen CM, Wang JY, Hsu LC (2021) An interactive test dashboard with diagnosis and feedback mechanisms to facilitate learning performance. Comput Educ: Artif Intell 2

Chen KZ, Li SC (2021) Sequential, typological, and academic dynamics of self-regulated learners: learning analytics of an undergraduate chemistry online course. Comput Educ: Artif Intell

Chen X, Xie H, Hwang GJ (2020) A multi-perspective study on artificial intelligence in education: grants, conferences, journals, software tools, institutions, and researchers. Comput Educ: Artif Intell

Chen X, Xie H, Zou D, Hwang GJ (2020) Application and theory gaps during the rise of artificial intelligence in education. Comput Educ: Artif Intell 1

Chen X, Zou D, Xie H, Cheng G (2021) Twenty years of personalized language learning. Educ Technol Soc 24(1):205–222

Cheuk T (2021) Can AI be racist? Color-evasiveness in the application of machine learning to science assessments. Sci Educ

Chih-Ming C, Ying-You L (2020) Developing a computer-mediated communication competence forecasting model based on learning behavior features. Comput Educ: Artif Intell 1

Chin DB, Dohmen IM, Cheng BH, Oppezzo MA, Chase CC, Schwartz DL (2010) Preparing students for future learning with teachable agents. Educ Technol Res Dev 58(6):649–669

Chin DB, Dohmen IM, Schwartz DL (2013) Young children can learn scientific reasoning with teachable agents. IEEE Trans Learn Technol 6(3):248–257

Chiu CK, Tseng JC (2021) A Bayesian classification network-based learning status management system in an intelligent classroom. Educ Technol Soc 24(3):256–267

Chiu TK, Chai CS (2020) Sustainable curriculum planning for artificial intelligence education: a self-determination theory perspective. Sustainability 12(14):5568

Chocarro R, Cortiñas M, Marcos-Matás G (2021) Teachers' attitudes towards chatbots in education: a technology acceptance model approach considering the effect of social language, bot proactiveness, and users' characteristics. Educ Stud 1–19

Chou CY, Chan TW, Lin CJ (2003) Redefining the learning companion: the past, present, and future of educational agents. Comput Educ 40(3):255–269

Chung JWY, So HCF, Choi MMT, Yan VCM, Wong TKS (2021) Artificial Intelligence in education: using heart rate variability (HRV) as a biomarker to assess emotions objectively. Comput Educ: Artif Intell 2

Conijn R, Martinez-Maldonado R, Knight S, Buckingham Shum S, Van Waes L, Van Zaanen M (2020) How to provide automated feedback on the writing process? A participatory approach to design writing analytics tools. Comput Assist Lang Learn 1–31

Cope B, Kalantzis M, Searsmith D (2020) Artificial intelligence for education: knowledge and its assessment in AI-enabled learning ecologies. Educ Philos Theory 1–17

Cox AM (2021) Exploring the impact of artificial intelligence and robots on higher education through literature-based design fictions. Int J Educ Technol High Educ 18(1):1–19

Cukurova M, Kent C, Luckin R (2019) Artificial intelligence and multimodal data in the service of human decision-making: a case study in debate tutoring. Br J Educ Technol 50(6):3032–3046

Dai Y, Chai CS, Lin PY, Jong MSY, Guo Y, Qin J (2020) Promoting students' well-being by developing their readiness for the artificial intelligence age. Sustainability 12(16)

Deo RC, Yaseen ZM, Al-Ansari N, Nguyen-Huy T, Langlands TAM, Galligan L (2020) Modern artificial intelligence model development for undergraduate student performance prediction: an investigation on engineering mathematics courses. IEEE Access 8:136,697–136,724

Devedžić V (2004) Web intelligence and artificial intelligence in education. Educ Technol Soc, 7(4):29–39

Divekar RR, Drozdal J, Chabot S, Zhou Y, Su H, Chen Y, … Braasch J (2021) Foreign language acquisition via artificial intelligence and extended reality: design and evaluation. Comput Assist Lang Learn 1–29

Dixon-Román E, Nichols TP, Nyame-Mensah A (2020) The racializing forces of/in AI educational technologies. Learn Media Technol 45(3):236–250

Du Boulay B (2019) Escape from the skinner box: the case for contemporary intelligent learning environments. Br J Educ Technol 50(6):2902–2919

Edwards BI, Cheok AD (2018) Why not robot teachers: artificial intelligence for addressing teacher shortage. Appl Artif Intell 32(4):345–360

El-Alfy ESM, Abdel-Aal RE (2008) Construction and analysis of educational tests using abductive machine learning. Comput Educ 51(1):1–16

Fachada N (2021) ColorShapeLinks: a board game AI competition for educators and students. Comput Educ: Artif Intell 2

Fu S, Gu H, Yang B (2020) The affordances of AI-enabled automatic scoring applications on learners' continuous learning intention: an empirical study in China. Br J Educ Technol 51(5):1674–1692

García P, Amandi A, Schiaffino S, Campo M (2007) Evaluating Bayesian networks' precision for detecting students' learning styles. Comput Educ 49(3):794–808

Gomede E, de Barros RM, de Souza Mendes L (2021) Deep auto encoders to adaptive e-learning recommender system. Comput Educ: Artif Intell 2

Han KW, Lee E, Lee Y (2009) The impact of a peer-learning agent based on pair programming in a programming course. IEEE Trans Educ 53(2):318–327

Hsieh YZ, Lin SS, Luo YC, Jeng YL, Tan SW, Chen CR, Chiang PY (2020) ARCS-assisted teaching robots based on anticipatory computing and emotional big data for improving sustainable learning efficiency and motivation. Sustainability 12(14)

Hsu TC, Abelson H, Lao N, Tseng YH, Lin YT (2021) Behavioral-pattern exploration and development of an instructional tool for young children to learn AI. Comput Educ: Artif Intell 2

Huang X (2021) Aims for cultivating students' key competencies based on artificial intelligence education in China. Educ Inf Technol 1–21

Hwang GJ, Sung HY, Chang SC, Huang XC (2020) A fuzzy expert system-based adaptive learning approach to improving students' learning performances by considering affective and cognitive factors. Comput Educ: Artif Intell 1

Hwang GJ, Xie H, Wah BW, Gašević D (2020) Vision, challenges, roles and research issues of artificial intelligence in education

Jalal A, Mahmood M (2019) Students' behavior mining in e-learning environment using cognitive processes with information technologies. Educ Inf Technol 24(5):2797–2821

Jonassen DH (2011) Ask systems: interrogative access to multiple ways of thinking. Educ Technol Res Dev 59(1):159–175

Kabudi T, Pappas I, Olsen DH (2021) AI-enabled adaptive learning systems: a systematic mapping of the literature. Comput Educ: Artif Intell 2

Kahn K, Winters N (2021) Constructionism and AI: a history and possible futures. Br J Educ Technol

Kaufmann E (2021) Algorithm appreciation or aversion? Comparing in-service and pre-service teachers' acceptance of computerized expert models. Comput Educ: Artif Intell

Kay J, Kummerfeld B (2019) From data to personal user models for life-long, life-wide learners. Br J Educ Technol 50(6):2871–2884

Kim K, Kim HS, Shim J, Park JS (2021) A study in the early prediction of ICT literacy ratings using sustainability in data mining techniques. Sustainability 13(4)

Kitto K, Knight S (2019) Practical ethics for building learning analytics. Br J Educ Technol 50(6):2855–2870

Knox J (2020) Artificial intelligence and education in China. Learn Media Technol 45(3):298–311

Koć-Januchta MM, Schönborn KJ, Tibell LA, Chaudhri VK, Heller HC (2020) Engaging with biology by asking questions: investigating students' interaction and learning with an artificial intelligence-enriched textbook. J Educ Comput Res 58(6):1190–1224

Kong SC, Cheung WMY, Zhang G (2021) Evaluation of an artificial intelligence literacy course for university students with diverse study backgrounds. Comput Educ: Artif Intell

Lai YH, Chen SY, Lai CF, Chang YC, Su YS (2021) Study on enhancing AIoT computational thinking skills by plot image-based VR. Interact Learn Environ 29(3):482–495

Latham A, Crockett K, McLean D (2014) An adaptation algorithm for an intelligent natural language tutoring system. Comput Educ 71:97–110

Lee HS, Lee J (2021) Applying artificial intelligence in physical education and future perspectives. Sustainability 13(1)

Lemay DJ, Baek C, Doleck T (2021) Comparison of learning analytics and educational data mining: a topic modeling approach. Comput Educ: Artif Intell 2

Li F, He Y, Xue Q (2021) Progress, challenges and countermeasures of adaptive learning. Educ Technol Soc 24(3):238–255

Lin CH, Yu CC, Shih PK, Wu LY (2021) STEM based artificial intelligence learning in general education for non-engineering undergraduate students. Educ Technol Soc 24(3):224–237

Lin HC, Tu YF, Hwang GJ, Huang H (2021) From precision education to precision medicine. Educ Technol Soc 24(1):123–137

Lin PY, Chai CS, Jong MSY, Dai Y, Guo Y, Qin J (2021) Modeling the structural relationship among primary students' motivation to learn artificial intelligence. Comput Educ: Artif Intell

Lin YS, Chen SY, Tsai CW, Lai YH (2021) Exploring computational thinking skills training through augmented reality and AIoT learning. Front Psychol 12

Liu M, Rus V, Liu L (2016) Automatic Chinese factual question generation. IEEE Trans Learn Technol 10(2):194–204

Loftus M, Madden MG (2020) A pedagogy of data and artificial intelligence for student subjectification. Teach High Educ 25(4):456–475

Luckin R, Cukurova M (2019) Designing educational technologies in the age of AI: a learning sciences-driven approach. Br J Educ Technol 50(6):2824–2838

Lu OH, Huang AY, Tsai DC, Yang SJ (2021) Expert-authored and machine-generated short-answer questions for assessing students learning performance. Educ Technol Soc 24(3):159–173

Lykourentzou I, Giannoukos I, Nikolopoulos V, Mpardis G, Loumos V (2009) Dropout prediction in e-learning courses through the combination of machine learning techniques. Comput Educ 53(3):950–965

McLaren BM, DeLeeuw KE, Mayer RE (2011) Polite web-based intelligent tutors: can they improve learning in classrooms? Comput Educ 56(3):574–584

McStay A (2020) Emotional AI and EdTech: serving the public good? Learn Media Technol 45(3):270–283

Mehmood R, Alam F, Albogami NN, Katib I, Albeshri A, Altowaijri SM (2017) UTiLearn: a personalised ubiquitous teaching and learning system for smart societies. IEEE Access 5:2615–2635

Mohammadreza E, Safabakhsh R (2021) Lecture quality assessment based on the audience reactions using machine learning and neural networks. Comput Educ: Artif Intell 2

Moridis CN, Economides AA (2009) Prediction of student's mood during an online test using formula-based and neural network-based method. Comput Educ 53(3):644–652

Mota-Valtierra G, Rodríguez-Reséndiz J, Herrera-Ruiz G (2019) Constructivism-based methodology for teaching artificial intelligence topics focused on sustainable development. Sustainability 11(17):4642

Orlando S, Gaudioso E, De La Paz F (2020) Supporting teachers to monitor student's learning progress in an educational environment with robotics activities. IEEE Access 8:48620–48631

Ortego RG, Sánchez IM (2019) Relevant parameters for the classification of reading books depending on the degree of textual readability in primary and compulsory secondary education (CSE) students. IEEE Access 7:79044–79055

Ouherrou N, Elhammoumi O, Benmarrakchi F, El Kafi J (2019) Comparative study on emotions analysis from facial expressions in children with and without learning disabilities in virtual learning environment. Educ Inf Technol 24(2):1777–1792

Ouyang F, Jiao P (2021) Artificial intelligence in education: the three paradigms. Comput Educ: Artif Intell 2

Paek S, Kim N (2021) Analysis of worldwide research trends on the impact of artificial intelligence in education. Sustainability 13(14)

Papadopoulos I, Lazzarino R, Miah S, Weaver T, Thomas B, Koulouglioti C (2020) A systematic review of the literature regarding socially assistive robots in pre-tertiary education. Comput Educ 155

Pereira HBDB, Zebende GF, Moret MA (2010) Learning computer programming: implementing a fractal in a Turing Machine. Comput Educ 55(2):767–776

Perrotta C, Selwyn N (2020) Deep learning goes to school: toward a relational understanding of AI in education. Learn Media Technol 45(3):251–269

Polito G, Temperini M (2021) A gamified web-based system for computer programming learning. Comput Educ: Artif Intell

Qin F, Li K, Yan J (2020) Understanding user trust in artificial intelligence-based educational systems: evidence from China. Br J Educ Technol 51(5):1693–1710

Renz A, Hilbig R (2020) Prerequisites for artificial intelligence in further education: identification of drivers barriers, and business models of educational technology companies. Int J Educ Technol High Educ 17(1):1–21

Richards D, Dignum V (2019) Supporting and challenging learners through pedagogical agents: addressing ethical issues through designing for values. Br J Educ Technol 50(6):2885–2901

Rico-Juan JR, Gallego AJ, Calvo-Zaragoza J (2019) Automatic detection of inconsistencies between numerical scores and textual feedback in peer-assessment processes with machine learning. Comput Educ 140

Rodríguez-Hernández CF, Musso M, Kyndt E, Cascallar E (2021) Artificial neural networks in academic performance prediction: systematic implementation and predictor evaluation. Comput Educ: Artif Intell

Rosé CP, McLaughlin EA, Liu R, Koedinger KR (2019) Explanatory learner models: why machine learning (alone) is not the answer. Br J Educ Technol 50(6):2943–2958

Samarakou M, Tsaganou G, Papadakis A (2018) An e-learning system for extracting text comprehension and learning style characteristics. J Educ Technol Soc 21(1):126–136

Shih PK, Lin CH, Wu LYY, Yu CC (2021) Learning ethics in AI—teaching non-engineering undergraduates through situated learning. Sustainability 13(7)

Smutny P, Schreiberova P (2020) Chatbots for learning: a review of educational chatbots for the facebook messenger. Comput Educ 151

Srinivasan V, Murthy H (2021) Improving reading and comprehension in K-12: evidence from a large-scale AI technology intervention in India. Comput Educ: Artif Intell 2

Standen PJ, Brown DJ, Taheri M, Galvez Trigo MJ, Boulton H, Burton A, … Hortal E (2020) An evaluation of an adaptive learning system based on multimodal affect recognition for learners with intellectual disabilities. Br J Educ Technol 51(5):1748–1765

Su JH, Liao YW, Xu JZ, Zhao YW (2021) A personality-driven recommender system for cross-domain learning based on Holland code assessments. Sustainability 13(7)

Sun M, Li Y (2020) Eco-environment construction of English teaching using artificial intelligence under big data environment. IEEE Access 8:193,955–193,965

Tan DY, Cheah CW (2021) Developing a gamified AI-enabled online learning application to improve students' perception to University Physics. Comput Educ: Artif Intell

Tanveer M, Hassan S, Bhaumik A (2020) Academic policy regarding sustainability and artificial intelligence (AI) sustainability 12(22):9435

Tlili A, Zhang J, Papamitsiou Z, Manske S, Huang R, Hoppe HU (2021) Towards utilising emerging technologies to address the challenges of using open educational resources: a vision of the future. Educ Technol Res Dev 69(2):515–532

Topal AD, Eren CD, Geçer AK (2021) Chatbot application in a 5th grade science course. Educ Inf Technol 1–25

Tsai SC, Chen CH, Shiao YT, Ciou JS, Wu TN (2020) Precision education with statistical learning and deep learning: a case study in Taiwan. Int J Educ Technol High Educ 17(1):1–13

Ulloa-Cazarez RL, López-Martín C, Abran A, Yáñez-Márquez C (2018) Prediction of online students performance by means of genetic programming. Appl Artif Intell 32(9–10):858–881

Vázquez-Cano E, Mengual-Andrés S, López-Meneses E (2021) Chatbot to improve learning punctuation in Spanish and to enhance open and flexible learning environments. Int J Educ Technol High Educ 18(1):1–20

Villegas-Ch W, Arias-Navarrete A, Palacios-Pacheco X (2020) Proposal of an architecture for the integration of a chatbot with artificial intelligence in a smart campus for the improvement of learning. Sustainability 12(4)

Wang HC, Chang CY, Li TY (2008) Assessing creative problem-solving with automated text grading. Comput Educ 51(4):1450–1466

Wang J, Hwang GH, Chang CY (2021) Directions of the 100 most cited chatbot-related human behavior research: a review of academic publications. Comput Educ: Artif Intell

Wang J, Xie H, Wang FL, Lee LK, Au OTS (2021) Top-N personalized recommendation with graph neural networks in MOOCs. Comput Educ: Artif Intell 2

Wang S, Christensen C, Cui W, Tong R, Yarnall L, Shear L, Feng M (2020) When adaptive learning is effective learning: comparison of an adaptive learning system to teacher-led instruction. Interact Learn Environ 1–11

Wang S, Yu H, Hu X, Li J (2020) Participant or spectator? Comprehending the willingness of faculty to use intelligent tutoring systems in the artificial intelligence era. Br J Educ Technol 51(5):1657–1673

Wang T, Park J (2021) Design and implementation of intelligent sports training system for college students' mental health education. Front Psychol 12

Wang Y, Liu C, Tu YF (2021) Factors affecting the adoption of AI-based applications in higher education. Educ Technol Soc 24(3):116–129

Wanichsan D, Panjaburee P, Chookaew S (2021) Enhancing knowledge integration from multiple experts to guiding personalized learning paths for testing and diagnostic systems. Comput Educ: Artif Intell 2

Webb ME, Fluck A, Magenheim J, Malyn-Smith J, Waters J, Deschênes M, Zagami J (2020) Machine learning for human learners: opportunities, issues, tensions and threats. Educ Technol Res Dev 1–22

Westera W, Dascalu M, Kurvers H, Ruseti S, Trausan-Matu S (2018) Automated essay scoring in applied games: reducing the teacher bandwidth problem in online training. Comput Educ 123:212–224

Wong LH, Looi CK (2012) Swarm intelligence: new techniques for adaptive systems to provide learning support. Interact Learn Environ 20(1):19–40

Wu JY, Hsiao YC, Nian MW (2020) Using supervised machine learning on large-scale online forums to classify course-related Facebook messages in predicting learning achievement within the personal learning environment. Interact Learn Environ 28(1):65–80

Wu JY (2021) Learning analytics on structured and unstructured heterogeneous data sources: perspectives from procrastination, help-seeking, and Machine-Learning defined cognitive engagement. Comput Educ 163

Xie H, Chu HC, Hwang GJ, Wang CC (2019) Trends and development in technology-enhanced adaptive/personalized learning: a systematic review of journal publications from 2007 to 2017. Comput Educ 140

Xie H, Hwang GJ, Wong TL (2021) Editorial note: from conventional AI to modern AI in education: reexamining AI and analytic techniques for teaching and learning. J Educ Technol Soc 24(3)

Xie T, Zheng Q, Zhang W, Qu H (2017) Modeling and predicting the active video-viewing time in a large-scale E-learning system. IEEE Access 5:11,490–11,504

Xu Y, Wang D, Collins P, Lee H, Warschauer M (2021) Same benefits, different communication patterns: comparing children's reading with a conversational agent vs. a human partner. Comput Educ 161

Yang AC, Chen IY, Flanagan B, Ogata H (2021) Automatic generation of cloze items for repeated testing to improve reading comprehension. Educ Technol Soc 24(3):147–158

Yang AC, Chen IY, Flanagan B, Ogata H (2021) From human grading to machine grading. Educ Technol Soc 24(1):164–175

Yang CC, Chen IY, Ogata H (2021) Toward precision education: educational data mining and learning analytics for identifying students' learning patterns with ebook systems. Educ Technol Soc 24(1):152–163

Yang C, Chiang FK, Cheng Q, Ji J (2021) Machine learning-based student modeling methodology for intelligent tutoring systems. J Educ Comput Res

Yang D, Oh ES, Wang Y (2020) Hybrid physical education teaching and curriculum design based on a voice interactive artificial intelligence educational robot. Sustainability 12(19)

Yang F, Wang M, Shen R, Han P (2007) Community-organizing agent: an artificial intelligent system for building learning communities among large numbers of learners. Comput Educ 49(2):131–147

Yang SJ (2021) Guest editorial: precision education-a new challenge for AI in education. J Educ Technol Soc 24(1)

Yang SJ, Ogata H, Matsui T, Chen NS (2021) Human-centered artificial intelligence in education: seeing the invisible through the visible. Comput Educ: Artif Intell 2

Yueh HP, Lin W, Wang SC, Fu LC (2020) Reading with robot and human companions in library literacy activities: a comparison study. Br J Educ Technol 51(5):1884–1900

Yu SJ, Hsueh YL, Sun JC. Y, Liu HZ (2021) Developing an intelligent virtual reality interactive system based on the ADDIE model for learning pour-over coffee brewing. Comput Educ: Artif Intell

Zhang K, Aslan AB (2021) AI technologies for education: recent research future directions. Comput Educ: Artif Intell

Zhu M, Liu OL, Lee HS (2020) The effect of automated feedback on revision behavior and learning gains in formative assessment of scientific argument writing. Comput Educ 143

Analysis of Research Progress in the Field of Educational Technology in China-Research on Knowledge Graph Based on CSSCI (2015–2020)

Yaru Zhao, Chaoqian Li, Xinyu Tian, Yunhong Chen, and Shuming Li

Abstract Select the 2015–2020 domestic higher education information technology scientific research key areas, and all the articles selected in the Chinese Social Science Journal Citation Index (CSSCI), make full use of the word frequency visual analysis method and the Cite Space visual analysis tool to draw 6,740 domestic education articles. The knowledge map of the C journal literature in the technical research field visually displays the research progress in the field of educational technology research in China, reveals the research trends, hotspots and trends in the field of educational technology research in China in the past six years, hoping to provide future educational technology research reference.

Keywords Educational technology · CSSCI source journals · Research progress · Visual analysis

1 Introduction

Exploring and in-depth analysis of the latest research results in the field of intellectual property rights of related disciplines can more accurately grasp the focus of tracking the current discipline development research, grasp the latest research trends of discipline evolution, predict the research direction of discipline development and further solve key research issues. This article uses CNKI statistical database as the main information source, based on the research and analysis method of academic literature quantitative measurement, and uses the data analysis function of Cite Space to conduct a visual quantitative statistical analysis of a total of 6,740 CSSCI journal articles from 2015 to 2020. The research content includes related knowledge structure diagrams such as the academic collaboration source network of the authors of the

Y. Zhao · C. Li · Y. Chen · S. Li (✉)
Hubei Normal University, Huangshi 435001, Hubei, China
e-mail: 545671848@qq.com

X. Tian
Huangshi Education Information Development Center, Huangshi 435001, Hubei, China

paper, the co-occurrence of keyword knowledge, and visual data analysis, presenting important experts and scholars, research field themes, research hotspots, and main research and practical methods. To provide reference for educational researchers and educators engaged in educational technology.

2 Data Sources and Research Methods

2.1 Sample Collection

The review of CSSCI journals is carried out almost every two years. The purpose is to screen out the most influential and authoritative professional academic journals in the past two years as its selected source journals. In order to fully ensure the authenticity and accuracy of the research sample data of the paper, the sample of this study uses "CSSCI source journals" as the retrieval standard of the CNKI database, and the period from January 2015 to December 2020 is selected for the period of January 2015 to December 2020. All articles are data sources (see Table 1 for details). From Table 1, it can be found that the seven journals in this field, "Research on Audio-visual Education", "Modern Educational Technology", "Research on Open Education", "Research on Modern Distance Education", "Journal of Distance Education" and "China Audio-visual Education" From 2015 to 2020, it has been selected as the source journal of CSSCI. The journal Comparative Education Research was listed as the source journal of CSSCI in 2015 and 2019 respectively. The journal Modern Distance Education was not selected as the source journal of CSSCI in 2019. The replacement of "China Distance Education" journals shows that the quality of journals has changed.

Table 1 12015–2020 CSSCI source journals

Year	CSSCI source journal name	Quantity
2015–2016	"International and Comparative Education", "e-Education Research", "Open Education Research", "Modern Educational Technology", "Modern Distance Education Research", "Modern Distance Education", "Journal of Distance Education", "China Educational Technology"	8
2017–2018	"e-Education Research", "Open Education Research", "Modern Educational Technology", "Modern Distance Education Research", "Modern Distance Education", "Journal of Distance Education", "China Educational Technology"	7
2019–2020	"International and Comparative Education", "e-Education Research", "Open Education Research", "Modern Educational Technology", "Modern Distance Education Research", "Journal of Distance Education", "China Educational Technology", "Distance Education in China"	8

Comparing the CSSCI journals every two years in the statistical Table 1, all the qualified journal articles published from January 2015 to December 2020. Perform secondary classification and reclassification screening to eliminate unqualified professional journal article information such as the first language, news, advertisements, notices and the latest developments of professional journals of similar publications. Finally, 6740 qualified journal article samples are obtained.

2.2   Research Methods

The main content of the literature research measurement method used in this study includes:

Using the literature data measurement method to carry out the statistical publication volume, publishing organization and all published articles of all the articles published in the national education technology research field selected in CSSCI journals from 2015 to 2020. The author's relevant statistical data analysis, summarizes the current research topic hotspots and development status of the current education information technology research field.

Through the citespace5.7.r2 software, using the LLR clustering method, clustering analysis of 6740 research samples, analysis and summary China's higher education science and technology literature research in the field of literature research development in the past six years.

Using content analysis methods to conduct in-depth analysis and discussion of the highly cited literature in the sample, and combine the above statistics and clustering results to summarize China's education Research progress, trends and research priorities in the field of technology.

3   Analysis of Research Results

3.1   Analysis of the Amount of Papers Issued

The documents in the statistical sample are classified according to the statistical year. The number of statistical documents and their changes are shown in Fig. 1. It can be clearly seen from Fig. 1 below that the average number of authors published in international authoritative academic journals in the field of higher education science and technology research in my country during the same period of 2015–2018 has shown a continuous decreasing trend year by year, and reached the lowest in the same period of 2018. Since 2018–2019, the annual number of articles published by various core academic journals in the fields of education and technical sciences has continued to rise and decline, and the number of articles published is basically the same as the amount of articles published in 2015 and 2016.

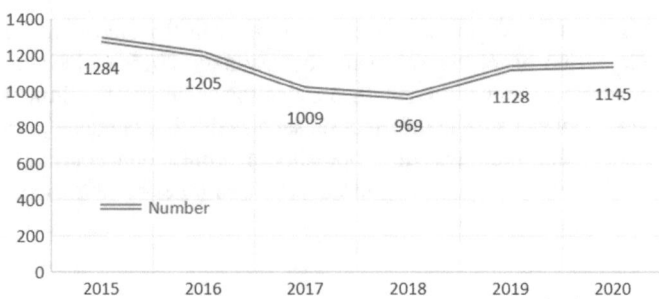

Fig. 1 The annual sample size change graph

In the field of education science and technology, a total of 8 titles of the new version of CSSCI source core journals were selected from 2015 to 2016 in September, a total of 7 titles of new version of CSSCI source core journals were selected from 2017 to September 2018, and a total of 7 titles of new CSSCI source core journals were selected from 2019 to September 2020. There are a total of 8 journals. From this we can see that during the 2017–2018 period, due to the fact that the field of education science and technology was a part of the smaller number of core academic journals published each year, the reason is also likely to be mainly due to the fact that CSSCI has removed the "Comparative Education Research" is a professional monthly academic journal.

At the same time, due to the fact that the core journals of the CSSCI source in the field of education science and technology in recent years have paid more attention to the publication quality of academic papers and the depth of research content [1], the average publication length of academic papers published by each journal. The total length has begun to show an upward and downward trend year by year, resulting in a significant decrease in the volume of C publications in the field of educational technology in 2016–2018.

3.2 Distribution of Academic Research Institutions

By counting the number of articles issued by various institutions, we can explore the core institutions in the field of educational technology research in my country. The analysis results are shown in Table 2. The results showed that the publishing organizations of 6,740 articles were mainly various universities and some primary and secondary schools, and some education research centers were also among them. From the comparison of Table 2, it can be clearly seen that the scientific research institutions with the top or the bottom in the total number of issued documents are named Beijing Normal University, and the research institute system under the same normal university is merged to obtain the documents issued in the same year. The number of scientific research institutions ranked in the top 10, a total of 8 normal

Table 2 Statistics on the number of articles by institutions in ET from 2015 to 2020

Number	Frequency	Centrality	Institution name
1	218	0.17	School of Educational Information Technology, Central China Normal University
2	191	0.12	South China Normal University Institute of Information Technology
3	185	0.06	East China Normal University Department of Information Technology
4	168	0.32	Beijing Normal University Faculty of Education
5	142	0.04	Nanjing Normal University Educational Science
6	122	0.05	Shaanxi Normal University Institute of Education
7	117	0.08	Beijing Normal University Institute of Technology
8	107	0.06	Northwest Normal University Institute of Technology
9	92	0.02	Beijing Normal University Institute of International and Comparative Education
10	91	0.07	National Digital Learning Engineering Research Center, Huazhong Normal University

undergraduate universities, 2 comprehensive undergraduate universities, indicating that most of the current domestic higher education vocational and technical colleges and research in my country are normal undergraduate universities, and there are a few comprehensive undergraduate universities. The University's Educational Technology Research Department has also made important contributions. The geographic locations of all the top ten institutions in the number of publications cover all regions in the east, west, south, and north of my country, indicating that the research field of educational technology in China is evenly distributed. Among them, there are three universities in Jiangsu Province, which are significantly higher than those in other provinces, indicating that there are more frequent teacher-student interactions in this field and more extensive research in this area.

Further analysis found that the top three institutions in the field of educational technology in my country belong to the School of Educational Information Technology, followed by the Department of Education, the School of Educational Science, and the School of Educational Technology. The disciplines are consistent. It is also found by consulting the data that the predecessors of the two non-normal universities in the top ten are normal universities. The predecessor of Southwest University was the famous "East Sichuan Normal School", and then for the further development of the school, it was divided into Southwest Normal University and Southwest Agricultural University. Due to the gradual increase in the strength of universities, it developed into the present Southwest University. It's known as "Sanjiang Normal School", after the merger of institutions, it became the present Jiangnan University. Therefore, it can be concluded that the teachers and students of various normal universities and educational departments have made major contributions to the research field of educational technology in my country.

3.3   Distribution of High-Output and High-Influencers

To a large extent, the core authors influence the research direction and trend of this subject field [2]. The authors of 6,740 samples are counted, and the Price formula is used as the quantitative standard for selecting core authors, where Nm is the number of articles published by the most productive authors, and the minimum standard Ni for the number of articles published by selected core authors is calculated, as in the formula (1) shown.

$$N_i = 0.749 \sqrt{Nm} \tag{1}$$

In this research, Professor Gu Xiaoqing of East China Normal University is the most prolific author, and 70 articles were published in C during 2015–2020. Substituting $N_m = 70$ into the formula (1) to calculate, $N_i = 6.27$, so the number of articles published by the selected core authors should be greater than or equal to 7 articles. The number of articles published by all authors in the statistical sample is 397 authors, of which 144 authors have published no less than 7 articles, accounting for 36.27%. Therefore, there are more than one-third of the authors in the field of educational technology research in my country from 2015 to 2020. The number of articles published is more than six, indicating that the core author's research is relatively stable. Table 3 shows the statistics of the top 20 authors who have published articles.

Taking the author as the node, using Cite Space to obtain the author's co-occurrence network map in the field of educational technology research in my country is shown in Fig. 2. A total of 397 nodes, 486 connections, and a density of 0.0062 are analyzed. In the knowledge graph of Cite Space, the more frequently the nodes appear, the larger the nodes. The color and thickness in the inner circle of the nodes indicate the frequency of occurrence in different time periods. The connection between the nodes represents the co-occurrence relationship, and its thickness indicates the strength of the co-occurrence. The color corresponds to the time when the node first co-occurs. The change in the color of the annual ring from the cold blue to the warm

Table 3 Statistics on the number of articles published in the field of education technology in China

Number	Author	Frequency	Centrality
1	Gu Xiaoqing	70	0.05
2	Zhu Zhiting	63	0.03
3	Yang Xianmin	61	0.12
4	Huang Ronghuai	55	0.2
5	Ren Youqun	54	0.12
6	Chen Li	53	0.06
7	Zhao Wei	50	0
8	Zheng Xudong	48	0.07
9	Liu Qingtang	48	0.07
10	Yu Shengquan	47	0.04

Fig. 2 A network map of authors' cooperation in the field of educational technology research in China

red color represents the change in time from early to recent [3]. It can be seen from the figure that the nodes of Yang Xianmin, Huang Ronghuai, Ren Youqun, Gu Xiaoqing, Zhu Zhiting, Li Yi, Yu Shengquan and others are larger. Among them, the red circles of Huang Ronghuai, Yang Xianmin, Wu Yu, and Ren Youqun nodes are thicker than other nodes, indicating that these nodes. The author has been most active in the field of educational technology research in the past two years.

Centrality is usually an important index used to measure the importance of a node in a wireless network. Generally, the higher the centralization index of the information intermediary enterprise of the key business node, the greater the important bridge role played in the entire information network center. The nodes marked with purple circles in the knowledge graph of Cite Space indicate that these nodes have greater centrality. As can be seen from Fig. 2, the six nodes of Huang Ronghuai, Yang Xianmin, Ren Youqun, Zhao Chengling, Wan Liyong, and Du Jing are marked with purple circles, indicating that the central role of these authors from 2015 to 2020 is greater. Other authors have more cooperation. The two color lines in the color of the knowledge thematic map represent the year of creation, the strength of the color line between the two small nodes represents a partnership between the two creators, and the strength of the color line can represent the author's cooperation. From this, the cooperative relationship network of the core authors can be analyzed. In this study, the colors from purple, blue, cyan, green, yellow, and red represent the years 2015–2020. As can be seen from the figure, Zhao Wei and Jiang Qiang collaborated more closely in 2015 and 2016, and together with other authors formed a close network map dominated by purple and blue; Liu Qingtang, Wu Linjing and Zhang Si formed a yellow The main triangle network; there is a darker red connection between the two nodes of Wang Yining and Zhang Hai, indicating that the two authors have more cooperation in 2020; Chen Li, Zheng Qinhua, and Zeng Haijun have also formed a red-based connection. Close cooperation network.

3.4 Research Hotspots and Topics

Keyword co-occurrence analysis. The frequency of keywords can be regarded as a theoretical basis for judging whether a research target has become a research hotspot in the research field at a certain level. Centrality is also another important indicator for us to measure research hotspots. Centrality indicates that there are different types of interconnected hubs between keywords and topics. By referring to these two indicators, we can judge the current and future research development hotspots. Taking keywords as nodes, using all sample data of Cite Space software for comprehensive analysis, the resulting word frequency statistics table and corresponding keyword co-occurrence maps are shown in Table 4 and Fig. 3.

The keywords are merged semantically, such as "MOOC" and "MOOC" are merged into "MOOC", and "big data" and "education big data" are merged into "big data", etc. It can be seen from Table 4: First, "MOOC", "Educational Information", "Artificial Intelligence", "Flip Class", and "Learning Analysis" have received

Table 4 2015–2020 keyword statistics in domestic educational technology field

Number	Frequency	Centrality	Keyword
1	226	0.07	MOOC
2	226	0.05	Educational information
3	216	0.06	Artificial intelligence
4	195	0.03	Flipped classroom
5	178	0.09	Learning analysis
6	168	0.08	Big data
7	145	0.06	Maker education
8	143	0.01	United States
9	137	0.06	Wisdom education
10	125	0.07	Information technology

Fig. 3 Co-occurrence map of keywords in the research field of educational technology in China

more attention, followed by "Big Data", "Maker Education", "America", "Smart Education", "Information Technology", "Online Teaching". The hot content of the research can be summarized into the following important aspects: First, in the field of online intelligent education, MOOC, smart education, online learning and in-depth interaction are paid more attention in China; second, in terms of education technology, The research fields of domestic educational technology disciplines pay more attention to artificial intelligence technology, learning analysis technology, information technology and teaching design; third, the choice of teaching methods has flipped the attention of education models such as classroom and maker education in my country. At most; finally, in terms of research objects, domestic researches on higher education are paid more attention. It is noteworthy that the word frequency of the keyword "United States" ranks eighth, indicating that my country's educational technology research field pays a lot of attention to American educational technology research.

Further statistics on the centrality ranking of keywords, the two keywords "learning analysis" and "Internet+" have the highest centrality, both at 0.09, followed by "big data", "online learning", "higher education" and "teaching" The centrality of the four keywords of "Design" is 0.08, and the centrality of the keywords of "MOOC" and "Information Technology" are both 0.07. It shows that the two keywords "learning analysis" and "Internet+" are at the center of the entire network structure, and they have a strong role in the field of educational technology research in my country. Combining the frequency of each keyword, the centrality and frequency of the keywords of "big data", "MOOC" and "information technology" are in the forefront, indicating that the application of big data for learning and analysis is now a hot spot in our country. Comparing the frequency ranking and centrality ranking of keywords, it is found that the centrality of the four keywords "Internet+", "online learning", "higher education" and "teaching design" is very high, but the frequency is after ten; The keywords of "Classroom" and "America" have high frequency, but the centrality is very low. This shows that some keywords such as "flipped classroom" and "America" with high frequency and low centrality have received widespread attention from scholars. However, due to their insufficient research depth, further research is needed; while some are highly central, Although the low-frequency "Internet + ", "online learning", "higher education" and "teaching design" have not yet become research hotspots, they can be found to be potential research directions and are worthy of in-depth research.

Keyword cluster analysis. Using LLR (Log Likelihood Ratio) algorithm for clustering, 5 topics are finally counted, namely #0 MOOC, #1 distance education, #2 online education, #3 artificial intelligence and #4 maker space. The main purpose of the time horizon parallel view is to focus on how to outline the different time span relationships between the clustered documents and each cluster, and the different time spans of all documents in the cluster. Each cluster node is arranged side by side on the same time horizontal line in the order of the length of the document time, so each document in the clustered document appears as if the string is arranged on the same horizontal time line. Shows a different timeline and span structure that outlines the cluster document and each cluster.

#0 MOOC clustering started in 2015 and achieved the most results in 2016 and 2017. The amount of literature in 2019 and 2020 is gradually decreasing. Related keywords include online learning and flipped classrooms that appeared in 2015. In 2016, they included augmented reality, smart classrooms and virtual reality. In 2017, learning investment, learning effects, and meta-analysis were increased. Related keywords in 2018 include learning mode and personality. In 2019, we will pay more attention to cognitive network analysis and online collaborative learning. The keywords in 2020 are cognitive investment, quality standards, and large-scale epidemics. In China, Professor Zeng Mingxing [4] has had a great influence on MOOC research. His published articles on flipped classroom teaching in the MOOC environment have been cited 577 times and downloaded more than 19,000 times. This article focuses on exploring how to apply the flipped classroom teaching model in the MOOC environment, and innovates three teaching models that integrate flipped classrooms and MOOCs, which has played a strong role in the development of MOOC in my country.

#1 Distance education clustering also started in 2015, with the most results in 2016, and subsequent researches are in a stable state. The keywords that appeared in 2015 were MOOC, higher education, and the United States. The keywords that appeared in 2016 were Australia, education reform and educational reform. The main keywords in 2017 were ICT, twenty-first century skills and Internet + education. The keywords include modern distance education, educational research and educational modernization. The keywords in 2019 are mainly Canada, and the Belt and Road Initiative. The keywords in 2020 include discipline construction, learning achievement certification, and China's education modernization. From this key word, we can clearly see the progress and historical development of my country's distance education research. The research focuses are as follows: ① Based on the distance education research of my country's Open University, for example, Professor Lv Jingjing has discussed the Open University of my country in depth. The new concept and connotation of mixed classroom teaching in China [5]; ② Research on distance education subject in the context of higher education reform; ③ Research on the use of distance education resources, such as Liu Jia and others studied "Live + education" in The new form and value in learning [6]; ④ The practice and application of modern distance education; ⑤ The development research of educational technology discipline construction.

#2 Online education clustering started to increase in 2017. Related keywords include development path, information literacy, evaluation indicators, online open courses, etc. The iconic documents include Zeng Mingxing and Li Guiping [7] who constructed a deep learning model based on SPOC, and Yang Fang, Zhang Huanrui and others [8] studied the mixed teaching model based on MOOC and Rain Classroom, all of which are researches on online teaching in my country is a big impact.

#3 Artificial intelligence clustering, related research results have increased significantly since 2015, and a certain amount of literature will be maintained every year thereafter. The research focuses from smart education, learning analysis, adaptive learning to education governance in the past two years, learning design, future schools, and extended reality. It can be seen that the research on artificial intelligence in the field of educational technology research in my country has already started from the analysis of learning. From research to education management research, technology applications are becoming more and more mature.

#4 Maker space clustering began in 2015, and the research results for four consecutive years have stabilized, and the popularity has gradually declined in 2019. Key words that appeared in 2015 were information technology, Internet + and maker education. The 2014 Higher Education Edition "American Horizon" research report predicts that students will change from ordinary knowledge acquisition to future creative knowledge [9]. Therefore, the term "maker" has been introduced into educational scientific research, especially in the United States. Yang Xianmin [10] pointed out that the environment, teachers, and education fairness and balance issues that my country faces under the emerging wave of maker education play a role in advancing the research of maker education in my country.

3.5 Research Trend Analysis

The mutation word detection in Cite Space can present the staged characteristics and evolutionary trends of the educational technology research field in my country in the past six years. Keywords that increase sharply in a certain period are called mutation words, which represent the frontiers of research in that period. Using the Citation Burst algorithm, the top ten keywords with mutation strength are shown in Table 5.

It can be seen from Table 5 that domestic research in the field of education technology closely follows the trend of the times. Among the top ten keywords of sudden change intensity, one-half of them are related to the Internet, including "MOOC", "Electronic Schoolbag", "Micro-course", "Maker Space" and "Micro-video". The second is the study of the teaching mode of "Flipped classroom" and "Case Study",

Table 5 2015–2020 keywords statistics with the strongest citation bursts (top 10)

Number	Keyword	Strength	Emergence time
1	MOOC	17.65	2015–2020
2	Flipped classroom	15.15	2015–2020
3	United States	9.05	2015–2020
4	Electronic book bag	6.31	2015–2020
5	Micro course	5.41	2015–2020
6	Maker space	5.06	2015–2020
7	Micro video	4.51	2015–2020
8	Informal learning	4.2	2015–2020
9	Case study	3.6	2015–2020
10	Primary mathematics	3.6	2015–2020

the study of "Informal learning" and the subject research of "Primary school mathematics". In addition, the emergence of these keywords is from 2015 to 2020, indicating that the Internet's application research in education, new teaching models, learning methods research and basic education research have been hotspots in the field of educational technology research in China in recent years.

4 Research Conclusions

This study visually analyzes all the CSSCI source journal documents in the field of educational technology in China from 2015 to 2020, and draws the following conclusions:

Combining the author's co-occurrence network map and the institutional cooperation co-occurrence map, it is found that the research field of educational technology in China has formed a research situation dominated by universities. There is more cooperation between universities and scholars have also carried out more cross-institutions. The cooperation between the two will help the research progress in the field of education technology in my country.

Combining the keyword co-occurrence network map, specific analysis of frequency and centrality, and discovery of learning analysis, MOOC, online learning, teaching design and other big data educational applications, it is a hot spot in the field of educational technology research in my country.

Combining the timeline view of keyword clustering and emergent words, discovering the research of distance education, maker space, and new educational models is the current research trend in the field of educational technology in my country.

Based on the above research results, this research believes that domestic educational technology research needs to be improved in the following aspects: First, theoretical research should be strengthened, theoretical innovation should be realized from an interdisciplinary perspective, and cognitive science research can be conducted from human psychology and behavior. Secondly, we should pay attention to the improvement of students' learning effects, apply big data, learning analysis and other network technologies to improve the quality of talent training in terms of promoting personalized learning and core skills training. Finally, Chinese scholars in the field of educational technology should strengthen international Exchanges and cooperation, strengthen knowledge sharing and experience exchanges with countries around the world, and help the common development of global education.

Acknowledgements Research and practice of information literacy of normal students under the action plan of teacher education revitalization: TED01 22-2019-10, Research on classroom quality evaluation of educational technology specialty under the background of normal specialty certification: 20210148.

References

1. Hu XY, Li LJ (2013) Visual analysis of research trends in the field of educational technology in my country. Mod Educ Technol 23(09):17–20+28
2. Zhao YX, Wang X, Jin K, Zhang P (2019) Research path and trend analysis of learning analysis technology in the domestic big data environment. J Mod Educ Technol 29(08):34–40
3. Chen C (2006) Cite space II: detecting and visualizing emerging trends and transient patterns in scientific literature. J Am Soc Inf Sci Technol 57(3):359–377
4. Zeng MX, Zhou QP, Cai GM, Wang XB, Chen SP, Huang Y, Dong JF (2015) Research on MOOC-based flipped classroom teaching mode. J China Audio-Visual Educ 2015(04):102–108
5. Lv JJ (2015) Research on the new connotation of open university blended teaching——Based on the enlightenment of SPOC. J Distance Educ 33(03):72–81
6. Liu J (2017) "Live + Education": the new form and value of "Internet + Learning." J Distance Educ 35(01):52–59
7. Zeng MX, Li GP, Zhou QP, Qin ZY, Xu HZ, Zhang BL, Huang Y, Guo X (2015) From MOOC to SPOC: a deep learning model construction. J China Audio-Visual Educ 2015(11):28–34+53
8. Yang F, Zhang HR, Zhang WX (2017) A preliminary study on the hybrid teaching based on MOOC and rain classroom——Taking the teaching practice of "Life English Listening and Speaking" MOOC and rain classroom as an example. J Mod Educ Technol 27(05):33–39
9. Zhu ZT, Sun YY (2015) Maker education: innovative education practice field enable by information technology. J China Audio-Visual Educ 2015(01):14–21
10. Yang XM, Li JH (2015) The value potential of maker education and its controversy. J Mod Distance Educ Res 2015(02):23–34

AI Technologies and Innovations

A Two-Stage NER Method for Outstanding Papers in MCM

Shuting Li◉ **and Lu Han**◉

Abstract In this paper, we introduce a two-stage NER method to recognize the entities about papers occurring in MCM. Firstly, we pick out the NN-labeled and the NP-labeled words from the pre-labeled dataset to generate the preprocessed word frequency table and labeled the method-related words with "MET". Secondly, $[-3, 3]$ windows are applied to identify the capital words and words occurring in the above frequency table around the MET-labeled words which we recognize in step 1 and refine the tags with "T-MET". After the above two steps, we use these recognized entities with "T-MET" tags to construct the document-by-method matrix for further clustering. We then have an exploratory data analysis of our dataset.

Keywords Named entity recognition · Vector space · Jaccard clustering

1 Introduction

Named entity recognition (NER) is a significant subtask for natural language processing. NER task aims to recognize some certain words' categories which are equipped with several common characteristics, in order to analyze sentences or texts more effectively.

Mathematical Contest in Modeling (MCM) is the only international mathematical modeling competition and the most influential one in the world, sponsored by the American Association for Mathematics and Applications. The questions in the contest involve many fields such as economy, management, environment, resources, ecology, medicine, security and so on. The types of awards include Outstanding Winner, Finalist, Meritorious Winner, Honorable Winner, etc., among which the Outstanding Winner is the highest award.

In the process of modeling, participants usually integrate multiple methods to obtain a more scientific and reasonable mathematical model, rather than only use a single method. However, contests in different years have different themes and fields,

S. Li · L. Han (✉)
Central University of Finance and Economics, Beijing 100081, China
e-mail: hanluivy@126.com

so the methods used will naturally vary among diverse years. However, among these methods, there are a number of approaches that are widely applicable and are often used in different types of modeling.

So in this paper, exploiting a two-stage NER approaches, we intend to explore the methods, algorithms, models, etc., which are more widely applied among those outstanding papers in the modeling competition, so as to provide some reference for the study of modeling methods in the future.

The main contributions of this paper are as follows:

1. We apply a preprocessed world frequency table for the first stage NER.
2. We explore a two-stage NER approach for identifying the most several applicable methods.
3. Considering that an outstanding paper in MCM usually includes more than one method or algorithm rather than single one, we construct a document-by-method matrix with binary value for each element.

The rest of our study is organized as follows: literature review which introduces the related work is established in Sect. 2. Section 3 consists of all our main idea and methods. Section 4 is writing to summarize our conclusion and to introduce our shortage and the future work.

2   Literature Review

The traditional named entity recognition (NER) methods emphasize entities that frequently occur in all fields such as person, location, organization and so on.

Jung [1] develops a method based Named Entity Recognition System for these three types of entities with tweets [1]. Liu and Zhou [2] propose two methods for extracting entities from the unstructured text with clustering using tweets [2, 3]. Van Cuong Tran et al. [4] introduce a NER approach for the structured data in free text using CRF technique for person, location, organization entities [4]. They propose a novel hierarchical two-stage NER method for tweets. At the first-stage, the tweets are pre-tagged based on linear CRF model; in the second period, tweets with similar contents are put in the clusters. And then, the enhanced CRF model can refine the tag of each tweet in each cluster using cluster-level information, traditional features and features obtained from the clustering of pre-tagged results.

Recently, many researchers are committed to develop named entity recognition methods to a wider range of fields and two-stage NER seems to be more popular [5]. Ritika and Satwinder [6] use text similarity to construct vector space, and then extract entities [6]. Mu et al. [7] propose a two-stage deep learning named entity recognition and relationship extraction method based on medical text. This NER approach is performed by combining bidirectional long-term short-term memory (BI-LSTM) and conditional random field (CRF) [7]. Goyal et al. [8] sum up the latest researches in named entity recognition and classification techniques [8].

3 Our Method

3.1 Descriptive Statistical Analysis

We collected the Conclusion parts of papers of outstanding winners from the 2010 to 2017 in MCM. At the beginning, we have a descriptive statistical analysis of it. Through descriptive statistical analysis, we obtained that this dataset contains 8,459 tokens and 2,403 different word types, with an average occurrence of 3.5 times for each word. We also count word frequency and find that the majority of the most frequent thirty words in our dataset are "the", "of", "to", "and", "we" and so on (Table 1) which provide quite a little information for our research and are often called "stop words". Obviously, it's necessary for us to eliminate them from the datasets for further study.

Then, we show the Zipf's law picture in Fig. 1. We intend to introduce Zipf's law briefly. Zipf's law can be depicted as, where stands for the frequency of a word and rank r is its position in the vocabulary list. In other words, Zipf's law proposes that there are a few very common words, a middling number of medium frequency words and many low frequency words. The graph shows rank on the x-axis versus frequency on the y-axis. The points correspond to the ranks and frequencies of the words in our dataset. The line represents the relationship between rank and frequency. Also, the graph in Fig. 1 shows that our text is fit for Zipf's law roughly.

3.2 Pre-processing

It's worth noticing that in Table 1, the majority of the words are what we call "stop words", and only a few words like "model", "algorithm", etc. provide information that we consider as important. As a result, the emphasis of pre-processing is stop words removing.

At this stage, we lower the words at the beginning of all sentences, stem the words which occur in different forms, remove all the punctuations, the most significantly, remove the stop words out of the dataset.

It's remarkable that we remain the capitalized words except for the beginning of all sentences, because capitalized words that appear in the middle of a sentence are usually those proper nouns, methods, or terms. Besides, we lower those words like "Model", "Algorithm", "Method", etc. which always stand for method, approaches used in the process of modeling.

Table 1 Word frequency table

Index	Word	Frequency	The proportion
1	the	662	7.825984159
2	of	302	3.570161958
3	to	259	3.061827639
4	and	256	3.026362454
5	we	216	2.553493321
6	a	198	2.340702211
7	in	134	1.584111597
8	is	118	1.394963944
9	model	118	1.394963944
10	for	86	1.016668637
11	that	81	0.957559995
12	on	68	0.803877527
13	by	68	0.803877527
14	water	61	0.721125429
15	our	61	0.721125429
16	with	57	0.673838515
17	this	56	0.662016787
18	as	51	0.602908145
19	are	47	0.555621232
20	which	43	0.508334318
21	based	39	0.461047405
22	two	36	0.42558222
23	from	31	0.366473578
24	algorithm	29	0.342830122
25	an	29	0.342830122
26	find	27	0.319186665
27	three	26	0.307364937
28	be	26	0.307364937
29	can	26	0.307364937
30	each	26	0.307364937

3.3 POS Labeling

After pre-processing our dataset, we have POS tags for words. The labels are "NN", "NP", "JJ", "DT", "VV" and so on. And then, based on the data labels, we draw a pie chart that revolves the composition ratio of the tags, as is shown in Fig. 2.

Obviously, the majority of words are labeled by "NN" and "NP"-77% approximately. As we all know, the tag "NP" are usually used when proper nouns, methods, or

Fig. 1 Zipf's law

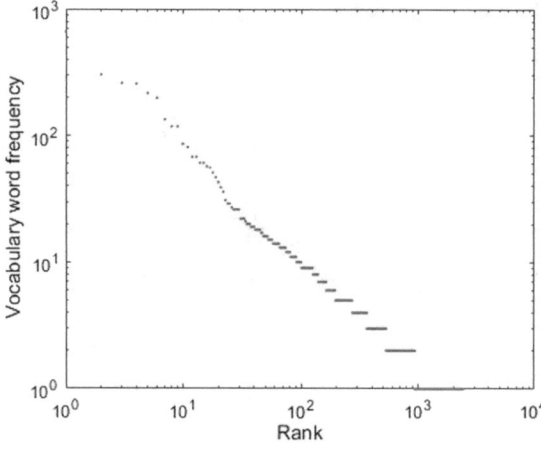

Fig. 2 The proportion of each tags

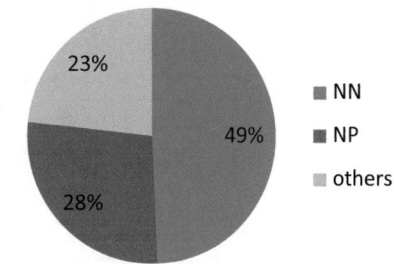

terms occur in a sentence. Meanwhile, the tag "NN" also stand for certain methods sometimes because some authors may prefer to use lowercase words for specific methods instead of capitalizing them. Consequently, these labels are chosen to be a feature for our following work.

After preprocessing and POS labeling, we pick out the NN and NP labeled words and re-counted the word frequency of the data set, and Table 2 was obtained. It's obvious that the most frequent words are more useful than the one before preprocessing.

3.4 Two-Stage NER

The First Stage: Pre-labeling. After we completed the POS labeling we introduce in the Sect. 3.3, the core of our research next is to searching for those widely applicable methods, algorithms, etc. among the "NN" or "NP" labeled words.

Firstly, based on NN-labeled words, we relabeled the words like model, method, algorithm, analysis, etc. which often co-occurrence with well-established method with "MET".

Table 2 Word frequency table after preprocessing

Index	Word	Frequency	The proportion
1	Model	51	11.48648649
2	Analysis	15	3.378378378
3	Method	8	1.801801802
4	Algorithm	6	1.351351351
5	Evaluation	5	1.126126126
6	Hierarchy	4	0.900900901
7	Linear	4	0.900900901
8	Fuzzy	4	0.900900901
9	Comprehensive	4	0.900900901
10	Pagerank	4	0.900900901
11	Network	3	0.675675676
12	Method	3	0.675675676
13	Grey	3	0.675675676
14	Programming	3	0.675675676
15	Component	3	0.675675676
16	Differential	3	0.675675676
17	Process	3	0.675675676
18	Weighted	3	0.675675676
19	ahp	2	0.45045045
20	Optimization	2	0.45045045
21	Genetic	2	0.45045045
22	Cluster	2	0.45045045
23	Neural	2	0.45045045
24	Classification	2	0.45045045
25	Geometric	2	0.45045045
26	Regression	2	0.45045045
27	Statistical	2	0.45045045
28	Sensitivity	2	0.45045045
29	Dynamic	2	0.45045045

The Second Stage: Searching for the Often-Used Methods. Secondly, according to those MET-labeled words, we adopt [-3, 3] windows model to search for the words around these recognized entities. If the word to the left and right side of the central word has a capital letter or has at least a word occurring in Table 2 among them, then the word before or after the central word, along with itself, is re-marked as "T-MET". Otherwise, we will leave it out of our consideration.

The process of the whole two stages is shown in Fig. 3 on the next page:

According to our two-stage method, we recognize the target entities-the methods used in modeling, after that we compare our NER outcome with manually-annotated

Fig. 3 The working
procedure of our method

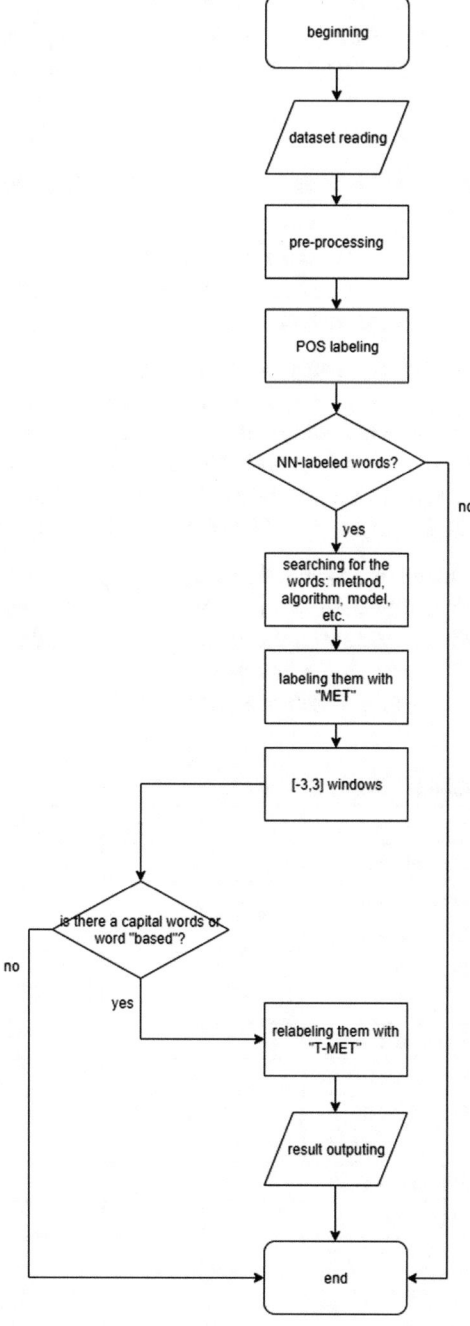

one using Precision. Supposing that manually-annotated approach can identify all target entities, i.e. an annotated dataset by human being is 100% accuracy rate, we finally get Precision 77.8%, a good performance relatively.

3.5 Two-Stage NER Method Applying

The purpose of our study is to statistic the most frequent methods which we usually exploit in modeling contests. So, after introduce our method, we are going to apply it to the papers completed in MCM recent years.

Statistical Analysis. After NER process, we listed all the identified entities and counted several commonly used modeling methods with the most frequent occurrences. Some statistical results are shown in Table 3.

The most frequently exploited methods are Analytic Hierarchy Process (AHP), Principal Component Analysis (PCA), Fuzzy Comprehensive/Synthetic Evaluation (FCE/FSE), Linear Model, Grey Model and so on.

Clustering-An Exploratory Analysis. To conduct the exploratory analysis, we try clustering the comments with different top-down hierarchical clustering methods which measure similarity between various documents with different distance metrics [9]. Because the document-by method matrix in Table 3 is sparse, and the value with 1 in the matrix is more significant than the value with 0, in other words, the

Table 3 Statistical results

	AHP	PCA	...	FCE/FSE	Grey model	SUM
Document 1	1					1
Document 2						1
Document 3						0
Document 4	1				1	3
...			...			...
Document 38						0
Document 39						0
Document 40		1				2
Document 41						0
Document 42						0
Document 43						1
Document 44						0
Document 45						0
Document 46						0
SUM	6	5		6	5	39

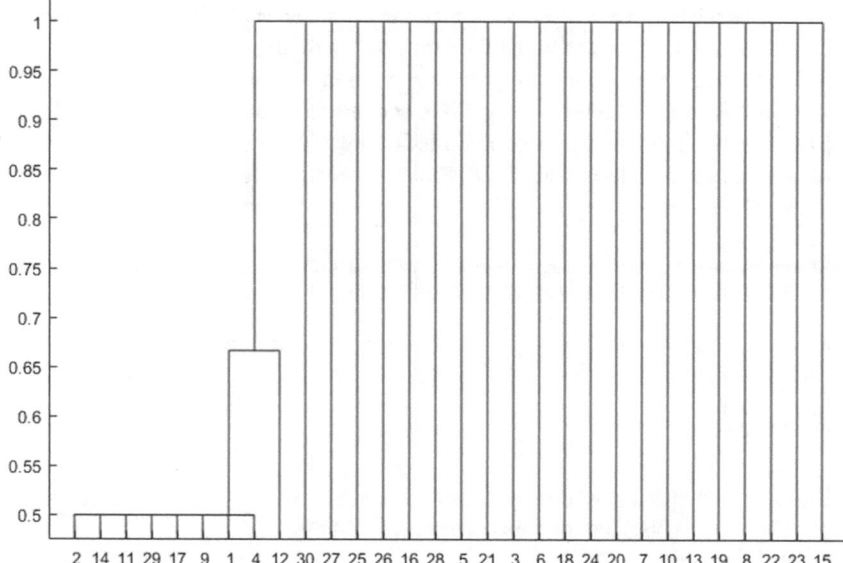

Fig. 4 Jaccard similarity

attributes of our research are all asymmetric, we adopt Jaccard method which is used for special purpose for asymmetric attributes to measure the similarity between different documents. The result is shown in Fig. 4.

It can be seen in Fig. 4 that there are 8 documents sharing 50% similarity, that is to say, these 8 documents share most common methods.

If we statistic all the recognized entities, we can obtain a more accurate matrix and a more precise result of exploratory clusters. And then, we can learn the most similar papers together to find the common thread. The performance will be better if documents share 2 or more methods in common.

4 Conclusion and Future Work

We produce a two-stage NER method for papers published in MCM, identifying the method-related entities based on the capitalized words and the preprocessed word frequency table which reflect the most frequent words. Furthermore, exploiting the above two-stage NER method, we explore the similarity of documents by structuring the document-by-method matrix. We use the vectors provided by the matrix space as an input for top-down hierarchical clustering algorithm.

The main contributions of this paper are as follows: 1. a preprocessed word frequency table for the second NER stage; 2. a successful NER method for papers published in MCM; 3. the construction of the vector space according to document-by-method matrix about the extracted methods.

In the future, we are going to improve our research from following aspects: 1. take more available and useful word feature into account to recognize as more target entities as possible; 2. search for more credible features more than word frequency table to achieve a better performance; 3. explore a more effective method for integrating all features together.

Acknowledgements The work was supported by the Education Reform Fund of Central University of Finance and Economics under Grant No. 2020XYJG26.

References

1. Jung JJ (2012) Online named entity recognition method for microtexts in social networking services: a case study of twitter. Expert Syst Appl 39(9):8066–8070
2. Liu X, Zhou M (2013) Two-stage NER for tweets with clustering. Inf Process Manag 49(1):264–273
3. Xiaohua L, Furu W, Shaodian Z, Ming Z (2013) Named entity recognition for tweets. ACM Trans Intell Syst Technol (TIST) 4(1)
4. Tran VC et al (2017) A combination of active learning and self-learning for named entity recognition on twitter using conditional random fields. Knowl-Based Syst 132:179–187
5. Erjing C, Enbo J (2017) Review of studies on text similarity measures. Data Anal Knowl Discov 1(06):1–11
6. Ritika S, Satwinder S (2020) Text similarity measures in news articles by vector space model using NLP. J Inst Eng (India): Ser B (Prepublish)
7. Mu X, Wang W, Xu A (2020) Incorporating token-level dictionary feature into neural model for named entity recognition. Neurocomputing 375
8. Goyal A, Gupta V, Kumar M (2018) Recent named entity recognition and classification techniques: a systematic review. Comput Sci Rev 29:21–43
9. Han L, Su Z, Lin J (2020) A hybrid KNN algorithm with Sugeno measure for the personal credit reference system in China. J Intell Fuzzy Syst 39:6993–7004

Comprehensibility Analysis and Assessment of Academic Texts

José Medardo Tapia-Téllez⬤, Aurelio López-López⬤, and Samuel González-López⬤

Abstract Written communication can be affected by low comprehensibility of documents. In this work, we present a computational-linguistic analysis, evaluation and feedback for academic documents based on three comprehensibility measures: Connectivity, Dispersion, and Comprehensibility Burden. These measures are explored and validated in a document collection of undergraduate and graduate levels, ranging from technician to doctoral. This preliminary exploration also provides parameters to assess new texts, i.e. writings in progress (drafts). A visualization feedback during the assessment process is proposed and illustrated, aimed to understand and resolve detected deficiencies. Reported results set the basis to reach an online comprehensibility assessment application.

Keywords e-Learning · Natural language processing · Comprehensibility · Automatic writing evaluation · Academic texts

1 Introduction

Communication in science is based on the exchange of ideas through academic texts. However, this communication can be affected due to the low comprehensibility of texts. Likewise, creating these texts is not an easy task for authors since they face several obstacles, which could create a vague text or, in the worst-case scenario, the complete abandonment of the task. To avoid this, authors receive support from

J. M. Tapia-Téllez · A. López-López (✉)
Instituto Nacional de Astrofísica Óptica y Electrónica (INAOE), Puebla, México
e-mail: allopez@inaoep.mx

S. González-López
Universidad Tecnológica de Nogales, Sonora, México
e-mail: sgonzalez@utnogales.edu.mx

reviewers and editors, but even they could benefit from receiving academic texts of a higher quality; and this would allow to focus more on research content rather than on writing issues.

The above mentioned problems are in essence of language, so is through natural language processing and analysis that can be tackled. Low comprehensibility, whether the challenge is simply to assess or, as in our case, improve it, is of great relevance. Utterly related to this are the tools that are needed in order to tackle these challenges. This will only be possible through research such as those in [1, 2]. So the research points to computational-linguistic techniques that contribute for more comprehensible texts with high quality.

Our work aims to analyse and assess comprehensibility based on previously proposed measures. These measures are explored on a text collection of different academic levels in Spanish language, which provides parameters to assess new texts, possibly writings in progress (drafts), to provide feedback on these text attributes. In consequence, the main contributions are: (a) a process to assess comprehensibility based on three measures validated on an academic text collection, and (b) visualization feedback for the evaluation of new writings (drafts).

The paper is organized as follows. Section 2 details related work to this research. The methodology employed is provided in Sect. 3, along the text representation and comprehensibility measures. Section 4 describes the document collection analysed, followed by results of the comprehensibility analysis of academic text at different levels in Sect. 5. In Sect. 6, new text assessment and feedback are presented and discussed. Conclusions and work in progress are included in Sect. 7.

2   Related Work

This work is focused on the creation of methods and techniques that obtain an evaluation based on comprehensibility measures of academic texts, providing feedback for drafts or writings in progress. This in the line as that of [3], that focus on several features for analysis and assessment.

To evaluate comprehensibility, we have to assess first the difficulty of these texts. A guide and background of methods and techniques related to automatic assessment of reading difficulty can be found in [4], that also provides a review on the state of the art for algorithms for predictive and automatic modeling of text difficulty.

We can even go further and evaluate theses drafts through other modern approaches such as automated essay scoring, which are utilized in [5]. Also, in [6], they assess the comprehensibility of 'raw' machine translation outputs by comparing it to humans, to obtain high-quality translations. However, the comprehension of the text is a complex task. In [7], the authors assess the relationship between text readability levels and systems based on language models. The results indicated a low correlation, which motivates us to work on comprehension and readability tasks.

Once we have an assessment of the difficulty of a text, our goal is help to improve it. In [2], the authors address this issue through a language checker; they semiautomatize the process of an intralingual translation (simplification of medical text) to later compare it against a non-automated, and evaluate if it had effects over the comprehensibility of the text. The results were positive, since the semi-automatized process indeed provided improved comprehensibility texts. Another example can be found in [8], where an scheme for a computational-linguistic evaluation is provided to measure the readability and comprehensibility of a patients information document, this in the context of projects regarding clinical research. Also, in [9], the result of the estimate analysis to text readability for L2 learners, considering the extraction of lexical, syntactic, and traditional characteristics (the number of sentences, average and maximum number of words by sentence, average number of characters by word, and average number of syllables per word) was encouraging. The best performance was the combination of the different features with an accuracy of 0.785.

Our research efforts aims to incorporate a feedback component. In [1], the authors do not properly create this feedback step, however, this research set a basis on how to develop them. Specifically, they developed a qualitative evaluation for eBooks through parameters such as Coverage, Readability, and Comprehensibility Burden (CB). These three measures become of special interest in our research, despite the specific representation they employed did not adapt well to our texts of interest. However, their evaluation did identify specific parts of the document that could be improved. So this work prompted us to formulate a component for user feedback.

The research in [10, 11] was intended to create fake documents by manipulating the comprehensibility of a given document. Three different measures for comprehensibility after manipulation were used in such work: CB, Connectivity and Dispersion. The document representation utilized inspired us to formulate the evaluation of academic texts or draft. Although our idea for using these measures is entirely different, we were able to apply them to academic texts, and through them, create a basis of evaluation values based on thesis and proposals of different academic levels.

3 Applied Methodology

Our evaluation and feedback method was planned in three stages. The first corresponds to the representation of each document in the collection in a unidirectional graph of concepts. The second stage defines the comprehensibility metrics and finally, a methodology for evaluation and feedback was established. Each of the stages of our method is detailed next.

3.1 Document Representation

A. Sequences of Sentences and Concepts. A document d consists of a finite sequence of K paragraphs: $d = (p_1, ..., p_K)$. The kth paragraph is composed of a finite sequence of sentences n_k, in such a way that $p_k = (s_1, ..., s_{n_k})$. The jth sentence of the paragraph p_k consists of a finite sequence of $m_{(j,k)}$ terms or words, such that $s_{j,k} = (w_{(1,j,k)}, ..., w_{m_{j,k},j,k})$.

A word $w_{(i,j,k)}$ or a phrase (sequence of words) in a sentence is defined as a concept c if it belongs to the set of selected regular expression patterns[1], predominant in the language. We identify a concept by extracting the vocabulary of d, tagging by part-of-speech (POS), and extracting the patterns, employing spaCy.[2]

Finally, a finite set of concepts represented by C and a finite sequence X of occurrences is defined as $O = (c_1, ..., c_X)$ where $c_x \in C$. A concept $c' \in C$ is called a *related-concept* of c if it co-occurs in more than one paragraph p_k. A set of related concepts of c is denoted by $R(c)$, each with frequency $f(c, p_k)$, and where each concept has its *key-paragraph* p_{k_c} where the concept c is explained.

B. Concept Graph. A document is represented as a unidirectional graph of concepts: $G = (C, E, w)$, where C is the finite set of concept nodes, $E \subseteq C \times C$ is the finite set of edges between the concepts, and $w : E \to \mathbb{N}$ is a weight function that is assigned to each edge $(c, c') \in E$, in such a way that is the total frequency of the co-occurrence of c and c' along the paragraphs in the document d. Once the representation for the document or text has been built, the comprehensibility measures can be computed.

3.2 Comprehensibility Measures

Comprehensibility Burden (Sequentiality Measure). If a concept c is utilized to explain a related concept c' before c is explained adequately, this adds a comprehensibility burden to the reader and thus reveals a deficiency in sequentiality. The comprehensibility burden is thus obtained as follows.

Significance. Given a document d that contains a concept c, its set of related concepts $R(c)$, a paragraph p_k, and the frequency of c in p_k as $f(c, p_k)$; the significance of c_{p_k} is defined as:

$$\lambda(c, p_k) = f(c, p_k) * |R(c)| \tag{1}$$

Comprehensibility Burden for a Specific Concept. The key paragraph p_{k_c} of c is defined as the first paragraph in d where the significance is maximum, and the comprehensibility burden for c in a respective paragraph p_k is defined as:

$$\psi(c, p_k) = \begin{cases} \lambda(c, p_{k_c}), & \text{if } k < k_c \\ 0, & \text{in other case} \end{cases} \tag{2}$$

Comprehensibility Burden for the Document. Finally, the comprehensibility burden for the entire document d is

---

[1] Set of regular expression patterns: N-N, N-A, N-[A]-Prep-N-[A], N-[A]-Prep-Art-N-[A], N-[A]-Prep-V-[N-[A]], according to [12].

[2] An open source Python library for natural language processing [13].

$$\psi(c) = \sum_{p_k \in D} \psi(c, p_k) \tag{3}$$

Connectivity. A document is easier to comprehend when its concepts are well-connected to each other, thus meaning that a document has to be written in a way that each of its concepts must have a relation with another in order to comprehend it correctly. Based on this, we employ a measure of connectivity, defined as the average strength of concept relations in a document based on the collocation of concepts. This is obtained through a graph-theoretic formulation.

Connectivity for a Specific Concept. Given the graph $G(C, E, w)$ of a document d, concepts $c, c' \in C$, and the connections of the edges for c in E, the connectivity of c is defined as:

$$\theta = \left(\sum_{(c,c') \in E} w(c, c') \right)^{\alpha} \tag{4}$$

Connectivity for the Document. The connectivity for the entire document d is defined as:

$$\text{Connectivity}(d) = \frac{\sum_{c \in C} \theta(c)}{|C|} \tag{5}$$

Dispersion. Focus is an essential part of a well written document, this is why the lack of it may cause an information burden on the reader and the need to remember more conceptual information. A possible way to measure this is through theoretical information formulation as describe next.

Dispersion for a Specific Paragraph. Given a document d, a concept $c \in C$, a paragraph p_k, and the probability of occurrence of a concept c in p_k denoted as $P(c, p_k)$, defined as the ratio of $f(c, p_k)$ and the total number of concepts in p_k, the dispersion of concepts in p_k, based on Shannon entropy, is computed as:

$$\theta(p_k) = -\sum_{c \in C} P(c, p_k) * log_2 P(c, p_k) \tag{6}$$

Dispersion for the Document. This value for document d is formulated as:

$$\text{Dispersion}(d) = \frac{\sum_{1 \leq k \leq K} \theta(p_k)}{K} \tag{7}$$

3.3 Methodology for Evaluation and Feedback

Evaluation Based on Previously Reviewed Texts Given that our purpose is to have an assessment of writings in progress and provide some sort of guidance to the author, we decided to obtain the measures on a collection of texts. These prior

Table 1 Collection data by academic level

	Complete	Selected	Problem Stmt	Justification	Results
ACT	227	202	80	104	102
Undergraduate (UG)	150	136	28	21	77
Master	269	254	92	81	179
Doctoral	66	64	21	13	43

analysis, besides validating the computing procedures on texts of different academic levels, can provide reference values to assess new texts. This evaluation is done on three specific lengthy sections for each document. These are: Problem Statement, Justification and Results.

Feedback Visualization For the evaluation of drafts or writings in progress, we foresee a visualization feedback from the document. This feedback is based on the document representation built to compute the comprehensibility measures. We employ three different visualization approaches: word cloud[3], graph, and bar plot. The word cloud allows the author to visualize the different discussed concepts and their importance in general, and those with higher CB and Connectivity value. The graph allows the author of the document to visualize concept relations, key paragraphs, and a complete Document Concept Graph (G). Finally, the bar plot visualization displays how the dispersion values of paragraph are with respect to average, hinting improvements to be done in those above.

4 Document Collection

In our research, we utilize the document collection reported in [14]. The data set consists of theses and proposals of different academic levels. These levels are: Advanced College-level Technician[4] (ACT), Undergraduate, Master, and Doctoral. As mentioned above, the following sections were extracted from each document: Problem Statement, Justification, and Results. Table 1 includes the number of the thesis of each academic level in the collection (Complete), along with the selected thesis (documents that include more than one paragraph where a paragraph was considered if it has more than two sentences), and the number of sections for each of the academic levels.

Word/token statistics for the three sections in document collection is given in Table 2, that includes average, larger text (Max) and shorter text (Min).

---

[3] A visual representation of the words used in a particular piece of text, with the size of each word indicating its relative frequency.

[4] A two year program offered in some countries.

Table 2 Statistics of tokens per section and academic level

	Problem statement			Justification			Results		
	Avg.	max.	min.	Avg.	max.	min.	Avg.	max.	min.
ACT	409.33	1708	75	385.07	1123	110	493.55	1883	138
UG	477.60	1253	114	342.33	820	109	508.92	2386	121
Master	399.39	1063	121	407.81	1612	113	656.65	5184	137
Doctoral	754.85	1643	310	433.84	802	247	840.27	3765	59

5 Comprehensibility Measures in Academic Levels

We report the results of comprehensibility measures for the three sections of interest and the different academic levels. The tables include the average, standard deviation, and percentage of texts used to compute the statistic values, with separate columns for each measure, i.e., Connectivity, Dispersion, and Comprehensibility Burden.

As one can observe in the tables, the dispersion was computed in all the sections. However, connectivity was computed in percentages of section texts ranging between 21 and 81. Comprehensibility Burden was computed in only a few texts since most of them did not show a deficiency in this aspect (null burden).

As explained above, these statistics also serve to determine thresholds to grade new texts. We expect to define five different grades: Excellent, Good, Regular, Can-be-improved, and Work-is-needed.

5.1 Problem Statement Section

The number of texts for this section was not the more abundant. We can notice in Table 3 that comprehensibility burden was quite high in the few ACT-level texts used to compute this measure, compared to other academic levels, followed by doctoral texts.

5.2 Justification Section

This section was the most scarce in the data set, except in the ACT level. Burden values are lower in the technician level but still the highest, as Table 4 shows. Notice also that given that there are few doctoral justifications (13), none of them reached a non zero comprehensibility burden.

5.3 Results Section

Table 5 includes the values obtained for the Results section and again the technical level texts reached the highest results. This section has the highest number of texts for three of the four academic levels.

5.4 Analysis of Measures in Academic Text Collection

From the statistics included in Tables 3, 4 and 5, we can notice that dispersion is the measure that shows a stable behavior across the different sections.

Table 3 Measures statistics for problem statement

	Connectivity			Dispersion			Comprehensibility burden		
	Avg.	SD	%Non-zero	Avg.	SD	%Non-zero	Avg.	SD	%Non-zero
ACT	1.02	±1.26	21.25	1.21	±0.53	100.00	1340.00	±1058.42	5.00
UG	5.51	±12.45	25.00	1.59	±0.85	100.00	54.00	±63.69	10.71
Master	1.97	±2.86	45.65	1.37	±0.53	100.00	48.60	±91.16	10.87
Doctoral	4.09	±6.07	80.95	1.27	±0.76	100.00	158.20	±185.97	23.81

Table 4 Measures statistics for justification

	Connectivity			Dispersion			Comprehensibility burden		
	Avg.	SD	%Non-zero	Avg.	SD	%Non-zero	Avg.	SD	%Non-zero
ACT	0.57	±0.65	30.77	1.18	±0.54	100.0	343.50	±272.75	7.69
UG	0.91	±0.89	38.10	1.39	±0.57	100.0	33.00	±27.00	9.52
Master	2.97	±5.91	30.86	1.50	±0.71	100.0	90.67	±105.59	3.70
Doctoral	2.34	±2.22	38.46	1.36	±0.61	100.0	0.0	±0.0	0.0

Table 5 Measures statistics for results

	Connectivity			Dispersion			Comprehensibility burden		
	Avg.	SD	%Non-zero	Avg.	SD	%Non-zero	Avg.	SD	%Non-zero
ACT	4.59	±23.39	35.29	1.02	±0.41	100.0	1433.25	±1920.65	7.84
UG	5.11	±9.22	22.08	1.29	±0.72	100.0	336.60	±372.63	6.49
Master	29.07	±257.22	58.10	1.07	±0.54	100.0	138.27	±237.50	12.29
Doctoral	18.72	±57.74	76.74	1.16	±0.79	100.0	270.25	±233.17	18.60

Regarding connectivity, this measure varies slightly, showing higher values for graduate academic levels, caused possibly by their length and being more elaborated writings.

Comprehensibility Burden is the measure that reflects an overall assessment of the text since evidences how the concepts are developed across the paragraphs of the text. This measure was computed in the least number of texts because requires an analysis at concepts, concept relations, paragraph, and entire document. We notice that ACT level texts obtained higher values for the three sections, possibly because several concepts are mentioned but not adequately introduced and sequenced.

6 New Texts Evaluation and Feedback

Since we aim to provide a resource to evaluate documents in progress (drafts) to determine its comprehensibility in terms of the three measures, some ways to report the evaluation to the author are proposed. During the different evaluation stages, some preliminary outcomes are worthy of reporting visually. These are presented and illustrated next for texts of technical and doctoral level.

6.1 Example of Feedback Visualization of an ACT-Level Text

The first step for the analysis is the identification of the concepts discussed in the document under evaluation (as detailed in Sect. 3.1A) to build the representation. Once this set of concepts is identified, a word cloud is generated and reported as Figure 1a illustrates, with central concepts in large size font. The concepts co-occur in the paragraphs and this relation among concepts is represented in a concept graph (see Sect. 3.1B). Figure 1b depicts the concept graph built for the text being evaluated.

Figure 2a includes a concept graph built for the evaluation of key paragraphs, showing a paragraph and associated concepts. Once the comprehensibility burden is

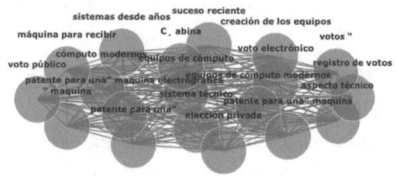

(a) Word cloud for Concepts   (b) Relation among Concepts

Fig. 1 Visual representation for concept frequencies and relation

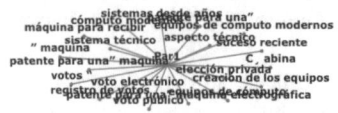

(a) A paragraph and related concepts

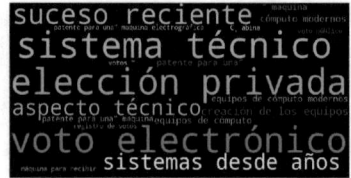

(b) Word cloud for CB measure of concepts

Fig. 2 Visual representation for key paragraphs and CB measure of concepts

(a) Document Concept Graph

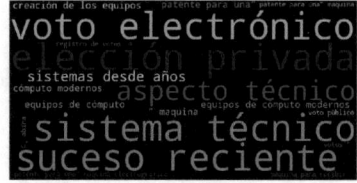

(b) Word cloud for Connectivity measure of concepts

Fig. 3 Visual representation for document concept graph and connectivity measure

computed for concepts, another word cloud can be displayed as illustrated in Fig. 2b. From concept CB measures, the whole document burden is got.

A Document Concept Graph built during the document analysis is depicted in Fig. 3a. After computing the connectivity measure of concepts, a connectivity word cloud of concepts can then be reported (Fig. 3b).

The results of calculating the Dispersion measure of paragraphs is reported with a bar plot as shown in Fig. 4a for a text of technical level. Here a reference threshold (red line) indicates those paragraphs with a density above the average for the academic level. Figure 4b illustrates the results of the evaluation of dispersion for a doctoral level text with more paragraphs.

6.2 Example of Feedback Visualization of Doctoral level Text

We provide here two illustrations of how visual feedback changes from short to larger documents, as a doctoral level thesis can be. For instance, from the beginning, the number of concepts and frequencies are higher for this kind of document as Fig. 5a shows. Also the key paragraphs and related concepts can show more elaborated graphs (Fig. 5b). In previous section, an example of dispersion bar plot for a doctoral text was given (Fig. 4b).

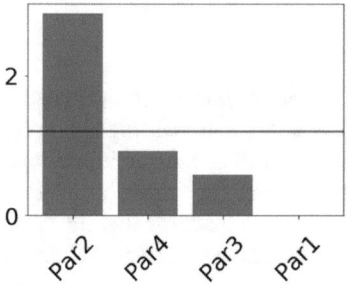

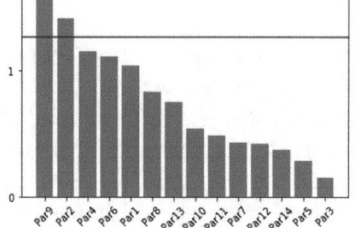

(a) Bar plot of Paragraph Dispersion measure (ACT-level text)

(b) Bar plot of Paragraph Dispersion measure (Doctoral-level text)

Fig. 4 Visual representation for dispersion measure (ACT and Doctoral texts)

6.3 Additional Comments

Word clouds displayed at a different stage of the analysis allow visualizing concepts under varied perspectives. Initially, most mentioned extracted concepts are displayed and later reflecting their burden and connectivity.

Concept graphs displayed also show how the comprehensibility analysis progresses, going from how concepts are related to more elaborated graphs indicating key concepts in all the paragraphs, and the complete document graph. It is important to point out that these graphs are generated with an interactive tool where the user can zoom and look in detail different concepts and their interactions. This is illustrated with a box in Fig. 5b, displaying key paragraphs.

Bar plots are used to display the evaluation of dispersion of paragraphs, sorting them in decreasing order, so that those with higher values are clearly identified. This along the reference line reveal clearly those needing improvement.

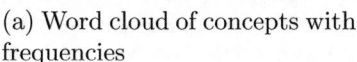

(a) Word cloud of concepts with frequencies

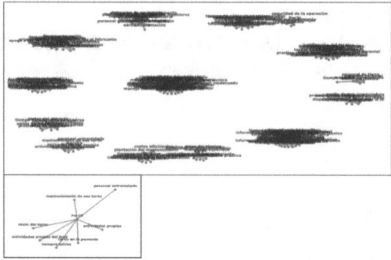

(b) Graph of Key Paragraphs and related concepts

Fig. 5 Visual representation of concepts and key paragraphs for a doctoral document

7 Conclusions

A set of measures to assess comprehensibility was presented and evaluated on a collection of academic texts at different levels. A computational-linguistic process to evaluate and provide feedback for a new writing was detailed and illustrated.

Dispersion, a measure to reveal whether the concepts of a document are adequately focused, provides the most reliable measure since it can be computed for any text, as showed when applied in the text collection. The range of values did not show variation to the academic level of texts.

Connectivity, a measure indicating that concepts in a text are properly connected, was the second most reliable metric. Contrary to dispersion, this measure tended to reach higher values with more elaborated texts, i.e. according to the academic level.

Comprehensibility Burden is a measure that focuses on determining an inappropriate sequence of concepts in a document, demanding an effort on the reader to understand the text. We found CB as the least reliable metric by two reasons; first, only a few documents in the collection produced a non-null CB value, and in consequence, its statistics showed a high variation. Nevertheless, a tendency can be observed where the lower the academic level, the higher the comprehensibility burden value.

We expect that visualization feedback during the stages of comprehensibility analysis can help the text author to understand how the measures are computed and ameliorate the deficiencies detected. But this can only be confirmed with further studies.

Despite the fact that our experimental document collection and results are for Spanish, the assessment can be brought to other language with minor changes.

We plan to bring these analysis and feedback techniques to an online application. So a student can submit his/her writing in progress for comprehensibility assessment. Afterwards, a pilot study can be set to evaluate the benefits of the application. Additionally, as future work we would like to increase the text collection with the aim of balancing it, so that the number of texts at undergraduate and doctoral level increases. In this way, the reference statistics values can be more reliable.

Acknowledgements The first author thanks the support provided through scholarship number 1009285 by Conacyt, México. The second and third authors were partially supported by SNI, México.

References

1. Relan M, Khurana S, Singh VK (2013) Qualitative evaluation and improvement suggestions for ebooks using text analytics algorithms. In: Proceedings of second international conference on eco-friendly computing and communication systems, Solan, India
2. Rossetti A (2019) Intralingual translation and cascading crises: evaluating the impact of semi-automation on the readability and comprehensibility of health content, 1st edn. Routledge

3. González-López S (2015) Linguistic analysis of undergraduate research drafts. Doctoral dissertation, Instituto Nacional de Astrofísica, Óptica y Electrónica
4. Collins-Thompson K (2014) Computational assessment of text readability: a survey of current and future research. ITL-Int J Appl Linguist 165(2):97–135
5. Zhang H, Litman D (2020) Automated topical component extraction using neural network attention scores from source-based essay scoring. arXiv preprint arXiv:2008.01809
6. Macken L, Fonteyne M, Tezcan A, Daems J (2020) Assessing the comprehensibility of automatic translations (ArisToCAT). In: EAMT2020, pp 485–486
7. Benzahra M, Yvon F (2019) Measuring text readability with machine comprehension: a pilot study. In: ACL BEA2019, pp 412–422
8. Glaser I, Bonczek G, Landthaler J, Matthes F (2019) Towards computer-aided analysis of readability and comprehensibility of patient information in the context of clinical research projects. In: Proceedings of the seventeenth international conference on artificial intelligence and law, pp 260–261
9. Xia M, Kochmar E, Briscoe T (2016) Text readability assessment for second language learners. In: ACL BEA2016, pp. 12–22
10. Karuna P (2019) Manipulating comprehensibility of text: an automated approach to generate deceptive documents for cyber defense. Doctoral dissertation, George Mason University
11. Karuna P, Purohit H, Jajodia S, Ganesan R, Uzuner O (2020) Fake document generation for cyber deception by manipulating text comprehensibility. IEEE Syst J
12. Peñas A, Verdejo F, Gonzalo J (2011) Corpus-based terminology extraction applied to information access. In: Proceedings of corpus linguistics, p 458
13. Honnibal M, Montani I (2017) spaCy 2: natural language understanding with Bloom embeddings, convolutional neural networks and incremental parsing
14. González-López S, López-López A (2015) Colección de tesis y propuesta de investigación en tics: un recurso para su análisis y estudio. In: XIII Congreso Nacional de Investigación Educativa, pp 1–15

Deep Learning Techniques for Automatic Short Answer Grading: Predicting Scores for English and German Answers

Jörg Sawatzki, Tim Schlippe, and Marian Benner-Wickner

Abstract We investigate and compare state-of-the-art deep learning techniques for *Automatic Short Answer Grading*. Our experiments demonstrate that systems based on the *Bidirectional Encoder Representations from Transformers* (BERT) [1] performed best for English and German. Our system achieves a Pearson correlation coefficient of 0.73 and a Mean Absolute Error of 0.4 points on the Short Answer Grading data set of the University of North Texas [2]. On our German data set we report a Pearson correlation coefficient of 0.78 and a Mean Absolute Error of 1.2 points. Our approach has the potential to greatly simplify the life of proofreaders and to be used for learning systems that prepare students for exams: 31% of the student answers are correctly graded and in 40% the system deviates on average by only 1 point out of 6, 8 and 10 points.

Keywords Automatic short answer grading · Artificial intelligence in education · Natural language processing · Deep learning

1 Introduction

The research area "AI in Education" addresses the application and evaluation of Artificial Intelligence (AI) methods in the context of education and training [3]. One of the main focuses of this research is to analyze and improve teaching and learning processes. Many educational institutions–public and private–already conduct their courses and examinations online. This means that student examinations and their assessments are already available in digital, machine readable form, offering a wide range of analysis options. An exam typically consists of multiple choice and free text questions. While answers to multiple choice questions can easily be evaluated by machines, the evaluation of free text answers still requires tedious manual work by the examiners.

J. Sawatzki · T. Schlippe (✉) · M. Benner-Wickner
IU International University of Applied Sciences, XXX, Germany
e-mail: tim.schlippe@iu.org

© The Author(s), under exclusive license to Springer Nature Singapore Pte Ltd. 2022 65
E. C. K. Cheng et al. (eds.), *Artificial Intelligence in Education: Emerging Technologies,
Models and Applications*, Lecture Notes on Data Engineering and Communications
Technologies 104, https://doi.org/10.1007/978-981-16-7527-0_5

The focus of our paper is on Automatic Short Answer Grading (ASAG) with deep learning, i.e., the evaluation and further development of various deep learning state-of-the-art approaches to automatically evaluate free text answers. We investigate the following two architectures to compare a student answer with a given model answer and to predict an evaluation in the form of a score:

- Our *feature extraction architecture* uses general pre-trained sentence embeddings in a high-dimensional semantic vector space as input values and predicts a score using a linear classifier.
- Our *fine-tuning architecture* is based on a pretrained deep learning model, which supplemented by a linear layer is adapted to the specific task of ASAG. In contrast to our *feature extraction architecture*, the parameters of the embeddings are tuned as well.

A data set with manually graded exams and sample solutions from the area *Business Administration* of a German bachelor's program serves to optimize models and evaluate quality. To evaluate and classify the findings in the context of current research, an English data set from the University of North Texas with questions from the undergraduate studies of *Computer Science* is also used.

In the next section, we present the latest approaches of other researchers for ASAG. Section 3 describes our experimental setup. Section 4 characterizes the models which we evaluate and compare. Our experiments and results are outlined in Sect. 5. We conclude our work in Sect. 6 and suggest further steps.

2   Related Work

A good overview of rule-based and statistical-based approaches in ASAG before the deep learning era is given in [4]. Newer publications are based on *bag-of-words*, a procedure based on term frequencies [5, 6]. The latest trend which has proven to outperform traditional approaches is to use neural network-based embeddings, such as *Word2vec* [7]. [8] have developed *Ans2vec*, a *feature extraction architecture*-based approach. It is based on combine-skip sentence embedding [9] and logistic regression. They evaluated their concept with a non-public data set from the University of Cairo, the SciEntsBank data set [10], and the English data set of the University of North Texas [2]. This English data set of the University of North Texas–like the German data set of our university–contains scored student answers. Consequently, it is a comparative data set which is also evaluated in the experiments of this paper. Like us, [8] use the Pearson correlation coefficient for evaluation. They report a best value of 0.63 on the data of the University of North Texas. [11] use a data set from the Hewlett Foundation[1] to compare different approaches based on deep learning models, including a *fine-tuning architecture* based on the *Bidirectional Encoder Representations from Transformers (BERT)* [1]. [12] and [13] deal in their work exclusively

---

[1] https://www.kaggle.com/c/asap-sas.

with *BERT fine-tuning architectures.* In their work, answers are categorized into 3 classes–there is no point-based grading. *BERT* also provides the basis for our *fine-tuning architecture,* but we focus on point-based grading. [14] and [15] developed systems which classify German student answers as "correct" and "wrong"—based on traditional features such as lemmas.

Our contributions are the analysis of deep learning architectures for transfer learning on an English and a German data set. This includes the investigation of multilingual deep learning models. We are the first to examine point based ASAG of questions with variable maximum score.

3 Experimental Setup

In this section we describe our evaluation metrics and corpora.

3.1 Evaluation Metrics

As in related literature, we evaluate our results with the *Pearson correlation coefficient* [16], the *Mean Absolute Error* and the *Root Mean Square Error (RMSE).*

Pearson Correlation Coefficient. The Pearson correlation coefficient (*Pearson*) indicates how strong the linear relationship between predictions and target values is. If the value is close to 0, there is no correlation; if it is close to 1, there is a strong correlation. The Pearson correlation coefficient is the normalized covariance. Therefore, it is independent of the scaling used in the data.

Mean Absolute Error. The Mean Absolute Error *(MAE)* is calculated from the average deviations of the prediction from the target value. It depends on the units and scaling in the evaluated data set.

Root Mean Square Error. Unlike Mean Absolute Error, in the Root Mean Square Error *(RMSE)* the error is squared, and the square root of the mean square deviation is considered. Squaring the error results in strong deviations being weighted more heavily than small ones.

3.2 Corpora

We evaluate our experiments with a German and an English data set which are compared in Table 1. To provide insights into both data sets,[2] Tables 2 and 3 indicate

---

[2] Questions and answers were modified for the German data set due to confidentiality.

Table 1 Information of the data sets

	German	English
Subject	Business administration	Data structures
#questions with model answer	233	87
#answers (total)	3,560	2,442
#answers per question	15.4	28.1
Ø length of answer (#words)	87.6	18.4
Maximum scores (in points)	6/8/10	5
Annotated model answer	yes	no

Table 2 Original sample question and answers from the English data set

Question	What is a variable?
Model answer	A location in memory that can store a value
Example: Answer 1	A variable is a location in memory where a value can be stored
Grading: Answer 2	5 of 5 points
Example: Answer 2	Variable can be an integer or a string in a program
Grading: Answer 2	2 of 5 points

Table 3 Modified and translated sample question and answers from the German data set

Question	• Explain: What is the role of models in business administration? • Explain how statements of a model can be distinguished from each other according to the completeness of the information
Model answer	In business administration, models are used to obtain, formulate, and test knowledge from the operational context (2 points). Statements with complete information are statements with certainty (3 points). Statements with incomplete information are statements under uncertainty or risk (3 points)
Example: Answer 1	(a) In business administration, models are used to explain, describe, forecast and design macro- and microeconomic phenomena (b) Complete information represents security. Incomplete information represents uncertainty and risk
Grading: Answer 2	8 of 8 points
Example: Answer 2	Models are used for information. Explanatory model: Explains reasons in the company, e.g., employee motivation. Descriptive model: describes business phenomena in the company, e.g., accounting which records the entire flow of money in the company. Decision model: here different information is combined with each other. For example, the optimal order quantity. This depends on various factors
Grading: Answer 2	2 of 8 points

for each data set, a typical question, the corresponding model answer and two student answers. One of them was given the full score, the other one is a rather weak answer.

3.3 English Short Answer Grading Data Set

The short answer grading data set of the University of North Texas [2] contains 87 questions with corresponding model answer and on average 28.1 evaluated answers per question about the topic *Data Structures* from the undergraduate studies. The questions are rather short and are not divided into sub-questions. They can usually be answered in only one sentence and no knowledge transfer is required.

3.4 German Short Answer Grading Data Set

The German data set is taken from an online exam system in the learning management system *Moodle*.[3] It contains 233 questions with corresponding model answer and on average 15.4 evaluated answers per question from the bachelor module *Business Administration*. A special feature of the German data set is that the maximum achievable score varies from question to question. Depending on the question a maximum of 6, 8 or 10 points can be achieved. Another feature is that the model answers include annotations with the criteria for grading performance. Many model answers contain only short hints for the corrector, so that in many cases additional background knowledge is needed for correction in addition to the model answer. A question usually consists of several sub-questions on a common topic and, in addition to the pure reproduction of knowledge. In many cases knowledge transfer is expected from the students.

4 Techniques

This paper describes and compares the following two architectures for transfer learning: A *feature extraction architecture* and a *fine-tuning architecture*.

4.1 Feature Extraction Architecture

This architecture is based on the *Ans2vec* approach described by [8]: The model answer and the student answer are first converted into the two embedding vectors

---

[3] https://moodle.org.

MA and *SA*. Then the dot product and the absolute difference of the two embedding vectors are calculated and concatenated. The result of the concatenation is the input vector for a linear model to predict the score.

Ans2vec-Skip-Logit-Baseline. [8] use combine-skip vectors for the embeddings and logistic regression as a classifier (*Ans2vec-Skip-Logit-Baseline*). Since no combine-skip embeddings are available for German, we evaluated this model only on English.

Ans2vec-MUSE-Logit. *Ans2vec-MUSE-Logit* refers to a model that corresponds to *Ans2vec-Skip-Logit-Baseline*, but for sentence embedding the *Multilingual Universal Sentence Encoder* (*MUSE*) [17] is used.

Ans2vec-Skip-SVM. *Ans2vec-Skip-SVM* refers to a model that corresponds to *Ans2vec-Skip-Logit-Baseline*, but for the classification a *Support Vector Machine* (*SVM*) is used [18]. Due to the lack of German combine-skip embeddings, we also evaluated this model only on the English data set.

Ans2vec-MUSE-SVM. *Ans2vec-MUSE-SVM* refers to a model that corresponds to *Ans2vec-Skip-Logit-Baseline*, but for the sentence embedding the *MUSE* is used and for the classification an *SVM*.

4.2   Fine-Tuning Architecture

This architecture is based on *BERT* [1] from the family of transformer models. We supplemented *BERT* with a linear regression layer that provides a prediction of the score given an answer. The model takes the model answer and the student answer without prior embedding as input, separates the model answer and the student answer with a *separator token* and performs a tokenization into *word pieces*. Since in our German and English data sets the scores are not only discrete integer values, this approach uses regression instead of classification. Our evaluated *BERT* models are characterized in the following sections.

BERT-EN. *BERT-EN* refers to the English BERT[4] published by [1].

BERT-DE-Deepset. *BERT-DE-Deepset* refers to a German *BERT* model provided by Deepset GmbH. The model is trained on Wikipedia and Open Legal Data.[5]

BERT-Multilingual. *BERT-Multilingual* refers to a multilingual *BERT* model[6] which is published by Google, supports 104 languages and is trained on Wikipedia.

---

[4] https://github.com/google-research/bert.

[5] https://deepset.ai/german-bert.

[6] https://github.com/google-research/bert/blob/master/multilingual.md.

5 Experiments and Results

Randomization and splitting of the data sets into training, validation and test data using 5-fold cross-validation is performed in all experiments to determine most accurate models for the German and English data as shown in Table 4. After the most accurate models for the German and English data set were determined, we evaluated them on the held-out test set.

5.1 English Automatic Short Answer Grading

The results of the experiments with the English data set and their relative improvements compared to *Ans2vec-Skip-Logit-Baseline* are shown in Table 5. For comparison, the first line also contains the values published by [8]. Since they do not provide further details on the implementation, parameters, and the procedure for evaluating the model, the reasons for deviation cannot be further analyzed. If instead of the combine-skip embeddings the *MUSE* embeddings are used, the results improve, and the training effort is reduced considerably. With only 512 dimensions, the *MUSE* embeddings are significantly more compact than the combine-skip vectors with 4,800 dimensions. The *Ans2vec* model also provides better predictions if the linear regression is replaced by an *SVM* classifier. However, the *BERT* models provide the best

Table 4 Preparation of data sets for cross-validation

Data set	Portion	German	English
Total	100%	3,560	2,442
Cross validation	80%	2,848	1,953
Test (held-out)	20%	712	489
Cross validation	100%	2,848	1,953
Training	80%	2,278	1,953
Validation	20%	570	391

Table 5 Basic experiments with the English data set

Model	Pearson	RMSE	MAE
[7]	0.63	0.91	–
Ans2vec-Skip-Baseline	0.33 (+0.0%)	1.27 (+0.0%)	0.73 (+0.0%)
Ans2vec-Skip-SVM	0.49 (+48.6%)	1.09 (−14.0%)	0.60 (−16.8%
Ans2vec-MUSE-Logit	0.38 (+13.6%)	1.24 (−2.2%)	0.69 (−4.5%)
Ans2vec-MUSE-SVM	0.56 (+67.5%)	1.02 (−19.1%)	0.56 (−23.7%)
BERT-EN	**0.79** (+138.5%)	**0.69** (−45.3%)	**0.41** (−43.3%)
BERT-Multilingual	0.79 (+137.1%)	0.70 (−44,6%)	0.43 (−44,6%)

results. *BERT-EN* is the best of all models, but *BERT-Multilingual* provides only slightly worse numbers. With a *Pearson* of 0.79, we see that the scores predicted by the *BERT-EN* model have a strong linear relationship with the scores decided by the human corrector. On average, the evaluation of an answer by the model deviates by 0.41 points (see *MAE*). The *RMSE*, which weighs more strong deviations, also has the lowest number in this model. Compared to the numbers published by [8], the *BERT-EN* model achieves a relative improvement of more than 25% in *Pearson* and 23% in *RMSE*.

5.2 German Automatic Short Answer Grading

The results of the experiments with the German data set are shown in Table 6. Comparing these results with those of the evaluation of the English data set, only *Pearson* may be used. *MAE* and *RMSE* depend on the scaling of the score, which is different for both data sets. Looking at the linear correlation, one will notice that *Ans2vec-MUSE-Logit* performs slightly better on the German data set, while *Ans2vec-MUSE-SVM* performs slightly better on the English data set. The results of *BERT* on the German data set are only slightly worse than on the English data, even if the German exam questions are considerably more complex and extensive.

The German data set also demonstrates that the *BERT fine-tuning architecture* produces significantly better results. Again, the monolingual *BERT* model—in this case the *BERT-DE-DBMDZ*—is slightly better than the multilingual model. The best model is *BERT-DE-DBMDZ* which—with a *Pearson* of 0.75—shows a strong linear relationship between prediction and actual scores. The model's prediction deviates by 1.30 points from the human corrector's grading (see *MAE*). Compared to *Ans2vec-MUSE-Logit*, *Pearson* could be improved by 90% relative. The *RMSE* and *MAE* are almost 31% lower than the numbers of the *Ans2vec-MUSE-Logit* model.

Table 6 Basic experiments with the German data set

Model	Pearson	RMSE	MAE
Ans2vec-MUSE-Logit	0.39	2.52	1.87
Ans2vec-MUSE-SVM	0.44 (+11,2%)	2.68 (+6,6%)	1.90 (+1,9%)
BERT-DE-Deepset	0.74 (+89.2%)	1.76 ($-$30.3%)	1.31 ($-$29.7%)
BERT-DE-DBMDZ	**0.75** (+90.2%)	**1.75** ($-$30.6%)	**1.30** ($-$30.6%)
BERT-multilingual	0.71 (+81.4%)	1.83 ($-$27.4%)	1.38 ($-$26.3%)

Table 7 Further experiments with the German data set

Model	Pearson	RMSE	MAE
BERT-DE-DBMDZ	0.75	1.75	1.30
Annotations removed	0.74 (−1.4%)	1.78 (+2.0%)	1.31 (+1.3%)
Max. score annotated	**0.75** (+0.8%)	**1.73** (−0.9%)	**1.28**(−1.7%)

Table 8 Final results on the unseen text sets

Language	Pearson	RMSE	MAE
English	**0.73** (+15.5%)	**0.72** (−20.9%)	**0.42**
German	**0.78**	**1.62**	**1.19**

5.3 Experiments with Removed and Added Annotations on the German Data Set

We removed the annotations with the criteria for grading (e.g., "(2 points)", see Table 3) and re-evaluated the best model *BERT-DE-DBMDZ*. Additionally, we added annotations for the maximum achievable score of each question to the model answers (e.g., "maximum 8 points"). The results of the further experiments compared to *BERT-DE-DBMDZ* can be found in Table 7. Removing the annotations reduces the quality of the model's predictions, adding the annotations slightly improves them.

5.4 Final Results

After the most accurate models for the German and English data set were determined, we evaluated them on the unseen *test set*. As shown in Tables 5 and 7, *BERT-EN* is the best model on the English data set and *BERT-DE-DBMDZ (Maximum score annotated)* on the German data set. The results of the final evaluation are illustrated in Table 8. For English, the relative improvement compared to [8] is also shown.

6 Experiments and Results

We investigated and compared state-of-the-art deep learning techniques for *Automatic Short Answer Grading*. With our *BERT* models we achieved a significant performance improvement compared to our baseline system and related work. Our system achieves a Pearson correlation coefficient of 0.73 and a Mean Absolute Error of 0.4 points on the Short Answer Grading data set of the University of North Texas [2]. On our German data set we report a Pearson correlation coefficient of 0.78 and a Mean Absolute Error of 1.2 points. The result on our English and German data sets were comparable even though the German data set contains more complex questions and has variable maximum scores.

Future work will include an analysis of what types of questions and answers the system still has issues and how we can tackle them. For example, examination questions often contain sub-questions. It could be evaluated whether their separation into individual questions leads to better predictions. We also plan to analyze to what extent the quality of the model improves by training the model on further subject-specific corpora such as lecture notes or textbooks as suggested by [19]. The developed model could also be used as a warning system: The system detects when a human corrector's grading significantly deviates from the model (e.g., by a defined threshold) and initiates further steps, e.g., the transfer of the relevant student answer and correction to a further review process. Furthermore, we plan to investigate the time savings by our automation. For example, our first more detailed analyses of the results of the German data set indicate that in 31% of all cases the score can be just accepted. In 39.6% of the cases the suggested score only needs to be corrected 1 point up or down out of 6, 8 and 10 points. This means that in 70.6% the total score does not have to be corrected at all or only by 1 point, which could lead to significant time savings in the correction process and to be used for learning systems that prepare students for exams [20].

References

1. Devlin J, Chang M-W, Lee K, Toutanova K (2019) BERT: Pre-training of Deep Bidirectional Transformers for Language Understanding. In: Proceedings of the 2019 Conference of the North American Chapter of the Association for Computational Linguistics: Human Language Technologies, vol 1 (Long and Short Papers). Association for Computational Linguistics, Minneapolis, Minnesota, pp 4171–4186
2. Mohler M, Bunescu R, Mihalcea R (2011) Learning to Grade Short Answer Questions Using Semantic Similarity Measures and Dependency Graph Alignments. In Proceedings of the 49th Annual Meeting of the Association for Computational Linguistics: Human Language Technologies. Association for Computational Linguistics, Portland, Oregon, USA, pp 752–762
3. Libbrecht P, Declerck T, Schlippe T, Mandl T, Schiffner D (2020) NLP for Student and Teacher: Concept for an AI based Information Literacy Tutoring System. In: The 29th ACM International Conference on Information and Knowledge Management (CIKM2020), Galway, Ireland
4. Burrows S, Gurevych I, Stein B (2015) The Eras and Trends of Aautomatic Short Answer Grading. Int J Artif Intell Educ 25(1):60–117
5. Süzen N, Gorban AN, Levesley J, Mirkes EM (2020) Automatic Short Answer Grading and Feedback Using Text Mining Methods. Proedia Comput Sci 169:726–743
6. Zehner F (2016) Automatic Processing of Text Responses in Large-scale Assessments. Ph.D. thesis, TU München
7. Mikolov T, Chen K, Corrado G, Dean J (2013) Efficient Estimation of Word Representations in Vector Space. In: 1st International Conference on Learning Representations, ICLR 2013, Workshop Track Proceedings, Scottsdale, Arizona, USA
8. Gomaa WH, Fahmy AA (2019) Ans2vec: A Scoring System for Short Answers. In: The International Conference on Advanced Machine Learning Technologies and Applications (AMLTA2019). Springer International Publishing, Cham, pp 586–595

9. Kiros R, Zhu Y, Salakhutdinov R, Zemel RS, Torralba A, Urtasun R, Fidler S (2015) Skip-Thought Vectors. In: Proceedings of the 28th International Conference on Neural Information Processing Systems, NIPS'15, vol 2. MIT Press, Cambridge, MA, USA, pp 3294–3302

10. Dzikovska M, Nielsen R, Brew C, Leacock C, Giampiccolo D, Bentivogli L, Clark P, Dagan I, Dang HT (2013) SemEval-2013 Task 7: The Joint Student Response Analysis and 8th Recognizing Textual Entailment Challenge. In: Second Joint Conference on Lexical and Computational Semantics (*SEM), Proceedings of the Seventh International Workshop on Semantic Evaluation (SemEval 2013), vol 2, pp 263–274. Association for Computational Linguistics, Atlanta, Georgia, USA

11. Krishnamurthy S, Gayakwad E, Kailasanathan N (2019) Deep Learning for Short Answer Scoring. Int J Recent Technol Eng 7:1712–1715

12. Sung C, Dhamecha T, Mukhi N (2019) Improving Short Answer Grading using Transformer-Based Pre-Training. Artif Intell Educ 469–481

13. Camus L, Filighera A (2020) Investigating Transformers for Automatic Short Answer Grading. Artif Intell Educ 12164:43–48

14. Meurers D, Ziai R, Ott N, Kopp J (2011) Evaluating Answers to Reading Comprehension Questions in Context: Results for German and the Role of Information Structure. In: Proceedings of the TextInfer 2011 Workshop on Textual Entailment. Association for Computational Linguistics, Edinburgh, Scottland, UK, pp 1–9

15. Pado U, Kiefer C (2015) Short Answer Grading: When Sorting Helps and when It Doesn't. In: Proceedings of the 4th Workshop on NLP for Computer Assisted Language Learning, NODALIDA 2015, Linköping Electronic Conference Proceedings. LiU Electronic Press and ACL Anthology, Wilna, pp 42–50

16. Pearson K (1895) Note on Regression and Inheritance in the Case of Two Parents. Proc R Soc Lond 58:240–242

17. Yang Y, Cer D, Ahmad A, Guo M, Law J, Constant N, Abrego GH, Yuan S, Tar C, Sung Y-H, Strope B, Kurzweil R (2019) Multilingual Universal Sentence Encoder for Semantic Retrieval. arXiv:1907.04307

18. Evgeniou T, Pontil M (2001) Support Vector Machines: Theory and Applications. Mach Learn Appl: Adv Lect 2049:249–257

19. Wölfel M (2021) Towards the Automatic Generation of Pedagogical Conversational Agents from Lecture Slides. In: 3rd EAI International Conference on Multimedia Technology and Enhanced Learning (EAI ICMTEL 2021). Cyberspace

20. Schlippe T, Sawatzki J (2021) AI-based Multilingual Interactive Exam Preparation. In: The Learning Ideas Conference 2021 (14th annual conference). ALICE—Special Conference Track on Adaptive Learning via Interactive, Collaborative and Emotional Approaches. New York, New York, USA

Solving Chemistry Problems Involving Some Isomers of Benzene Ring

Xinguo Yu, Lina Yan, Hao Meng, and Rao Peng

Abstract This paper presents an algorithm for understanding and solving the chemistry problems involving some isomers. The research into this problem involves three aspects. The first part is to design the method of understanding the problem text. This method adopts the syntactic-semantic model and it first constructs a pool of syntactic-semantic models. The second part is to design the method of extracting knowledge items from the problem figure and the method of combining the knowledge items from text and figure. The third part is to develop a solving procedure with the aid of a knowledge base. Based on the solution provided by the proposed algorithm, a diagnostic procedure is to judge whether the handwritten answer is correct. The experimental results show that the proposed algorithm has a good performance.

Keywords Automatic solution · Isomers with benzene ring · Image recognition\and automatic diagnosis

1 Introduction

1.1 Research Background

The automatic solving problems in basic education has been an active research problem since Artificial Intelligence (AI) appeared in 1950s [3]. The automatic solution can provide convenience, save time and improve efficiency for students

---

X. Yu (✉) · L. Yan · H. Meng · R. Peng
Central China Normal University, Wuhan 430079, China
e-mail: xgyu@mail.ccnu.edu.cn

L. Yan
e-mail: 13836754565@163.com

H. Meng
e-mail: menghao@mails.ccnu.edu.cn

R. Peng
e-mail: 41002093770@qq.com

and teachers. It includes natural language processing, intelligent reasoning, machine answering, and system diagnosis. More researchers participate in the field of automatic solutions, and promote forward this field. The significance of this paper about the automatic solution of organic isomers is as follows: the system mainly includes the elementary school mathematics application question, the plane geometry question [5], the semantic understanding question, the physical circuit question, etc. Generally speaking, the isomers of organic compounds are more comprehensive, and the strong logical ability to test the depth and breadth of students' understanding of organic chemistry.

1.2 Research Problem and Challenge

The research is about the automatic solution of isomers with a benzene ring. For the given structure of organic compounds, we can solve the isomers related to it: the organic compounds with the same molecular formula but different kinds of structural formula. The text and picture in the problem (with benzene ring organic compound structure simplified form) and the options in the answer are as the input, take the useful information in the title as the keyword extraction, the information and structure simplified form as two judgment standard. From the knowledge base to find the content that meets the conditions. Finally, the output of all the effective content compares with the number of the answers.

To solve and diagnose the results in organic chemistry automatically, which is composed of organic compounds' structural and text. The topic's key information is extracted by matching the syntactic-semantic model, and the structural simplification is transformed into the molecular formula pattern. The automatic solution to this type of question is proposed for the first time [8]. This research problem aims to propose an approach to solve problem and can generate a process which the primary school students can understand. This paper aims to output a knowledge base about organic isomers, which can be used in the related research of organic isomers. Although for the fields of isomers that have not yet been involved, preliminary solutions have been formed. More and more people pay attention to them and gradually put them into the research of related fields, resulting the automated organic chemistry solutions, providing phase information. The structure of the identification information, organic chemistry topic information for others to continue the in-depth study.

2 Literature Review

Throughout the field of the automatic solution, mathematics, physics, Chinese, geography, and chemistry are involved, but the related research of organic chemistry is less [2]. For the recognition of chemical structure, the chemical structure figure represented by molecular design Limited (MDL) mole is used as the input of optical structure recognition [3]. The MDL is divided into two parts, a series of different atoms, molecules, and the connection table of the definition graph. Through the steps of scanning, extraction, structural analysis, detection of signs, clearing up errors, structural identification, creating connection table, combination, and verification, the structural simplification of organic compounds is identified.

Vision recognition means figure mining. The organic chemical structure figure is used as input, and the output of the system is an SDF chemical file in standard format [1]. The system consists of five modules: pretreatment module, OCR module, vector module, reconstruction module, and chemical knowledge module. The optical structure recognition technology is used to process the figure and recover the chemical structural formula. OSRA can read more than 90 kinds of figure representation formats and automatically identify and extract information [4]. It can express the chemical structure formula into smiles string or SD form.

In the rule knowledge acquisition of automatic problem-solving for geography course, to obtain the text and the semantic relations between texts, these semantic reasoning relations are the key to solve the problem [7]. The syntactic semantic model is used to do relationship matching, which is composed of keywords, part of speech patterns, and the relationship between entities; The key process of the solution is to extract the quantitative relation of the question to construct the solution equation, to find the string with part of speech annotation and the implicit relation, and to replace the relation R with the defined variable when the clause matches the inherent model. In the implementation, the syntactic-semantic model S2[9] is mainly used to complete text analysis, extract relation, and instantiation. S2 model is a four tuple form, which is represented by $M = (K, P, Q, l)$. It is represented by keyword structure, part of speech tag mode, mathematical relationship, and corresponding list respectively.

In this paper, a mathematical framework of plane geometry theorem and arithmetic application problem is proposed, which is a relation extraction problem [9]. The model is used to describe the mapping of grammatical patterns to semantic patterns. S2 model pool is used to extract the display mathematical relation, and the framework is created. For the mathematical framework, including problem understanding and problem solving, the main process is to obtain the text, problem understanding, problem relation set, problem solving, and solution [6]. The step of identifying atomic regions is a key step of constructing equations. The remaining paper is organized as follows. Section 2 gives the outline of the proposed solver. Section 3 presents the detail of our algorithm. In Sect. 4, data set and evaluation are shown and discussed.

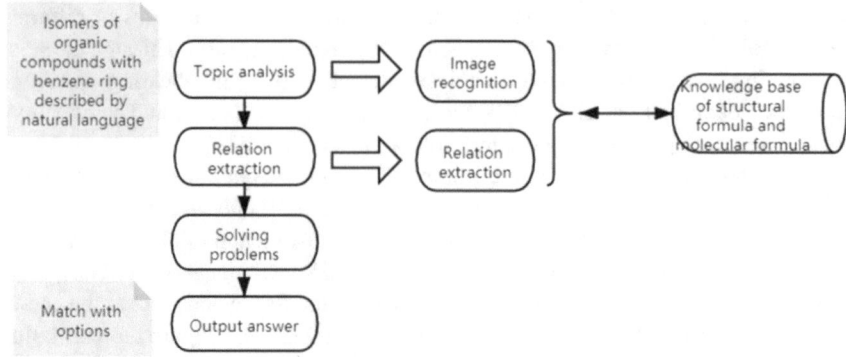

Fig. 1 The framework of the proposed solver

3 Automatic Solution Based on the Syntactic-Semantic Mode

The framework of the proposed solver is presented in Fig. 1. The syntactic and semantic model [9] for the automatic solution of the problem of isomers of organic compounds with benzene ring is a four tuple form $M = (K, P, Q, L)$. K represents the valid information of keywords, P represents the form with labels, Q is the extracted chemical text relationship, and L is a corresponding list. For different types of questions, different syntactic and semantic models need to be constructed. For the automatic solution of isomers of organic compounds with a benzene ring, different models represent different chemical text relations and different sentences and then match to extract the effective information of keywords. Because the text in the title has a sequence relationship, it will also have a certain sequence relationship afterword segmentation and part of speech tagging in Fig. 2.

4 Automatic Solution Isomers with Benzene Ring

When the text is used as input, the corresponding model matching means that the text has a certain syntactic order. At the same time, the corresponding syntactic and semantic model is searched. If the matching is successful, the text relationship is extracted to get the keywords. The specific process is as follows.

Input: Model pool $M = (K, P, Q, L)$;

The title of organic isomers with benzene ring described by natural language is named $A = 1...j$ afterword segmentation and part of speech tagging, which contains both syntactic and semantic parts. Output: chemical relationship or chemical valid information set.

Model	Applicable sentences	Extraction of chemical valid information
(NNP PRP NN VBZ, a=b, NNP NN)	Organic compounds A,..., the simple structure formula is	A=simple structure formula
(NNS TO NNS CC TO NNS, a and b,NNS NNS)	belongs to phenols, and there are () compounds which also belongs to esters	phenols and esters
(VBP TO NNS CC : NN, a and b, NNS NN)	belong to phenols and have - CH3	Phenols and CH3
(VBP TO NNS, a, NNS)	its isomers belong to common aromatic_alcohols	aromatic_alcohols

Fig. 2 NLTK part of speech tagging

The automatic answer to the problem of isomers of organic compounds with benzene ring is an important part of Compulsory Chemistry five, as well as the key and difficult point of organic chemistry. Most of these types of questions are multiple-choice questions. The text and the figure of organic structure in the questions are used as the input. The automatic answer to this kind of question needs basic text understanding, the storage of organic chemistry knowledge (different types of structure), the recognition and transformation of chemical structure. The key to the automatic answer is text processing and understanding.

Algorithm 1 flow.

There are six steps to automatically answer the questions of isomers with benzene ring: text processing, relation extraction, figure recognition and transformation, knowledge base matching, question solving and answer generation. Automatic solution of isomers with benzene ring.

Algorithm1: Text processing and understanding

Input: Organic isomers with benzene ring in natural language
Output: the process of solving the problem of isomers of organic compounds with benzene ring
Step 1: (text processing) text processing includes word segmentation, part of speech tagging and some proper noun replacement;
Step 2: (relation extraction) for the isomer topic, the syntactic semantic model is used to extract the direct relation and effective information expressed in the topic;
Step 3: (figure recognition and transformation) recognize the structural formula of organic compounds and transform it into molecular formula;
Step 4: (knowledge base matching) use the content extracted in steps 3 and 4 to match with the knowledge base;
Step 5: (output answer) sort out the content and options to match and output answer.

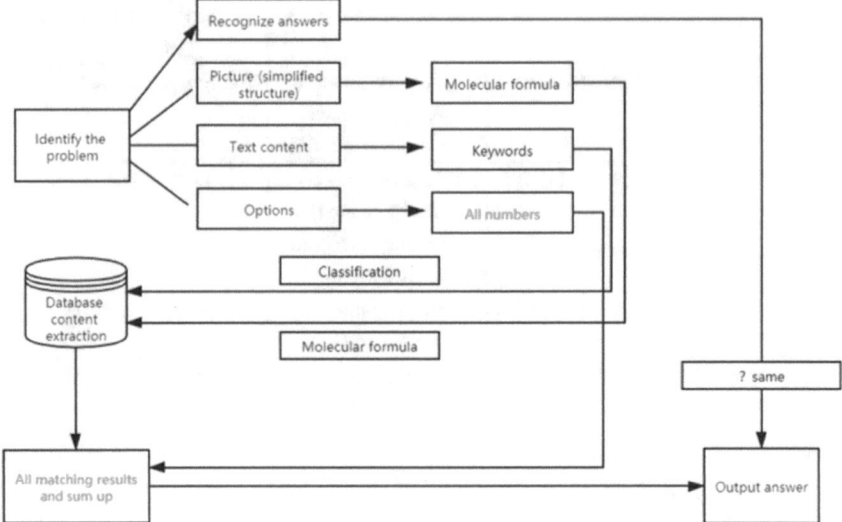

Fig. 3 Automatic solution flow chart

4.1 Text Processing

As the Fig. 3 shows that the first step of text processing is text preprocessing, the second step is proper noun replacement, and the third step is word segmentation and part of speech tagging. Since the automatic answer of isomers of organic compounds with benzene ring is a multiple-choice question, it is necessary to preprocess the text when understanding the text. First, only the words other than the options are recognized, that is, the content of the title. The proper noun (e.g. isomer) is replaced by a word, which is conducive to word segmentation and part of speech tagging. If it is not replaced, the isomer of a word will be divided into two or more words, and the part of speech will change accordingly, leading to errors. Because the topic is more monotonous and neater, there is no need for too much operation.

4.2 Figure Recognition and Transformation

For the automatic answering of isomers of organic matter with a benzene ring text understanding is one of the key points, and the other key point is the identification and transformation of organic matter structural formulas. The current more mature technology for chemical structure figure recognition is to use the optical structure for structure recognition. OSRA is a very useful tool that can convert the structure on the picture into InChI, In ChIkey, SMILES, SDF data [7]. OSRA can obtain a variety of formats, automatically recognize and extract graphical information that represents chemical structures, and generate SMILES or SD representations of molecular structure figures encountered. In general, for chemical figure recognition, the first step is OSRA recognition to extract the figure, and the second step uses the SMILES string to express the obtained structural formula. The SMILES string contains the numbers C and O.

5 Knowledge Base Matching

For many automatic answering questions, the support of a knowledge base is needed. To solve the problem of rate, choose to use the form of database to build this knowledge base. The database itself is a tool that can improve search efficiency. The title of organic compounds with a benzene ring and isomers in organic chemistry is sorted out. It is concluded that there are five kinds of functional groups involved in the solution of isomers, which are alcohols, esters, acids, phenols, and aldehydes. Their structural formulas are $-OH$ (not directly connected with a benzene ring), $-COOR$, $-COOH$, $-OH$ (directly connected with a benzene ring), $-CHO$. Pay attention to distinguish alcohols and phenols, phenols The functional groups need to be directly connected with the benzene ring. According to the input and output of knowledge base matching, input molecular formula and output organic compound structural formula with a benzene ring. Therefore, when building the database, we need to include the organic compound structural formula and molecular formula and need a structural formula ID as the prime Key to avoid repetition. For the only structural formula identified by the molecular formula and functional group, the database should also include the category field of organic compounds. The fields of the database table are as follows Table 1.

For the storage and search of database, it is similar to the form of decision tree. First, search downward with one condition, and then search downward with another condition until all the contents that meet the condition are found. It is similar to the search of decision tree. As shown in the figure below, the molecular formula and type

Table 1 Database table

Field name	Data type	Is it empty	Default value	Explaination
Structure ID	Integer		Identity	The structure of organic compounds
Molecular formula	Varchar (20)	Null		Molecular formula
Number C	Integer	Null		The total number of C
Number H	Integer	Null		The total number of H
Number O	Integer	Null		The total number of O
Classification	Varchar (20)	Null		Classification of organic matter
Simple structure	Varchar (80)	Null		Organic structure

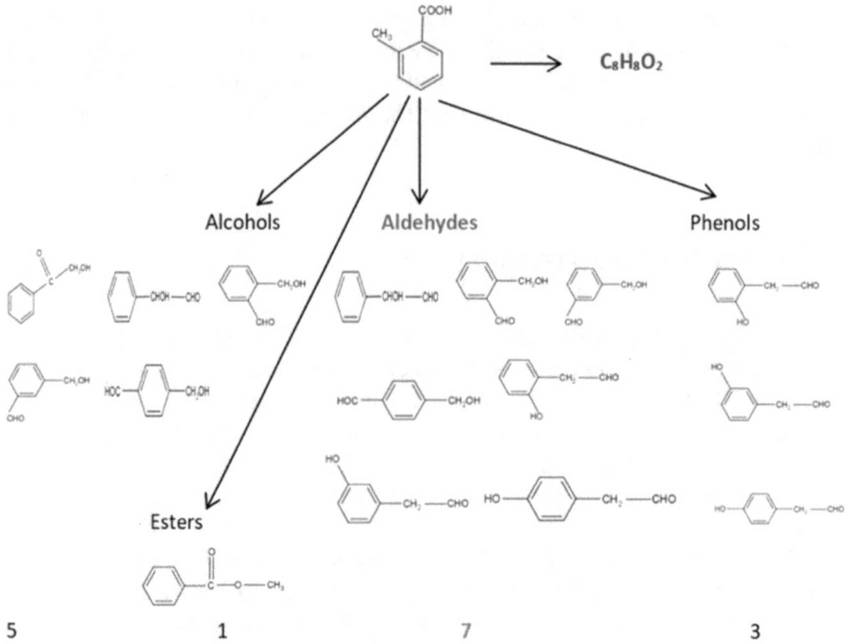

Fig. 4 Content extracted from the database

are the known search conditions. Taking $C_8H_8O_2$ as an example, the data is stored as follows: This Fig. 4 represents all the structural formulas formed by $C_8H_8O_2$. There are 16 in total, including four types, namely five alcohols, seven aldehydes, three phenols and one ester. After inputting the structural formula, it is converted into a molecular formula. Query all the structural formulas contained in the molecular formula in the database, and then search according to the type, and output the answer.

The storage and search of the database, it is similar to the form of the decision tree. First, search downward with one condition, and then search downward with another condition until all the contents that meet the condition are found. It is similar to the search of the decision tree. As shown in the figure below, the molecular formula and type are the known search conditions. Taking $C_8H_8O_2$ as an example, the data is stored as follows: This table represents all the structural formulas formed by $C_8H_8O_2$. There are 16 in total, including four types, namely five alcohols, seven aldehydes, three phenols, and one ester. After inputting the structural formula, it is converted into a molecular formula. Query all the structural formulas contained in the molecular formula in the database, and then search according to the type, and output of the answer. According to the effective input information, five kinds of structure formulas are extracted from the data, and then output and sum to 5. Based on the above information, we can find the suitable structure in the knowledge base and sum it statistically, then compare it with the numbers in the options and output the answers.

6 Automatic Diagnosis

The solution to the problem of organic isomers can not only complete the automatic answer to the problem but also can diagnose the problem automatically. The meaning of automatic diagnosis is that the machine can judge whether the answer to a problem can be correct or not. Automatic diagnosis is an important aspect of the education system and a technology needed at present in Fig. 5.

This type of question is a multiple-choice question, and the key to automatic diagnosis is the recognition of handwritten letters. For the test of handwritten letter recognition, there are some machine learning algorithms, such as KNN, K-means, SVM, and so on, but the efficiency and accuracy need to be improved.

This section describes the KNN algorithm for handwritten digit recognition. KNN algorithm: The samples in the dataset have corresponding labels, and each label represents a classification. After getting a new unlabeled data, we find the feature tag that is closest to the new data in the dataset. The principle of KNN is to judge which category x belongs to according to the category of the nearest k points when predicting a new value X. So the choice of K value is very important. Through the study of benzene ring isomers, it is found that there are at most eight kinds of isomers and at least 0 kinds of isomers. Therefore, data recognition is set between 0 and 8; that is, handwritten digits between 0 and 8 are recognized. The data is divided into training data and test data. The training data set is set to 50 pieces of each number, a total of 450 pieces; the test data set is set to 15 pieces of each number, a total of 135 pieces. InPython, firstly, the PNG format image is converted into a txt text containing only 0 and 1, the contour of the number is drawn, and the data are trained in the model. Then the test data set is used to test and classify each number, and the accuracy is calculated. The following figure shows the txt file, which is classified by number.

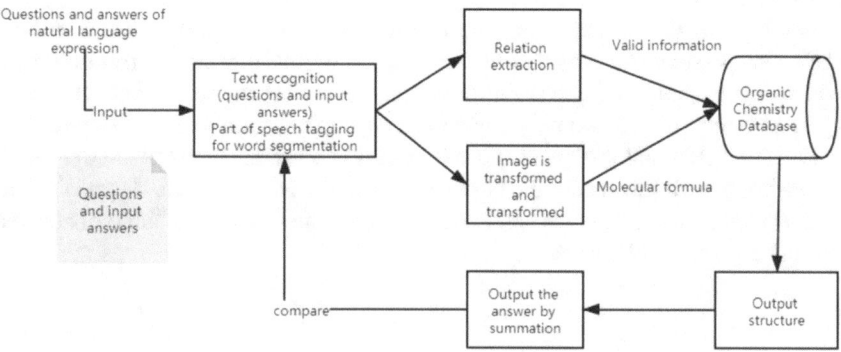

Fig. 5 Automatic diagnosis process

Table 2 Accuracy of different parts

Name	Total numbers	Correct numbers	Accuracy (%)
Syntactic-semantic model	63	59	93.65
OSRA figure recognition	60	57	95.00
Automatic solution result	63	55	87.30
Recognize handwritten letters	60	56	93.3
Recognize handwritten numbers	135	127	94.07

7 Experiment

This paper chooses 63 benzene ring organic isomers as the dataset for the automatic solution questions. The questions are all multiple-choice questions, and the questions contain the text and organic chemistry figures expressed in natural language. The questions come from the people's education press organic chemistry elective five, exercise book five-year high school entrance examination three-year simulation, real questions, and questions with answers on the Internet. The automatic solution to questions includes text understanding, syntactic-semantic model relation extraction, content matching, and output answers. In the syntactic and semantic model part, the model includes syntactic part and semantic part, representing a type of sentence, and one model matches a relationship to extract key information. For chemistry figures, 60 different organic compounds with benzene ring are selected as a dataset, which is characterized by various types of functional groups including carboxyl, hydroxyl, hydrocarbon, and ester groups are used to identify and test the accuracy of OSRA. The automatic diagnosis includes two algorithms, 200images are selected for recognizing handwritten letters, and the training dataset and test data set are divided into 7:3; 450 images are selected for recognizing handwritten numbers, and the data set and test data set are also divided into 7:3.

In this paper, the accuracy rate is used as the standard of the evaluation algorithm. The syntactic-semantic model is used to extract useful information from the text. When a text contains multiple key information, it is correct after it is extracted; OSRA figure recognition is correct only when the complete recognition of the benzene ring and all the functional groups connected with it is correct; when the answer result is consistent with the standard answer, it is correct. For automatic diagnosis, two methods are used to recognize handwritten letters. The following a Table 2 show the relevant experimental results.

8 Conclusion

This paper has developed an algorithm for understanding and solving the chemistry problems involving some isomers. This is the first algorithm for solving chemistry problems involving some isomers. It built on multiple new methods, which are the S2 method of understanding the problem text, the method of extracting knowledge items from problem figures and the method of combining the knowledge items from text and figure, and the inference method with the aid of knowledge base. The proposed algorithm has a good performance on the prepared dataset. The main contributions of this paper are as follows:

(1) A relation extraction method based on syntax and semantics is pro-posed to find the effective information or related relations in problem text. A pool of S2 models is constructed, which can extract the desired relations from text effectively.

(2) A database about the isomers of benzene ring organic compounds is established, which is used as a knowledge base to extract the content that meets the conditions.

(3) A diagnosis method is proposed based on the solution provided by the proposed algorithm.

There are multiple future jobs. First, the S2 models for isomers of organic compounds with Benzene ring needs to build the sophisticated pool. Second, we need to seek a better algorithm for understanding figures in chemistry problems. Third, we need to develop a new knowledge base to better serve the solving procedure.

References

1. Algorri M-E, Zimmermann M, Hofmann-Apitius M (2007) Automatic recognition of chemical images. In: Eighth Mexican international conference on current trends in computer science (ENC 2007). IEEE, pp 41–46
2. Andrenucci A, Sneiders E (2005) Automated question answering: review of the main approaches. In: Third International conference on information technology and applications (ICITA'05), vol 1. IEEE, pp 514–519
3. Casey R, Boyer S, Healey P, Miller A, Oudot B, Zilles K (1993) Optical recognition of chemical graphics. In: Proceedings of 2nd international conference on document analysis and recognition (ICDAR'93). IEEE, pp 627–631
4. Filippov IV, Nicklaus MC (2009) Optical structure recognition software to recover chemical information: Osra, an open source solution
5. Jiang Y, Tian F, Wang H, Zhang X, Wang X, Dai G (2010) Intelligent understanding of handwritten geometry theorem proving. In: Proceedings of the 15th international conference on Intelligent user interfaces, pp 119–128
6. Maoz S, Ringert JO (2018) A framework for relating syntactic and semantic model differences. Softw Syst Model 17(3):753–777
7. Shi S, Wang Y, Lin C-Y, Liu X, Rui Y (2015) Automatically solving number word problems by semantic parsing and reasoning. In: Proceedings of the 2015 conference on empirical methods in natural language processing, pp 1132–1142
8. Sneiders E (2002) Automated question answering using question templates that cover the conceptual model of the database. In: International conference on application of natural language to information systems. Springer, pp 235–239
9. Yu X, Wang M, Gan W, He B, Ye N (2019) A framework for solving explicit arithmetic word problems and proving plane geometry theorems. Int J Pattern Recognit Artif Intell 33(07):1940005

Automatic Question Answering System for Semantic Similarity Calculation

MinChuan Huang⑩, Ke Chen⑩, XingTong Zhu⑩, and GuoQuan Wang⑩

Abstract The automatic question answering system based on semantic similarity calculation includes three modules: word segmentation module, question understanding module and FAQ database module. Jieba, an open-source tool, is used in the word segmentation module. The problem understanding module can be further divided into problem classification, keyword extraction, and keyword expansion. The hierarchical classification method based on self-learning rules is used for problem classification. The common question database module distinguishes sentence similarity calculation and question matching. Sentence similarity calculation is based on the How Net semantic dictionary. The core algorithm is the rule algorithm design based on the corpus. The system relies heavily on each module, so it is difficult to establish a more perfect test scheme. Therefore, we only test the sentence similarity calculation which ultimately determines the accuracy of the problem matching, and finally realize the function of each module, and test and evaluate each module. The test results can be summarized that the sentence segmentation is relatively short, the part of speech contains less, and the similarity judgment is relatively concentrated, which is caused by the absence of specified parts of speech in both sentences. According to the part of speech coverage specified by the system, the more comprehensive the coverage, the more accurate the similarity calculation.

Keywords Machine learning · Chinese word segmentation · Question classification · Semantic similarity · Question base

1 Introduction

This is a newly developed Chinese question answering system [1–3], covering computer linguistics, machine learning [4–6] and natural language processing technology, code implementation, algorithm principle of the system, artificial intelligence and information retrieval [7]. System and design of the thinking scheme,

M. Huang · K. Chen · X. Zhu (✉) · G. Wang
Guangdong University of Petrochemical Technology, Guangdong 525000, China
e-mail: 305299282@qq.com

natural language questions, semantic similarity analysis, question answering system can automatically answer. In order to improve the efficiency of information retrieval when answering questions, semantic similarity calculation is needed to accurately match a large number of existing questions in the question database, find out the appropriate answers, and give feedback to users.

The way to retrieve information on the Internet is to type keywords through the search engine. The traditional search engine mode is that users return a large number of hyperlink indexes of relevant pages by typing keywords. Users filter relevant pages through the introduction of the page and the matching of keywords, and then enter the relevant pages to browse and further filter the information they want [8].

Advanced automatic question answering system mode, the user can directly input questions in natural language, and the system can return the user's desired answers in natural language by understanding the questions. As a more advanced information retrieval technology, automatic question answering system has been put forward and received widespread attention. Users can ask questions in natural language and get accurate answers quickly through automatic question answering system [9].

The main research institutions of Chinese retrieval information are Institute of computer science, Chinese Academy of Sciences, Fudan University, Harbin Institute of technology, Beijing Language and Culture University, etc. The Chinese Academy of Sciences has developed National Knowledge Infrastructure (NKI) question and answer system. The knowledge base includes more than ten knowledge bases, such as geographic knowledge, weather forecast, character knowledge, etc. In terms of Chinese word segmentation technology, there are nlpir of Chinese Academy of Sciences, LTP of Harbin Institute of technology, and thulac of Tsinghua University.

2 System Analysis and Theoretical Conception

The theoretical basis of Chinese automatic question answering system is different from English grammatical structure [10]. Chinese sentence structure is highly complex. It is difficult in natural language processing technology. How to make word segmentation fast and accurate is the primary problem of Chinese automatic question answering system. There is also the problem classification, which needs a deeper analysis of sentences. Sentence similarity calculation and question matching are based on the result of sentence similarity calculation. If the similarity is higher than a certain value, the matching is considered successful. There are also two problems, ambiguity resolution and unknown word recognition.

2.1 Chinese Word Segmentation Tool

Among the existing Chinese word segmentation tools [11], the Natural Language processing & Information Retrieval Sharing Platform (NLPIRSP) [3, 12] of the

Institute of computing technology, Chinese Academy of Sciences can fully meet the processing needs of users, including network crawling, text extraction, Chinese word segmentation, word tagging and so on. It is the most comprehensive text processing tool in the current word segmentation tools. At present, among the Chinese word segmentation tools, the most comprehensive text processing tool is the Natural Language Processing & Information Retrieval Sharing Platform (NLPIRSP) of Institute of computing technology, Chinese Academy of Sciences. NLPIRSP can meet the needs of users for big data text processing from all aspects, including the complete technical chain of big data: network capture, text extraction, text retrieval, text retrieval, etc. Chinese and English word segmentation, part of speech tagging, entity extraction, word frequency statistics, keyword extraction, semantic information extraction, text classification, emotion analysis, semantic depth expansion, complex and simple coding conversion, automatic phonetic notation, text clustering, etc. NLPIR provides rich open API, which can be integrated into all kinds of complex operating systems, and can also be called by all kinds of mainstream development languages.

2.2 Chinese Word Segmentation Algorithm

The segmentation algorithm of Chinese dictionary is based on the matching and cutting of Chinese sentences and existing dictionaries. According to the length of matching words, maximum matching and minimum matching are distinguished. Forward matching and reverse matching. Chinese word segmentation faces two major problems, ambiguity resolution and unknown word recognition. The causes of ambiguity can be divided into three types: intersection ambiguity, combination ambiguity and true ambiguity. The identification of unknown words is not included in the dictionary, which can be divided into proper nouns and non-proper nouns. With the change of social life, there will be more and more categories of unlisted words.

The main purpose of question classification is to classify according to the types of questions and improve the efficiency and accuracy of question answering system. Question classification can effectively simplify the classification of candidate answers. Bayes theorem formula, $P(H|x) = (P(x|h) P(H))/P(x)$, where p (H|x) belongs to a posteriori probability, which refers to the probability of h under condition X. P (H), P (x) and P (x|h) belong to a priori probability, and the former two refer to the probability of X and H events. The latter refers to the probability of X under H condition. Naive Bayes algorithm is used to adjust Bayes classification method, and its simple formula is shown in (1) below [13, 14].

$$P(X|C_i) = \prod_{n}^{k=1} P(x_k|C_i) \tag{1}$$

3 System Module Design

Chinese automatic question answering system is divided into user view interface and question processing background. The user view interface is used to give the user feedback of typing questions and getting answers [15, 16].

The background of question processing includes three parts: word segmentation module, question understanding module and quick Q & A library module. The structure of Chinese automatic question answering system is shown in Fig. 1.

3.1 Chinese Word Segmentation System

Chinese word segmentation adopts Jieba word segmentation [17, 18] of Python open source tool. Jieba word segmentation tool provides user dictionary interface, which is conducive to the expansion of subsequent professional vocabulary, and realizes more accurate word segmentation of professional vocabulary. Jieba word segmentation is based on trie tree structure to achieve efficient word graph scanning and generate all possible Directed Acidic Graph (DAG). The word segmentation tool has a dictionary, which contains more than 20,000 words. The word structure includes the word itself, word frequency and part of speech, and word frequency statistics. Before word segmentation, the dictionary is loaded in advance and loaded into a Trie tree to improve the search speed.

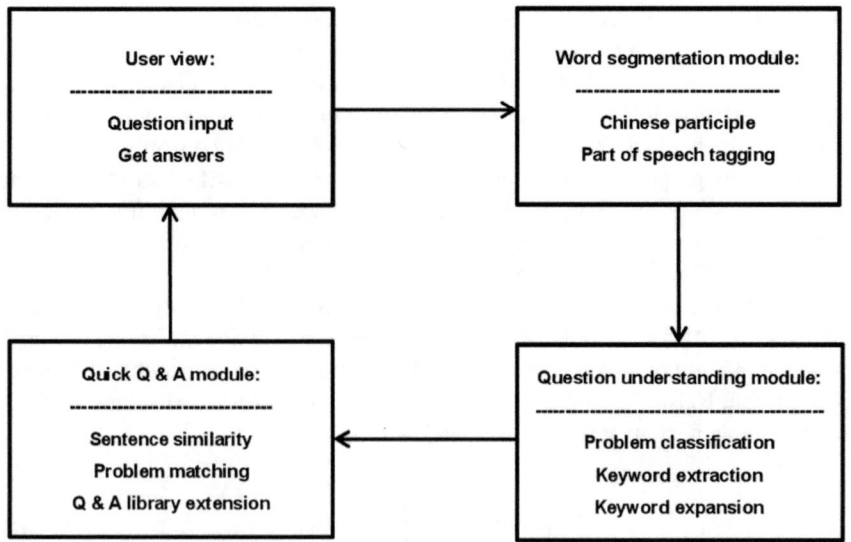

Fig. 1 The structure of Chinese automatic question answering system

Table 1 Styles available in the word templet

Code	Noun	Annotation
a	Adjective	The first letter
n	Noun	The first letter
v	Verb	The first letter
d	Adverb	The second letter
m	Numeral	The third letter
nr	Name	The first letter

For the recognition of the non-logins, the Jieba participle uses Viterbi algorithm based on the Hidden Markov Model (HMM) model to realize the prediction segmentation. HMM is a mathematical statistical model, which is used to describe the HMM process with unknown parameters. Through the word frequency record in the dictionary, the Jieba segmentation uses dynamic programming to find the maximum probability path, finds the maximum partition combination, and solves the problem of ambiguity resolution. The words are divided into the best, the frequency of words is determined, the maximum path is determined according to the dynamic programming method, and the maximum probability of segmentation combination is obtained.

Part of speech tagging, Jieba word segmentation tool supports part of speech tagging at the same time. Part of speech tagging adopts the part of speech tagging set of Peking University Institute of computing. Part of speech tagging is helpful to understand the structure of questions, and it is the main parameter for the subsequent problem classification and sentence similarity processing. See Table 1 for the code, noun and annotation of part of speech tagging.

3.2 Problem Classification and Keyword Generation

The hierarchical classification method based on machine learning rules is used to realize Chinese problem classification [19]. The training result of QCR and QHCR rule matching is incomplete, so it is not used. Bayesian classifier is selected to improve the classification accuracy, keyword extraction and keyword expansion. Figure 2 flow chart of problem classification.

Keyword extraction, input question segmentation, need to extract the key words in the question. We need to filter some words with low information content, extract the words with high information content, and finally use the remaining keywords for subsequent similarity calculation.

In order to identify synonyms and many uncommon professional words, the system needs to expand keywords.

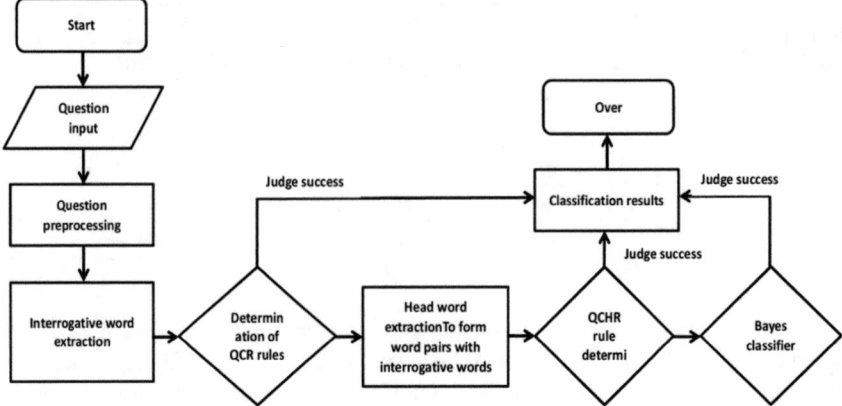

Fig. 2 Flow chart of problem classification

3.3 Sentence Similarity Calculation

It is divided into sememe similarity calculation, word semantic similarity calculation and sentence similarity calculation. The word similarity calculation based on HowNet is adopted, and the relevant sememes are extracted as the basis of similarity matching through the corresponding knowledge structure of each word in HowNet dictionary.

Sememe Similarity Algorithm

For the synonymous primitive tree, according to the characteristics of the tree structure, the distance between the two sememe nodes is calculated, and then the similarity value of the sememe is calculated through the formula. As shown in Fig. 3, Y1 and Y2 traverse upward to find the common cross node Y7, and the sum of their steps is the distance between the two primitive nodes.

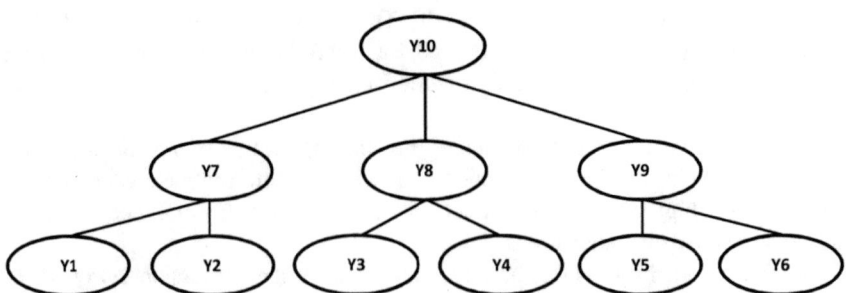

Fig. 3 The tree structures

Word Similarity

In China National Knowledge Infrastructure (CNKI) dictionary, every dictionary will have a sememe description, including one or more sememes. The similarity of each sememe is calculated by rules, and the word similarity is calculated by weighted conversion.

Sentence Similarity

The calculation of sentence similarity is based on the calculation of word similarity. Through the word segmentation module, the cut set of words is obtained. According to the importance of words in the sentence, the weighted calculation is carried out to get the sentence similarity value.

Figure Descriptions

Every figure should have a figure description unless it is purely decorative. These descriptions convey what's in the image to someone who cannot see it. They are also used by search engine crawlers for indexing images, and when images cannot be loaded.

Problem Matching

The method of question classification and sentence similarity calculation matches the question set in the fast question answering database, and the answers are responded to the users. Questions in question answering database will be classified and keywords extracted in advance, and the results will be loaded into memory to improve the efficiency of question matching.

4   Core Algorithm

The system uses Python language which is open source on the Internet to develop Jieba word segmentation component of Chinese word segmentation module. The function of Jieba word segmentation module includes word segmentation, part of speech tagging and keyword extraction. The keyword extraction of Jieba word segmentation is based on the TF-IDF algorithm and text rank algorithm, which calculates the weight of words in a large number of texts. The corpus of question answering corpus of this system mainly exists in the form of short questions, and there is no data base of a large number of texts. Therefore, the key word extraction function of Jieba word segmentation is not used [20, 21].

4.1 Word Segmentation Module

There are styles for block quotations, which should be used for quotes that are separated from in-line text. Below is an example.

Jieba word segmentation function and part of speech tagging function to achieve the design of word segmentation module. According to the question and answer corpus of this system, all the existing questions are directly segmented with part of speech, and the segmentation results are analyzed.

Jieba word segmentation is based on its own vocabulary. The parts of speech of punctuation and unmarked words are marked with X; There are no English words in the dictionary, and the English part of speech is marked as end, which is compatible with non-Chinese words and symbols in the text. Jieba word segmentation component is enough to meet the basic needs of daily sentence segmentation. The system is a question answering system for the information field. All questions contain words in the information field, which is different from the results of daily word segmentation.

4.2 Problem Classification

To classify problems, we need to determine the type system of problems. There is no unified classification system for Chinese problem classification. This system establishes its own problem classification system based on the characteristics of information field. According to the UIUC classification system standard and the characteristics of Chinese problems, Harbin Institute of technology adopts the hierarchical classification method to formulate the Chinese problem classification system.

This classification system is directly used as the basis of the system problem classification system to ensure that the system can better expand the problem classification system. To classify problems, we must first establish the type system of problems.

The question and answer set in information field focuses on DES (description class). Des (description class) includes reason, relation, effect, form, behavior, rules, comment, purpose, trait, type and advantage. See Table 2 problem classification system.

After establishing the classification system, we need to establish the problem classification rules. The hierarchical classification method based on self-learning rules can reduce the workload of rule establishment and the subjectivity of human operation in classification. To establish self-learning rules, we need to extract interrogative words and head words first.

Interrogative words include general interrogative words and special interrogative words. Special interrogative words are used to classify problems, which can be realized by one interrogative word. General interrogative words are used to classify problems when there are not enough interrogative words. When the question contains general interrogative words, it needs the auxiliary judgment of head words.

Table 2 Problem classification system

Classification	Details
DES (description class)	Adventage, Behavior, Comment, Charact, Effect, Form, Purpose, Reason, Relation, Rules, Type, Other
HUM (people)	Description, Organization, Person, Other
LOC (location)	Address, Country, Other
NUM (digital)	Area, Code, Count, Frequency, Percent, Range, Speed, Temperature, Other
OBJ (entity class)	Academic, Event, Food, Instrument, Language, Substance, Other
Time	Era, Time, Other
Unknown	

The head word reflects the essence of the problem and does not have ambiguity. The combination of general interrogative words and head words can classify questions.

To determine the role of interrogative words and head words in problem classification, it is necessary to design self-learning rules for problem classification. Self-learning rules can be divided into two parts, one is interrogative word type self-learning rule (QCR rule), the other is interrogative word head type self-learning rule (QHCR rule).

Self-learning Rules of Interrogative Words

In order to facilitate the expansion and adjustment of the interrogative words, the interrogative words are extracted from the self-learning rules, and the way of establishing the interrogative words list is to extract the interrogative words from the sentences. The purpose of learning rules is to determine the association rules of question words and question types, extract question types, and realize the manual division and annotation of question sentences in training corpus.

Self-learning Rules of Interrogative Head Words

The self-learning rule and the self-learning rule of interrogative word type add a parameter to match the head word. It is necessary to establish this rule to match the general interrogative words that cannot be matched. The extraction of head words is more difficult than the extraction of interrogative words. We can build an interrogative word list to match interrogative words, but we can't build a head word list because the number of head words is more than that of interrogative words, and there are many possibilities of matching with general interrogative words. The existing algorithms for the extraction of head words include syntactic dependency structure analysis. The implementation of syntactic dependency structure is complex and needs a large number of training corpus, so it is not used.

Improved Progressive Bayesian Classification Method

Self-learning rules can improve the accuracy of problem classification to a limited extent, but the rules extracted from the head words cannot cover all types of problems, which is caused by the complexity and diversity of Chinese question structure. The solution is to avoid the problems that cannot be covered by self-learning rules cannot be classified, import the third level classification rules, and improve the Bayesian classification method.

The first step of the system is Chinese word segmentation. The cut words are filtered by stop word list, and the remaining words are calculated as eigenvalues. The eigenvalues of Bayesian model should be independent as far as possible. Assuming that the problems in the problem set conform to the bag of words model, the remaining words after word segmentation are independent of each other, the position and order are not related, and some structural and semantic information will be lost. However, it can improve the algorithm implementation of Bayesian model. Set 0.5 as the zeroing factor, n represents the total number of problem types, and the value of n is 34. Combined with TF-IDF for weight processing, it needs to reduce the weight of words manually. X is the number of questions contained in the related question type, and Y is the number of times that the word appears in the related type.

$$P_2(qc, Q_1) = P_1(qc_i, Q_i) \times \log\left(\frac{N + 0.1}{N + 0.1}\right) \tag{2}$$

4.3 Keyword Extraction

Questions are segmented and keywords are extracted. The speed and accuracy of question matching can be improved by eliminating the low information words that hinder sentence similarity matching. The stop words database developed by Harbin Institute of technology is directly used as the basis for the establishment of stop words list. The low information words are filtered and the stop words list is expanded and adjusted.

The key points of the system include noun as part of speech, extended vocabulary of noun, adjective, verb and restrictive adverb. This system can't carry out this kind of division, and give different weights according to the part of speech for sentence similarity matching. Math statements should have the "Statement" style applied.

4.4   Word Similarity Calculation

After keyword extraction, sentence similarity matching becomes the similarity calculation of keyword set. Every computing unit is in the form of words, the first solution is the word similarity calculation.

This paper analyzes the structure of sememe description and divides it into four parts: the first independent sememe description, other independent sememe description, relational sememe description and symbolic sememe description. The word similarity calculation is divided into four aspects.

The first independent sememe description is calculated and its similarity is recorded as Sim1 (Y1, Y2). The similarity of other independent sememe descriptors was recorded as sim2 (Y1, Y2). All the independent sememes (except the first one) of two expressions are arbitrarily paired, and the sememe similarity of all possible pairings is calculated. The one with the highest similarity is selected, and they are grouped into one group. This process is repeated until all the independent sememes are grouped. The similarity is recorded as sim3 (Y1, Y2). And then calculate their similarity. The similarity of sememe description is calculated and recorded as sim4 (Y1, Y2). The calculation method of symbolic sememe description is consistent with that of relational sememe description. Through the above four parts of similarity calculation, weighted calculation can get the final similarity value of the two words. So far, the word similarity calculation is realized, which lays the foundation for the subsequent sentence similarity calculation.

$$\text{sim}(w1, w2) = \sum_{i=1}^{4} \beta_i \text{sim}_i(Y1, Y2) \tag{3}$$

4.5   Sentence Similarity Calculation

The final result of sentence similarity calculation affects the accuracy of problem recognition and matching. The following rules should be established to improve the efficiency and accuracy of sentence similarity calculation, and finally calculate the similarity value of sentences by combining the two.

The establishment of professional vocabulary list can improve the accuracy of the recognition of professional vocabulary by Jieba participators, and mark the word quality of professional words. The segmentation of Jieba has the ability to mark the word character. Only two words with the same word quality are calculated to improve the recognition efficiency of the system.

After the word and keyword extraction, we can get the key words set of two sentences. The first rule only calculates the similarity of words consistent. Match the words that meet the conditions, construct the matching matrix, calculate all similarity values, and take the maximum value for each line of the matrix, which is the similarity value of a keyword in the question to be matched.

When one word exists in one question and the other does not exist, the similarity value is defined as a small constant.

To determine the similarity value of key words in a sentence, we should make weighted calculation to obtain the final sentence similarity value. The formula is established: sentence similarity = noun similarity *a1 + verb similarity *a2 + adjective similarity *a3 + adverb similarity *a4 + other words *a5. After repeated statistics, the final value of each parameter is a1 = 0.3, a2 = 0.3, a3 = 0.18, a4 = 0.18, a5 = 0.04.

4.6 System Implementation Results

After the key words are extracted from the quick Q & A database module, we need to combine the word similarity method of HowNet and the weighted formula to calculate the sentence similarity. For the functional test of sentence similarity, the system test parameters are: the weight of noun part of speech is 0.3, the weight of verb part of speech is 0.3, the weight of adjective part of speech is 0.18, the weight of adverb part of speech is 0.18, and the weight of other parts of speech is 0.04, The similarity of the part of speech is 0.70. If one sentence has a specified part of speech and another sentence does not, the similarity of the part of speech is 0.1.

The test results show that the sentence structure is relatively short, the part of speech contains less, and the similarity judgment is relatively concentrated, which is caused by the system does not have the specified part of speech for both sentences. According to the part of speech coverage specified by the system, the more comprehensive the coverage, the more accurate the similarity calculation.

As shown in Table 3, in the last test case, the query content includes "rules", "abbreviations" and "hieroglyphs", and the artificial questions only contain "hieroglyphs". The system should be able to realize question matching, and the actual similarity value is less than 0.75, so the question matching should be realized by the way of question simplification. From the results, there are still some problems in sentence similarity calculation, which need to be further studied.

What is the content of computer security? What's the difference between e-mail protocol and e-mail protocol? The accuracy of Q & A similarity reached the highest of 0.88. How to construct open intranet? The similarity accuracy is the lowest, only 0.42.

Table 3 Sentence similarity test results

Original question	Artificial judgment	Similarity	Dissimilar sentences	Similarity
What is computer security?	What is the content of computer security?	0.88	What is hardware description language?	0.69
Why check the electrical rules?	What is the purpose of electrical rule detection?	0.80	Why design green chip?	0.61
What are the common email protocols?	What is the email protocol?	0.88	How do secure electronic transactions work?	0.54
How to realize informatization in enterprises?	How should a company realize informatization?	0.74	How to construct open intranet?	0.42
How to see a doctor through telemedicine system?	How can I use telemedicine to see a doctor?	0.82	How to set up campus network?	0.48
How to understand the rules, abbreviations and hieroglyphs of e-mail?	How to understand hieroglyphs?	0.61	How to use search engine?	0.53

5 Conclusions

Based on the core corpus of "the new 100,000 whys of computer and information science", this study constructs a common question database, and establishes a question system of more than 30 categories according to the characteristics of the corpus. Chinese word segmentation and part of speech tagging are realized by using Jieba word segmentation tool. Semantic understanding and similarity calculation are realized through the existing research results of HowNet dictionary. Hierarchical classification method improves the accuracy of rule classification. Python scripting language features, build functional module division, through the way of file construction to reduce the impact of module compatibility.

The development environment of the Chinese automatic question answering system is based on Windows system, combined with PyCharm development environment, using Python 3.6 programming language for development, using QT IDE of QT creator cross platform as interface and user interface (UI) for development. Python language is developed in the form of script. The core algorithm of the system includes word similarity calculation based on HowNet, sentence similarity calculation method based on word similarity, hierarchical classification method based on self-learning rules, etc. it is modular organization based on the characteristics of Python script language.

The success or failure of automatic question answering system depends on the integration of module functions, and its error will affect the decline of system accuracy.

What is the content of computer security? What's the difference between e-mail protocol and e-mail protocol? The highest similarity of Q & A was 0.88. How can I use telemedicine system to see a doctor? 82. What is the purpose of electrical rule testing? The similarity reached 0.80. How should a company realize informatization? The similarity reached 0.74. How to construct open intranet? The accuracy of similarity is the lowest, only 0.42.

This follow-up study can be extended to the wide application of university education administration. For example, when freshmen enter the University, they do not know the procedures and format templates of application documents for the relevant provisions of the student manual. At this time, the automatic question and answer app system of student manual is required to effectively guide the freshmen to handle relevant business. For example, the graduation design (Thesis) automatic question and answer web or app system of fresh graduates can assist the fourth-year students to inquire about the relevant provisions of the thesis format, how to download the relevant templates and templates, and how to handle the graduation and departure procedures.

Acknowledgements We would like to thank the School of Computer Science, Guangdong University of Petrochemical Technology (12440000727040230G). Project Number: 2019rc076, 2019rc078. Natural Science Foundation of Guangdong Province: 2018A030307032, 2018A030307038. Key research platform and project of universities in Guangdong Province: 2020zdzx3038.

References

1. Noraset T, Lowphansirikul L, Tuarob S (2021) WabiQA: a wikipedia-based thai question-answering system. Inf Process Manag 58:102431. https://doi.org/10.1016/j.ipm.2020.102431
2. Zihayat M, Etwaroo R (2021) A non-factoid question answering system for prior art search. Expert Syst Appl 177:114910. https://doi.org/10.1016/j.eswa.2021.114910
3. Pesquita C, Faria D, Falcão AO, Lord P, Couto FM (2009) Semantic similarity in biomedical ontologies. PLoS Comput Biol 5:e1000443. https://doi.org/10.1371/journal.pcbi.1000443
4. Chan H-Y, Tsai M-H (2019) Question-answering dialogue system for emergency operations. Int J Disaster Risk Reduct 41:101313. https://doi.org/10.1016/j.ijdrr.2019.101313
5. Zhu S, Cheng X, Su S (2020) Knowledge-based question answering by tree-to-sequence learning. Neurocomputing 372:64–72. https://doi.org/10.1016/j.neucom.2019.09.003
6. Zhao L, Zhang A, Liu Y, Fei H (2020) Encoding multi-granularity structural information for joint Chinese word segmentation and POS tagging. Pattern Recogn Lett 138:163–169. https://doi.org/10.1016/j.patrec.2020.07.017
7. Mohasseb A, Bader-El-Den M, Cocea M (2018) Question categorization and classification using grammar based approach. Inf Process Manag 54:1228–1243. https://doi.org/10.1016/j.ipm.2018.05.001
8. Xiong H, Wang S, Tang M, Wang L, Lin X (2021) Knowledge graph question answering with semantic oriented fusion model. Knowl-Based Syst 221:106954. https://doi.org/10.1016/j.knosys.2021.106954

9. Liu J, Wu F, Wu C, Huang Y, Xie X (2019) Neural Chinese word segmentation with dictionary. Neurocomputing 338:46–54. https://doi.org/10.1016/j.neucom.2019.01.085

10. Yuan Z, Liu Y, Yin Q, Li B, Feng X, Zhang G, Yu S (2020) Unsupervised multi-granular Chinese word segmentation and term discovery via graph partition. J Biomed Inform 110:103542. https://doi.org/10.1016/j.jbi.2020.103542

11. Ferreira JD, Couto FM (20) Multi-domain semantic similarity in biomedical research. BMC Bioinform 20:246. https://doi.org/10.1186/s12859-019-2810-9

12. Natural Language Processing & Information Retrieval Sharing Platform. NLPIRSP Homepage. http://www.nlpir.org/wordpress. Accessed 16 June 2021

13. Ayala J, García-Torres M, Noguera JLV, Gómez-Vela F, Divina F (2021) Technical analysis strategy optimization using a machine learning approach in stock market indices. Knowl-Based Syst 225:107119. https://doi.org/10.1016/j.knosys.2021.107119

14. Yu J, Zhu Z, Wang Y, Zhang W, Hu Y, Tan J (2020) Cross-modal knowledge reasoning for knowledge-based visual question answering. Pattern Recogn 108:107563. https://doi.org/10.1016/j.patcog.2020.107563

15. Zhao S, Wu Y, Tsang Y-K, Sui X, Zhu Z (2021) Morpho-semantic analysis of ambiguous morphemes in Chinese compound word recognition: an fMRI study. Neuropsychologia 157:107862. https://doi.org/10.1016/j.neuropsychologia.2021.107862

16. Li M, Li Y, Chen Y, Xu Y (2021) Batch recommendation of experts to questions in community-based question-answering with a sailfish optimizer. Expert Syst Appl 169:114484. https://doi.org/10.1016/j.eswa.2020.114484

17. Zhenqiu L (2012) Design of automatic question answering system base on CBR. Procedia Eng 29:981–985. https://doi.org/10.1016/j.proeng.2012.01.075

18. Zhang L, Lin C, Zhou D, He Y, Zhang M (2021) A Bayesian end-to-end model with esti-matuncertainties for simple question answering over knowledge bases. Comput Speech Lang 66:101167. https://doi.org/10.1016/j.csl.2020.101167

19. Zafar H, Dubey M, Lehmann J, Demidova E (2020) IQA: interactive query construction in semantic question answering systems. J Web Semant 64:100586. https://doi.org/10.1016/j.websem.2020.100586

20. Vanam MK, Amirali Jiwani B, Swathi A, Madhavi V (2021) High performance machine learning and data science based implementation using Weka. Mater Today: Proc S2214785321005617. https://doi.org/10.1016/j.matpr.2021.01.470

21. Xu Y, Zhou Y, Sekula P, Ding L (2021) Machine learning in construction: from shallow to deep learning. In: Developments in the built environment, vol 6, p 100045. https://doi.org/10.1016/j.dibe.2021.100045

Solving Shaded Area Problems by Constructing Equations

Zihan Feng, Xinguo Yu, Qilin Li, and Huihui Sun

Abstract This paper presents a method for solving shaded area problems that can generate the readable solution because it can adopt the way of constructing equations. Solving shaded area problems is a challenge research problem because of the expression diversity of the problems and the complicated relation between the shaded area and the other areas. Due to the diversity of expression of the problems, it is hard to design algorithms for this research problem. Some efforts have been made on this research problem; however, the proposed methods have some demerits. This paper proposes a novel way, whose main idea lies in constructing equations. Its main step is to acquire a system of equations. A batch of methods is proposed to construct the equations from the inputs. Thanks to the method of constructing equations, the proposed algorithm is concise and the solving process is understandable. The experimental results show that the proposed method is more accurate than the baseline one.

Keywords Problem solver · Shaded area problem · Equation construction

1 Introduction

Solving exercise problems in basic education has been an active research problem since Artificial Intelligence (AI) appeared in 1950s [1]. Solving shaded area problems is an interesting branch problem of solving exercise problems because shaded area problems are popular exercise problems, which is a kind of geometry reasoning problems that find the area of shaded region from a diagram with some constraints

Z. Feng · X. Yu (✉) · Q. Li · H. Sun
Central China Normal University, Wuhan 430079, China
e-mail: xgyu@mail.ccnu.edu.cn

Z. Feng
e-mail: zihanfeng@mails.ccnu.edu.cn

H. Sun
e-mail: 305863610@qq.com

[2]. Solving shaded area problems is to design algorithms that can generate solutions that students can understand and learn from them. This research problem aims to propose an approach to solve shaded area problem and can generate a process which the primary school students can understand. It is not considered as a solution if one uses a college-level method to find the area of the shaded region. For example, the following method is not acceptable. Simply mapping the problem into Cartesian coordinates and using finite integral to find the area enclosed by function curves is beyond the knowledge of the primary student. As a result, such a solving method cannot train students in thinking and deductive ability. Alvin et al. [2] worked on solving shaded area problems and proposed a hypergraph method. However, this method cannot promise the solving ability, which can only solving the problems with limited answer patterns. To build the auto tutorial systems, this paper is to propose a new method for solving shaded area problems.

The related work is in solving exercise problems in basic education. There were a batch of papers in solving math problem, like solving geometric proof problem [1, 3, 4], math word problem [5–8], etc. Most of these papers are not directly related with this research topic, but this paper will adopt the problem understanding methods developed by them. In particular, this paper develops a problem understanding method based on the method proposed in [9], which a batch of papers have demonstrated its wide application [6, 7]. And for problem solving framework, this work adopts the framework proposed in [5] which is a kind or relation extraction approach. Alvin et al. [2] is the most related paper to our research in solving shaded area problem. Alvin et al. [2] solved the shaded area problem by building an analysis hypergraph that targets to include all facts that can be derived of the figure, which probably is a hard task. In addition, their heuristic algorithm of constructing the hypergraph is complicated. The hypergraph construction is not efficient due to their hypergraph method which contains many steps. This paper proposes a new method, which adopts the popular techniques of constructing equations and Gaussian Elimination to overcome their limitation of hypergraph construction.

The step of identifying atomic regions is a key step of constructing equations.

The remaining paper is organized as follows. Section 2 gives the outline of the proposed solver. Section 3 presents the detail of our algorithm. In Sect. 4, data set and evaluation are shown and discussed.

2   The Outline of the Proposed Solver

2.1   The Shaded Area Problem

A typical shaded area problem consists of the text in natural language with diagram. In the problem scope, we follow the problem definition in [2] which ignore the situation that the area problem with no diagram. In [2], a shaded area geometry problem consists with three parts, a geometric figure, a set of assumption about the figure, and the shaded region, whose area is required to be found. The set of assumption contains the facts and constraints related the entities in the figure. For example, the length of side is 8, the area of triangle is 10, two lines are parallel to each other, are the constraints in the problem described by Fig. 2. The goal is to find the area of shaded region.

2.2   The Framework of the Auto Solver

This paper proposes a new method for solving the shaded area problems, which comprises three steps. The first step is called as problem understanding which is to extract the constraints and geometric facts from the problem. The second step is to find the answer from the extracted constraints and geometric facts. The third step is to generate the readable solution. The workflow is depicted in Fig. 1, in the top of this figure is a sample shaded area problem.

To extract the constraints from problem text, the text is first annotated by the annotation software. Then the constraints are extracted from the annotated text. As for the diagram, it is converted to a graph, whose vertices are intersection points in the figure and whose edges are segments which arise from non-crossing lines and arcs of

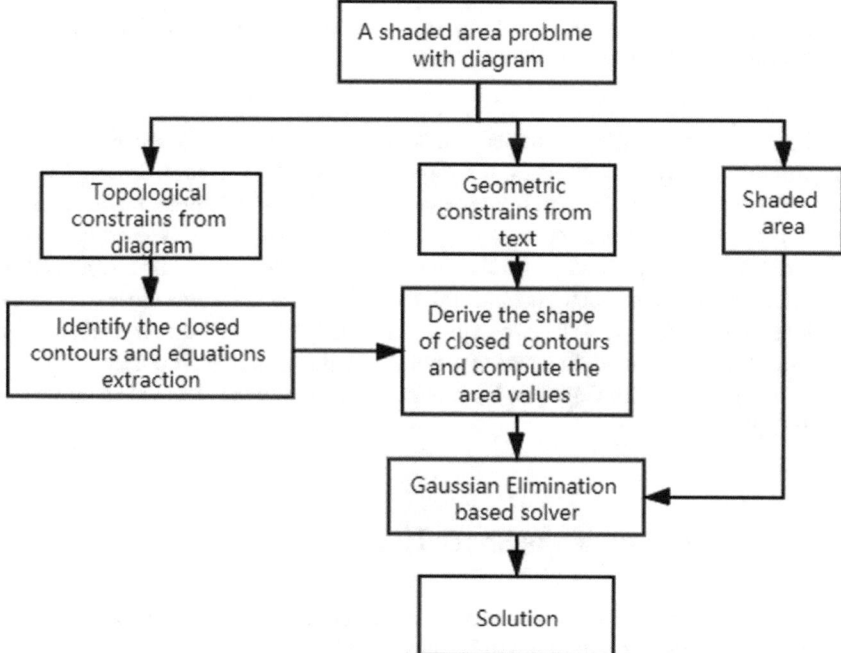

Fig. 1 The framework of the proposed solver

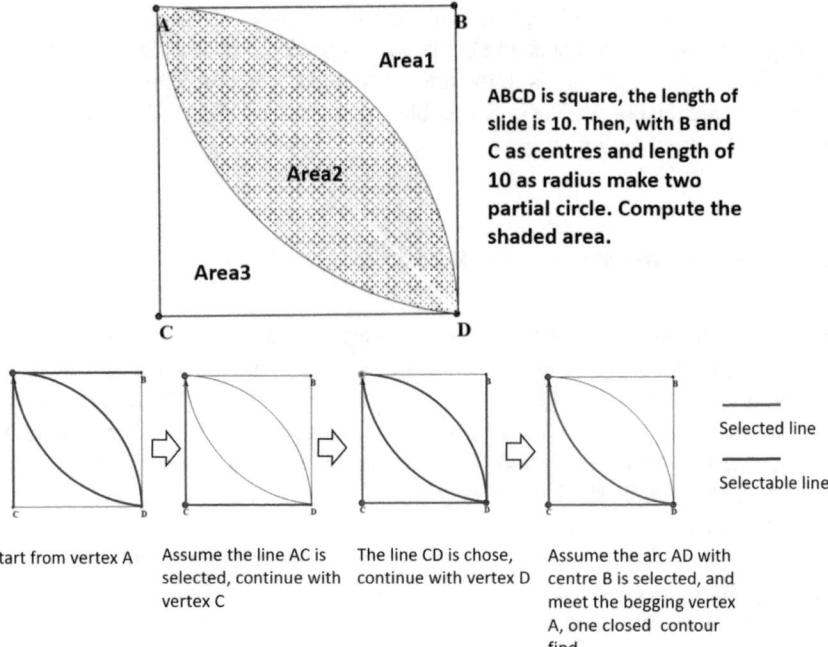

Fig. 2 The top part of the figure is a simple shaded area problem that contains both the text and the diagram, and the target area is shaded. The bottom part of the figure is an example to illustrates the iterations in closed contour searching algorithm

the figure. One challenge is to understanding the diagram which is essential to solve the problem. Many constraints do not directly express by text in real case. Consequently, we had to extract them from diagram. We define the diagram understanding as extracted the equations set from the relations appearing in the closed contours. We proposed a new method that can solve this task efficiently and be introduced in Sect. 3. After the closed contours and equations be extracted, some closed contours (regions) with certain shape (e.g., square, circle, triangle) could be computed, but for most shaded area problem that region is not the goal region, the proposed method to solve this problem with Gaussian Elimination which take the inference step which will be talked in Sect. 3.

3 Solver for Finding the Shaded Area

In this section, both the method to extract the relation from geometric diagram and the method to generate solution will be introduced.

3.1 Equation Construction from Geometric Diagram

A diagram consists of multiple regions. The key step of solving shaded area problems is to acquire the quantity relations among the areas of those regions. A region is called as an enclosure, which is defined by its closed contour in the geometric diagram. The closed contours are divided into two types, atom closed contour, abbreviated as ACC and composite closed contour, abbreviated as CCC. An ACC is the undivided region in the diagram and a CCC can be constructed by multiple ACCs. As shown is Fig. 2, Area1, Area2 and Area3 are three ACCs, and the square and the sectors are CCCs. Based on this fact, the diagram understanding is modelled as the problem of con-structing equations between ACCs and CCC, which is essential step in solving shaded area problem in many cases. This step of the proposed method replaces the hyper-graph construction in [2]. The procedure of our solver is more concise and understandable because it is based on relations and equations.

In order to construct the equations, we first have to identify all the ACCs among all the closed contours. At first, a DFS algorithm is used to find all the closed contours in the diagram (DFS stands for Depth First Search). The procedure for one iteration is illustrated in Fig. 2 using an example. This algorithm is modified from the classic DFS tree traversal algorithm.

After the searching procedure terminates, all the closed contours are divided into ACCs and CCCs according to the criterion that the ACC encloses no other enclosed regions. In other word, if one closed contour contain other closed contours that means this closed contour is CCC. And it is easy to find all the ACC when all the CCC are found. And in this process, the equations among the ACC and CCC can be obtained at the same time.

Using this method to identify the ACC is better than the method proposed in [2] by Alvin in two aspects. The first one is that this method needs less process for the nature geometric diagram. In their method, they used the algorithm in [10] to identify the ACC (They call this as the atomic region identification algorithm). Actually, the ACC is the facets in the planner graph. But for shaded area problem the ambiguities arise when arc exist in some case, and the delicate method is used to avoid the ambiguities by adding additional points. The second one is that this method is more concise than their method. This is due to the following facts. Their relations among ACCs and CCCs are derived in solution generation step. On the contrast, the proposed method can acquire these relations according to how a CCC encloses ACCs.

3.2 Constructing the Linear System and Problem Solving

The key idea of our solving method is that the target can often be computed by the linear combination of other directly calculated area. Hence, solving this problem can be converted into a problem of constructing a linear system. The constructed system is denoted as $MX = t$, where M is a $m * n$ matrix, the n is the number the ACCs in this problem and m is the number of directly computed region. The X is a vector which is the answer of this linear system. And the t is the vectorized representation of target. In the following part, the problem described in Fig. 2 will be used as example to illustrate our method.

The problem in Fig. 2, there are three regular areas (Sector (BAD), Sector (CAD), Square (ABCD)), and its corresponding area value is V_1, V_2, V_3 respectively. Combine the equations extracted from Sect. 3.1, the following equation set can be derived:

$$\text{Sector (BAD): Area1} + \text{Area2} = V_1$$

$$\text{Sector (CAD): Area2} + \text{Area3} = V_2 \tag{1}$$

$$\text{Square (ABCD): Area1} + \text{Area2} + \text{Area3} = V_3$$

To computer the Area2(target area), we try to find a set of X equal to $(x_1, x_2, x_3)^T$, which make

$$x_1 * V_1 + x_2 * V_2 + x_3 V_3 = \text{Area2}. \tag{2}$$

Then Eq. (2) can transform into the following Eq. (3).

$$(x_1 + x_3) * \text{Area1} + (x_2 + x_3) * \text{Area2} + (x_1 + x_2 + x_3) * \text{Area3} = \text{Area2}. \tag{3}$$

By further transforming Eq. 2. Area2 can be represented as [Area1, Area2, Area3]$*[0, 1, 0]^T$, where the t is $[0, 1, 0]^T$. For the problem in Fig. 2, the $MX = t$ is shown as follow:

$$\begin{bmatrix} 1 & 0 & 1 \\ 0 & 1 & 1 \\ 1 & 1 & 1 \end{bmatrix} \begin{bmatrix} x_1 \\ x_2 \\ x_3 \end{bmatrix} = \begin{bmatrix} 0 \\ 1 \\ 0 \end{bmatrix} \tag{4}$$

Then it is able to find out the values of x_1, x_2, x_3 by Gaussian Elimination. However, the proposed method does not use the Gaussian Elimination to compute the value of each ACC in the problem. Instead, the method is to finds a linear combination of the areas of ACCs that are computed by formula to represent the area of shaded region (target). For the problem in Fig. 2, it is able to compute each ACC in this problem with Gaussian Elimination, but this may not work for other problems due to the equations (known area) may not enough.

For example, in some case m will be larger than n, which will make this linear system cannot be solved with Gaussian Elimination (the number of equations is less than the number of variables). As shown in Fig. 3a, 8 regions are regular, which are 3 rectangles, 3 triangles, 1 sector and 1 trapezoid. The areas of these 8 regions can be directly computed. Since there are only 5 ACCs, then for this problem, the matrix

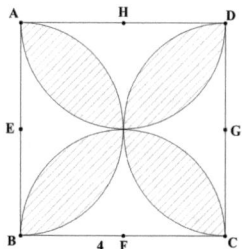 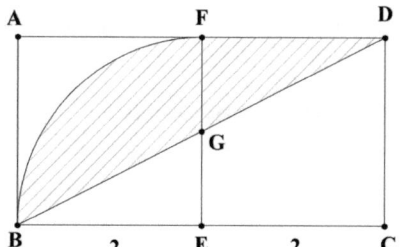

ABCD is a rectangle, and E,F is the middle point of the AD and BC. Compute the shaded area.

(a)

ABCD is a square, with E,F,G,H as centers make partial circle with half length of AB as radius. Compute the shaded area.

(b)

Fig. 3 Two shade area problems with text and diagram

M is a 5 * 8 matrix. This linear system is under-determined and according to the Rouché-Capelli theorem, there are indeed an infinitude of solutions since this linear system is consist. To handle this situation, for each time, we only select r identified regions instead of all identified regions to construct the linear system, where the r is the rank of the M. In each iteration we will get a M' where is $n * r$ matrix, and M' will be feed to the GE algorithm to get the solution. The solving algorithm be summarized as Algorithm1.

Algorithm1: GE based solving algorithm

Input: Extracted constraints from raw input
Output: The solution of given problem
1. Identify the shape of detected closed contours and compute the values
2. Construct the equation set S based on ACCs
3. Construct the linear system L by the method of undetermined coefficients
4. Check the solution situation of the L
5. Generate the solution

In some case, the matrix M can also be over-determined. For example, in Fig. 3b, there are 8 ACCs in total, and 5 regions (4 half circles and 1 square), can be directly computed, which means the matrix M is an 8 * 5 matrix, for this situation we directly feed the matrix M to the GE algorithm, because for this case if the solution exist in this linear system, there are some of rows of matrix M can be represented by other rows.

The other merit of the proposed method is that it is easy to know the problem can be solved or not by checking the order of matrix M and its augmented matrix $M^* = [M|t]$).

The solution situation be summarized as follow:

$$\left[\begin{array}{l} (1) \quad \text{Rank(M)=Rank(M*)} \begin{cases} \text{n>m} \\ \text{n}\leq\text{m} \end{cases} \qquad \begin{array}{l} \textbf{With multiple solution} \\ \text{Only one solution} \end{array} \\ (2) \quad \textbf{Rank}(\textbf{M}) \neq \textbf{Rank}(\textbf{M}*) \qquad\qquad\qquad \text{No solution} \end{array}\right.$$

4 Experimental Result and Evaluation

This section presents dataset preparation and experimental results. The problems in the dataset are from popular textbooks, reference books, examination papers, and from competition papers. There are totally 192 good quality shaded area problems, and 74 problems can be directly solved by our method. The remaining problems cannot be solved by the proposed method due to two main reasons. The first reason is that some problems need adding auxiliary lines. The dataset contains 56 such problems. The second reason is that some problems need geometric inference to find some facts. The dataset contains 62 such problems. These two parts are not in the scope of this paper. The more discussion on these two parts can be found in [1, 8, 11, 12].

The experimental results show that our method is time efficient. For the solved problem, the totally time for 74 problem is only 7.41 s. The algorithm runs on Intel Core i5-1035G4 CPU at 1.5 GHz with 16G RAM on 64-bit Windows 10 operating system.

The Fig. 4 is from Alvin's paper, the top part of this figure is a shaded area problem and the entire figure is shaded region. For their solving method, they calculated the regular region based on the formular which is similar to our method, and then

Fig. 4 An example from Alvin et al. [2]. The top part of the figure is a simple shaded area problem. The bottom part of the figure is an example to illustrate their solving method which they call it shape hierarchy

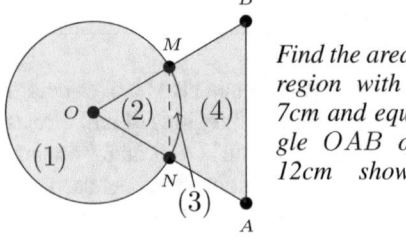

Find the area of the shaded region with circle radius 7cm and equilateral triangle OAB of side length 12cm shown at right.

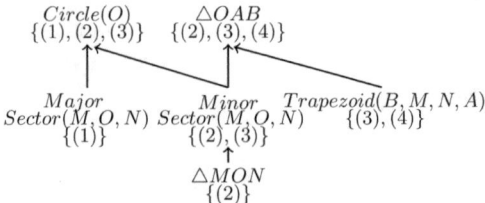

they organized those known regions into acyclic graph (they call this step as shape hierarchy), as shown in the bottom part of Fig. 4, the root is a root shape and each child of the node is fully contained in their parent. Then they take a heuristic method to derive more regions, they individually subtract each of its descendants to acquire a region and its associated area, if possible. For example, they can compute the area of region {(1), (3)} by taking the difference between Circle(O, OM) and Triangle(M, O, N): Area({(1), (3)}) = Area(Circle(O, OM)) − Area(Triangle(M, O, N)). Their last step is to exhaustively combine all regions and their associated areas. In each iteration, they only consider two regions, and if those two regions are disjoint, they compute the union of two region. In the case where one region is completely contained within the other region, they compute the area of the difference of the two regions. To analyze their method, for the last step, they take two kinds of opera and can be summarized to the pattern $A \pm B$, where the A and B are regions which come from shape hierarchy. There are two situations of the A and B, the first one is that the A and B are a shape that can be directly computed. The second one is that A and B is the difference of two shapes.

The solving pattern in their method certainly be covered by ours. It is because their method is also a kind of linear combination. And it is easy to know that this problem can be solved or not by using Rouché-Capelli theorem in our method, their method has to run all the procedures, then to know the problem can be solved or not. Hence, we only compare the problem that the time efficiency over the problems can be solved by our method.

To compare the solving ability, there are 74 problems totally, and 17 problems can not be solved by using their method. This is because their method not only limits the operator in the final expression but also limits the number of identified regions involved in the final expression. For example, the problem in Fig. 3a will not be solved by their method. The explanation is as follows. In their method, the maximum number of regions involved in the final expression is 4, and this problem can only be solved by taking the sum of 4 semi circle and then subtract the area of square which involved 5 regions. More detail of comparison be presented in Table 1.

To explain our method by the view of matrix analysis, each ACC in the diagram can be seen as the base of the vector space V, then each identified region is a vector e in V, basing on the problem input we can get a set of e, and target area is also a vector in V which is represented as t, then the problem solving can be seen to find the

Table 1 Experimental results

	Solved	Cannot solved due to the auxiliary lines	Cannot solved due to the lack of constraints	Unable to deduce the target area
The GE based method (ours)	74	56	62	0
The heuristics method in [2]	57	56	62	17

coordinate of t in the subspace V', where the e is the bases of the V', if this subspace covers the vector t, that means this problem can be solved.

5 Conclusion

The paper has proposed a new method to solve shaded area problems. The main step of the method is an equation construction step, which builds a set of linear equations among the areas of regions in the diagram. From this set of equations, the Gaussian Elimination is used to find a solution equation, its left side being the area of the shaded region and the right side being the linear combination of the areas of some ACCs that are computed by formula. The experimental results shows that the proposed method is better than the baseline method.

This paper has multiple contributions. First, it proposed an equation construction based method, which is an easy implementation method. Second, it proposed an equation construction method. Third, it proposed a solution synthesis method based on transforming the constructed system of equations.

The future work of this paper is to enhance the proposed method to solve more types of shaded area problems. The first job is to build the mathematics model to explain the proposed method. The second job is to integrate the geometric reasoning system to sole the shaded area problem that involves the geometric reasoning. The third job is to integrate adding auxiliary line method to solve more shaded area problems.

References

1. Zhang D et al (2019) The gap of semantic parsing: a survey on automatic math word problem solvers. IEEE Trans Pattern Anal Mach Intell
2. Alvin C et al (2017) Synthesis of solutions for shaded area geometry problems. In: FLAIRS conference
3. Mukherjee A, Garain U (2008) A review of methods for automatic understanding of natural language mathematical problems. Artif Intell Rev 29(2):93–122
4. Srihari RK (1994) Computational models for integrating linguistic and visual information: a survey. Artif Intell Rev 8(5–6):349–369
5. Yu X, Wang M, Gan W (2019) A framework for solving explicit algebra word problems and proving plane geometry theorems. Intl J Pattern Recognit Art Intell 1940005
6. Gan W, Yu X, Wang M (2019) Automatic understanding and formalization of plane geometry proving problems in natural language: a supervised approach. Int J Artif Intell Tools 28(04):1940003
7. He B, Yu X, Jian P, Zhang T (2020) A relation based algorithm for solving direct current circuit problems. Appl Intell
8. Chou S, Gao X, Zhang X, Zhang J (2000) A deductive database approach to automated geometry theorem proving and discovering. J Autom Reason
9. Yu X, Gan W, Wang M (2017) Understanding explicit arithmetic word problems and explicit plane geometry problems using syntax-semantics models 247–251. https://doi.org/10.1109/IALP.2017.8300590

10. Edelsbrunner H (1987) Algorithms in combinatorial geometry. Springer
11. Wang K, Su Z (2015) Automated geometry theorem proving for human-readable proofs. In: Proceedings of the 24th international conference on artificial intelligence (IJCAI'15) 1193–1199
12. Seo MJ et al (2014) Diagram understanding in geometry questions. In: Twenty-eighth AAAI conference on artificial intelligence

Cross-Lingual Automatic Short Answer Grading

Tim Schlippe and Jörg Sawatzki

Abstract Massive open online courses and other online study opportunities are providing easier access to education for more and more people around the world. However, one big challenge is still the language barrier: Most courses are available in English, but only 16% of the world's population speaks English [1]. The language challenge is especially evident in written exams, which are usually not provided in the student's native language. To overcome these inequities, we analyze AI-driven cross-lingual automatic short answer grading. Our system is based on a Multilingual Bidirectional Encoder Representations from Transformers model [2] and is able to fairly score free-text answers in 26 languages in a fully-automatic way with the potential to be extended to 104 languages. Augmenting training data with machine translated task-specific data for fine-tuning even improves performance. Our results are a first step to allow more international students to participate fairly in education.

Keywords Cross-lingual automatic short answer grading · Artificial intelligence in education · Natural language processing · Deep learning

1 Introduction

Access to education is one of people's most important assets and ensuring inclusive and equitable quality education is goal 4 of United Nation's Sustainable Development Goals [3]. Distance learning in particular can create education in areas where no educational institutions are located or in times of a pandemic. There are more and more offers for distance learning worldwide and challenges like the physical absence of the teacher and the classmates or the lack of motivation of the students are countered with technical solutions like videoconferencing systems [4] and gamification of learning [5]. The research area "AI in Education" addresses the application and evaluation of Artificial Intelligence (AI) methods in the context of education and training [6]. One of the main focuses of this research is to analyze and improve teaching and

T. Schlippe (✉) · J. Sawatzki
IU International University of Applied Sciences, XXX, Germany
e-mail: tim.schlippe@iu.org

E. C. K. Cheng et al. (eds.), *Artificial Intelligence in Education: Emerging Technologies, Models and Applications*, Lecture Notes on Data Engineering and Communications Technologies 104, https://doi.org/10.1007/978-981-16-7527-0_9

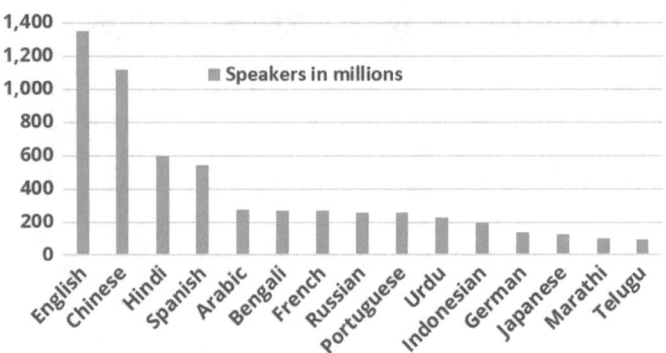

Fig. 1 The 15 most spoken *L1* and *L2* languages (based on [1])

learning processes. However, a major challenge is still the language barrier: Most courses are offered in English, but only 16% of the world population speaks English [1]. Figure 1 illustrates the 15 languages in the world which are spoken as first or second language. To reach the rest of the people with massive open online courses and other online study opportunities, courses would need to better support more languages. The linguistic challenge is especially evident in written exams, which are usually not provided in the student's native language.

To overcome these inequalities, we analyze AI-driven cross-lingual automatic short answer grading (ASAG). While the focus of related work in ASAG has been on the performance of a corpus in only 1 language–whether using monolingual or multilingual pre-trained natural language processing (NLP) models–the focus of this paper is on leveraging the benefits of a multilingual NLP model for the application on multiple languages in the context of cross-lingual transfer. The Multilingual Bidirectional Encoder Representations from Transformers model (*M-BERT*) [2] is such a multilingual NLP "model pre-trained from monolingual corpora in 104 languages" which can be adapted to a certain task with task-specific labeled text data in 1 or more languages (*transfer learning*) and then perform this learned task in other languages (*cross-lingual transfer*) [7].

Compared to separate monolingual ASAG systems, cross-lingual ASAG has the following advantages: First, only one model is required to cover many languages instead of separate models which saves storage space and is easier to maintain. Second, we do not need task-specific data in each target language for fine-tuning due to the cross-lingual transfer. To investigate cross-lingual ASAG, we compared the performances of three different approaches:

1. *M-BERT* fine-tuned on a single language
2. *M-BERT* jointly trained on 6 different languages and
3. Monolingual *BERT* models.

In the next section, we will present the latest approaches of other researchers for ASAG. Section 3 will describe the experimental setup for our study of cross-lingual ASAG with 26 languages. Our experiments and results are outlined in Sect. 4. We will conclude our work in Sect. 5 and suggest further steps.

2   Related Work

A good overview of approaches in ASAG before the deep learning era is given in [8]. Newer publications are based on *bag-of-words*, a procedure based on term frequencies [9, 10].

The latest trend which has proven to outperform traditional approaches is to use neural network-based embeddings, such as *Word2vec* [11]. [12] have developed *Ans2vec*, a *feature extraction architecture* based approach. They evaluated their concept with the English data set of the University of North Texas [13]. The advantage of this data set is that it contains scored student answers, while the answers of other short answer grading corpora, e.g., the SemEval-2013Task7 data sets [14] are only categorized into 3 classes–there is no point-based grading. [15] investigate and compare state-of-the-art deep learning techniques for ASAG and outperform [12] on the data set of the University of North Texas with a *fine-tuning architecture* based on the *Bidirectional Encoder Representations from Transformers (BERT)* [2] model. [16], [17] and [18] also deal in their work with *BERT* fine-tuning architectures. [18] report that their multilingual *RoBERTa* model [19] shows a stronger generalization across languages on English and German.

We extend their approach to 26 languages and use the smaller *M-BERT* [20] model to conduct a larger study concerning the cross-lingual transfer. While in most ASAG systems answers are categorized into only 3 classes, we focus on point-based grading. Our goal is to give a detailed analysis over the languages and investigate if cross-lingual ASAG allows students to write answers in exams in their native language and graders to rely on the scores of the system.

3   Experimental Setup

3.1   Evaluation Metrics

As in related literature, we evaluate our results with the Mean Absolute Error (*MAE*) which is calculated from the average deviations of the prediction from the target value. This metric provides an intuitive understanding in terms of the deviation of points which makes it possible to compare the systems' performance to human graders.

3.2   Data Set

The short answer grading data set of the University of North Texas [13] is used for our experiments. Table 1 summarizes the features of this data set.

Table 1 Information of the short answer grading data set of the University of North Texas

English	
Subject	Data structures
#questions with model answer	87
#answers (total)	2,442
#answers per question	28.1
Ø length of answer (#words)	18.4
Maximum scores (in points)	5

It contains 87 questions with corresponding model answer and on average 28.1 manually graded answers per question about the topic *Data Structures* from undergraduate studies. Each student answer received a score from 0 to 5 points from two independent graders. We used the average of these 2 scores as our prediction target. We randomly selected 80% of the ASAG data set (1.953 student answers) for training and the remaining 20% (489 student answers) for evaluation. Table 2 shows a typical question, the corresponding model answer and two student answers. One of them was given the full score, the other one is a rather weak answer. The structure of the questions is demonstrated in the example.

In order to produce artificial student and model answers for the adaptation, evaluation and comparison of the multilingual and the monolingual *BERT* models, we translated the English ASAG text data into 25 languages using Google's Neural Machine Translation System [21]. This procedure is also done by other researchers who experiment with multilingual NLP models [22] since this machine translation system comes close to the performance of professional translators [23–25]. An overview of the BLEU scores over languages is given in [24, 25]. The translation of the questions and answers in Table 1 into German and Chinese are demonstrated in Tables 3 and 4.

To get a first impression of how people evaluate our translations in particular, we had 33 German students evaluate the German translations. Most of them stated that the translations are linguistically correct and understandable, as shown in Fig. 2.

We produced ASAG data sets in the 26 languages that have the most Wikipedia articles [26]. These languages are spoken by more than 2.9 billion people (38% of

Table 2 Original sample question and answers from the English data set

Question	What is a variable?
Model answer	A location in memory that can store a value
Example: Answer 1	A variable is a location in memory where a value can be stored
Grading: Answer 1	5 of 5 points
Example: Answer 2	Variable can be an integer or a string in a program
Grading: Answer 2	2 of 5 points

Table 3 Machine-translated sample question and answers in German

Question	Was ist eine Variable?
Model answer	Eine Stelle im Speicher, die einen Wert speichern kann
Example: Answer 1	Eine Variable ist ein Ort im Speicher, an dem ein Wert gespeichert werden kann
Grading: Answer 1	5 of 5 points
Example: Answer 2	Eine Variable kann in einem Programm ein Integer oder ein String sein
Grading: Answer 2	2 of 5 points

Table 4 Machine-translated sample question and answers in Chinese

Question	什么是变量?
Model answer	内存中可以存储值的位置。
Example: Answer 1	变量是内存中可以存储值的位置。
Grading: Answer 1	5 of 5 points
Example: Answer 2	变量可以是整数, 也可以是程序中的字符串。
Grading: Answer 2	2 of 5 points

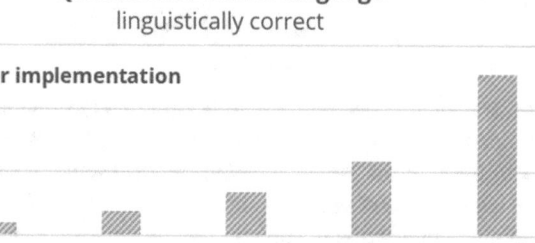

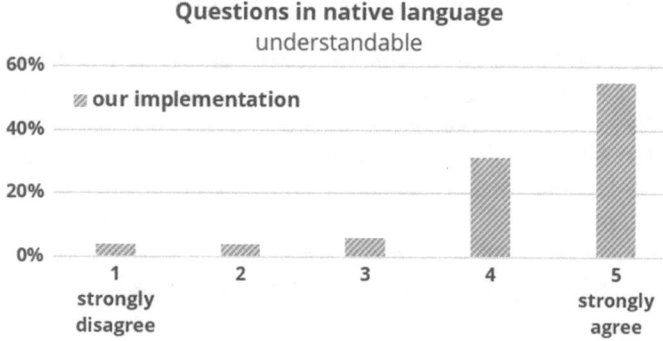

Fig. 2 Feedback of 33 students on the machine-translated ASAG data set in German

the world population) and cover the language families Indo-European, Austronesian, Austroasiatic, Japonic, Afroasiatic, Sino-Tibetan, Koreanic, and Uralic [26].

3.3   Natural Language Processing Models

Our goal was to analyze the performance of cross-lingual ASAG with the help of a multilingual model in comparison to monolingual ASAG.

To investigate cross-lingual ASAG for our languages, we experimented with NLP models based on *BERT* [2] since *BERT* models are small compared to other NLP models, e.g., *RoBERTa* [19], but still provide high performances on several NLP tasks [2]. Our evaluated NLP model *M-BERT* refers to a multilingual *BERT* which could support 104 languages [20].

To compare *M-BERT* to monolingual models from different languages families, we use the following 6 models:

- *bert-base-cased* (*en*),
- *bert-base-german-dbmdz-cased* (*de*),
- *bert-base-chinese* (*zh*),
- *wietsedv/bert-base-dutch-cased* (*nl*),
- *TurkuNLP/bert-base-finnish-cased-v1* (*fi*), and
- *cl-tohoku/bert-base-japanese-char* (*ja*).

The models were downloaded and fine-tuned through the *simpletransformers* library [27], which is based on the Transformers library [28]. We trained 6 epochs with a batch size of 8 using the AdamW optimizer [29] with an initial learning rate of 0.00004. We supplemented each fine-tuned *BERT* model with a linear regression layer that outputs a prediction of the *score* given an answer. The model expects the model answer and the student answer as input.

Figures 3, 4, and 5 demonstrate the training and testing procedures of our monolingual (*mono*) and multilingual ASAG systems: Our monolingual ASAG systems are exclusively fine-tuned with data from the target language, e.g., Chinese (*zh*) as shown in Fig. 3.

As illustrated in Fig. 4, the multilingual systems only need to be fine-tuned with ASAG data of 1 language, e.g., with the original English ASAG data (*Train ASAG en*). Then the multilingual *ASAG model* is able to receive a model answer together with a student answer in 1 of the other 103 languages and return a score in terms of points—without the need of fine-tuning with ASAG data in the other languages (*cross-lingual transfer*).

As shown in Fig. 5, we additionally investigated if adding translations of the ASAG data in more languages improves fine-tuning and performance, respectively.

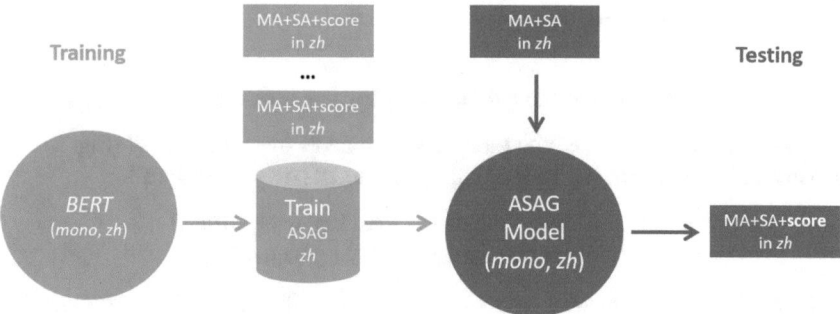

Fig. 3 Monolingual ASAG system (here trained with Chinese)

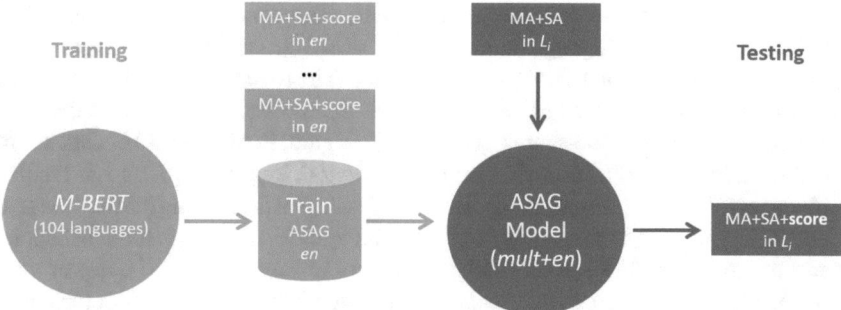

Fig. 4 Multilingual ASAG system with *cross-lingual transfer* (here trained with English)

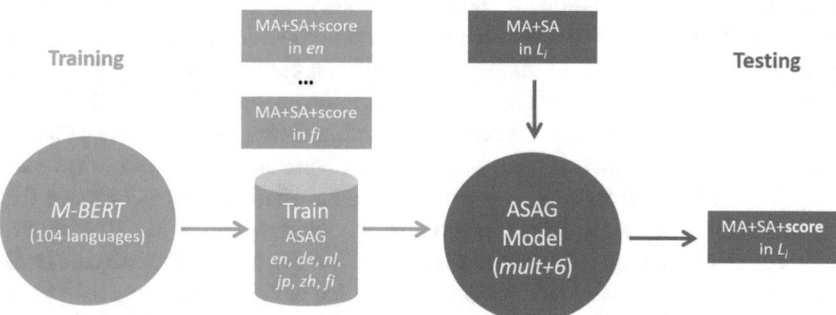

Fig. 5 Multilingual ASAG system with *cross-lingual transfer*
(trained with English, German, Dutch, Japanese, Chinese, and Finnish)

4 Experiments and Results

In our experiments we investigated the following research questions:

- How is the performance with monolingual models over languages? (Fig. 2)
- How is the performance with multilingual models over languages?

  - fine-tuned with task-specific data in target languages (Fig. 3)
  - fine-tuned with task-specific data in other languages (Fig. 4)

- How is the performance with monolingual/multilingual models compared to human graders?

Table 5 shows the deviation in context of the scoring scale from 0 to 5 points, represented by the Mean Absolute Error (*MAE*). The columns represent *M-BERT* fine-tuned on a single language (*multi + xx*), *M-BERT* trained on 6 languages (*multi + 6*) and our monolingual *BERT* models (*mono*). The rows represent the evaluation of the models in 26 languages. When we look at the results, we need to consider the grader variability: The scores given by the 2 graders of the ASAG data set of the University of North Texas differ on average by 0.75 points, which is a relative difference of 15% [13]. Our results in Table 5 indicate that fine-tuning the multilingual model *M-BERT* with task-specific data in 6 languages (*multi + 6*) is more beneficial than fine-tuning *M-BERT* with task-specific data in English (*multi + en*) or with task-specific data in 1 language (*multi + xx*)—even if the language *xx* is the target language (*multi + L*$_{target}$). If *xx* is not the target language, *multi + xx* performs worse than *multi + L*$_{target}$ but even fine-tuning with task-specific data from 1 other language results in *MAE*s lower than 0.86 points. This shows the strong effect of the cross-lingual transfer and is an impressive result considering that no data from the target language at all was used for training and that human graders differ by 0.75 points.

The ASAG performance of *multi + 6* shows only deviations between 0.41 points (Chinese (*zh*)) and 0.63 points (Cebuano (*ceb*)) which is 8% to 13% relative. Furthermore, Table 5 shows that *multi + L*$_{target}$ is more beneficial than *multi + en*: *multi + L*$_{target}$ achieves a cross-lingual performance with small deviations between 0.45 points (English (*en*)) and 0.52 points (Finnish (*fi*)) which is only 9% to 10% relative. However, if target language data is not available, fine-tuning with English data (*multi + en*) is sufficient since it comes only with marginal deviations between 0.45 points (English (*en*)) and 0.72 points (Arabic (*ar*)) which is only 9% to 14% relative.

The monolingual models (*mono*) slightly outperform *M-BERT* fine-tuned and evaluated on the same language (*multi + L*$_{target}$) with deviations between 0.43 points (English (*en*)) and 0.53 points (Japanese (*jp*)). However, *multi + 6* outperforms 4 of the 6 monolingual models demonstrating good overall performance and cross-lingual transfer.

Human graders deviate more (with 0.75 points, 15%) than the ASAG models which were cross-lingually adapted with English (worst *MAE*: 0.72), fine-tuned with the target language (worst *MAE*: 0.52), fine-tuned with our 6 languages (worst *MAE*: 0.63), and our monolingual models (worst *MAE*: 0.53).

Table 5 ASAG performance (MAE): multilingual and monolingual models

	multi + en	multi + de	multi + nl	multi + jp	multi + zh	multi + fi	multi + 6	mono
en	0.45	0.61	0.64	0.68	0.63	0.63	**0.43**	0.43
ceb	0.70	0.73	0.72	0.68	0.72	0.71	**0.63**	–
sv	0.63	0.67	0.68	0.73	0.72	0.68	**0.48**	–
de	0.64	0.51	0.67	0.70	0.70	0.65	0.46	**0.45**
fr	0.61	0.66	0.64	0.67	0.70	0.67	**0.54**	–
nl	0.62	0.64	0.52	0.70	0.73	0.67	**0.45**	0.47
ru	0.68	0.73	0.83	0.74	0.75	0.78	**0.52**	–
it	0.62	0.65	0.72	0.71	0.73	0.70	**0.52**	–
es	0.61	0.68	0.76	0.68	0.72	0.65	**0.49**	–
pl	0.62	0.71	0.77	0.69	0.72	0.68	**0.51**	–
vi	0.71	0.72	0.84	0.77	0.73	0.71	**0.52**	–
jp	0.66	0.70	0.73	0.49	0.63	0.71	**0.44**	0.53
zh	0.63	0.71	0.77	0.69	0.50	0.79	**0.41**	0.44
ar	0.72	0.78	0.85	0.78	0.76	0.76	**0.59**	–
uk	0.65	0.70	0.82	0.73	0.73	0.75	**0.54**	–
pt	0.59	0.67	0.75	0.69	0.73	0.69	**0.50**	–
fa	0.64	0.66	0.71	0.67	0.70	0.69	**0.56**	–
ca	0.64	0.70	0.74	0.70	0.76	0.67	**0.53**	–
sr	0.69	0.81	0.83	0.76	0.79	0.86	**0.56**	–
id	0.66	0.68	0.69	0.70	0.79	0.63	**0.49**	–
no	0.63	0.69	0.65	0.75	0.71	0.69	**0.45**	–
ko	0.70	0.70	0.76	0.66	0.66	0.67	**0.58**	–
fi	0.69	0.79	0.77	0.77	0.73	0.52	0.47	**0.45**
hu	0.69	0.76	0.81	0.72	0.76	0.69	**0.54**	–
cs	0.62	0.77	0.82	0.72	0.78	0.71	**0.51**	–
sh	0.66	0.77	0.79	0.74	0.78	0.79	**0.53**	–

Note: Human grader variability is **0.75** points

The significant improvements of *multi* + 6 over *multi* + *en* and *multi* + L_{target} for English (*en*), German (*de*), Dutch (*nl*), Japanese (*jp*), Chinese (*zh*), and Finnish (*fi*) are listed in Tables 6 and 7. With the wide range of English online study opportunities, in many cases ASAG data from English courses would be used for fine-tuning. However, in Table 6 we see that we can achieve up to 35% improvement by adding more languages. Even if we already have ASAG data in the target language, adding the 5 languages provides improvements of up to 18%, as demonstrated in Table 7.

Table 6 ASAG performance (*MAE*): *multi + en* versus *multi + 6*

	multi + en	multi + 6	Rel. improvement
en	0.45	0.43	4.4%
de	0.64	0.46	28.1%
nl	0.62	0.45	27.4%
jp	0.66	0.44	33.3%
zh	0.63	0.41	34.9%
fi	0.69	0.47	31.9%

Table 7 ASAG performance (*MAE*): *multi + L_{target}* versus *multi + 6*

	multi + L_{target}	multi + 6	Rel. improvement
en	0.45	0.43	4.4%
de	0.51	0.46	9.8%
nl	0.52	0.45	13.5%
jp	0.49	0.44	10.2%
zh	0.50	0.41	18.0%
fi	0.52	0.47	9.6%

5   Conclusion and Future Work

Our analysis on 26 languages demonstrated the potential of cross-lingual ASAG to allow students to write answers in exams in their native language and graders to rely on the scores of the system. With *MAEs* which are only between 0.41 and 0.72 points out of 5 points, our best models *multi + 6*, *multi + xx* and *mono* have even less discrepancy than the 2 graders, which is 0.75 points in our corpus. Augmenting training data with machine translated task-specific data for fine-tuning improves performance of multilingual models. We are aware that our results have to be considered experimentally. Depending on the domain and the language combination, we see challenges in achieving optimal quality in machine translation. Nevertheless, we are very confident and plan to investigate this augmentation with different combinations and numbers of languages. We hope that performance is in a similar range for further languages and intend to analyze this in the future. If this is true, with multilingual models, we do not need training data in the target language at all to reach human level. To enhance online and distance learning, our next step includes to analyze the integration and application for online exams on the one hand but on the other hand for interactive training programs to prepare students optimally for exams. Figure 6 demonstrates our visualization of a multilingual interactive conversational artificial intelligence tutoring system for exam preparation [30], where students can prepare for exams in their native language, e.g., Dutch, in a gamification approach and automatically receive points for their free text answers.

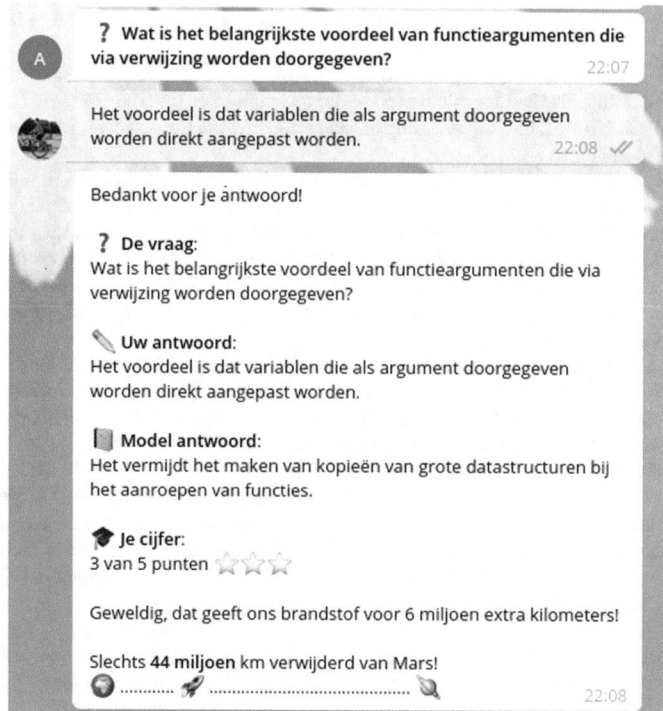

Fig. 6 Conversation with greeting, language selection, exam question, student answer, scoring, model answer and motivation

References

1. Statista: The Most Spoken Languages Worldwide in 2019 (2020). https://www.statista.com/statistics/266808/the-most-spoken-languages-worldwide
2. Devlin J, Chang MW, Lee K, Toutanova K (2019) BERT: Pre-training of Deep Bidirectional Transformers for Language Understanding. In: Proceedings of the 2019 Conference of the North American Chapter of the Association for Computational Linguistics: Human Language Technologies, vol 1 (Long and Short Papers). Association for Computational Linguistics, Minneapolis, Minnesota, USA, pp 4171–4186
3. United Nations: Sustainable Development Goals: 17 Goals to Transform our World (2021). https://www.un.org/sustainabledevelopment/sustainabledevelopment-goals
4. Correia AP, Liu C, Xu F (2020) Evaluating Videoconferencing Systems for the Quality of the Educational Experience. Distance Educ 41(4):429–452
5. Koravuna S, Surepally UK (2020) Educational Gamification and Artificial Intelligence for Promoting Digital Literacy. Association for Computing Machinery, New York, NY, USA
6. Libbrecht P, Declerck T, Schlippe T, Mandl T, Schiffner D (2020) NLP for Student and Teacher: Concept for an AI Based Information Literacy Tutoring System. In: The 29th ACM International Conference on Information and Knowledge Management (CIKM 2020). Galway, Ireland. Accessed 19–23 Oct 2020
7. Pires T, Schlinger E, Garrette D (2019) How Multilingual is Multilingual BERT? In: Proceedings of the 57th Annual Meeting of the Association for Computational Linguistics. Association for Computational Linguistics, Florence, Italy, pp 4996–5001

8. Burrows S, Gurevych I, Stein B (2014) The Eras and Trends of Automatic Short Answer Grading. Int J Artif Intell Educ 25:60–117
9. Süzen N, Gorban A, Levesley J, Mirkes E (2020) Automatic Short Answer Grading and Feedback Using Text Mining Methods. Procedia Comput Sci 169:726–743
10. Zehner F (2016) Automatic Processing of Text Responses in Large-Scale Assessments. Ph.D. thesis, TU München
11. Mikolov T, Chen K, Corrado G, Dean J (2013) Efficient Estimation of Word Representations in Vector Space. In: Bengio Y, LeCun Y (eds), Workshop Track Proceedings of the 1st International Conference on Learning Representations, ICLR 2013, Scottsdale, Arizona, USA
12. Gomaa WH, Fahmy AA (2019) Ans2vec: A Scoring System for Short Answers. In: Hassanien AE, Azar AT, Gaber T, Bhatnagar RF, Tolba M (eds) The International Conference on Advanced Machine Learning Technologies and Applications (AMLTA2019). Springer International Publishing, Cham, pp 586–595
13. Mohler M, Bunescu R, Mihalcea R (2011) Learning to Grade Short Answer Questions Using Semantic Similarity Measures and Dependency Graph Alignments. In: Proceedings of the 49th Annual Meeting of the Association for Computational Linguistics: Human Language Technologies. Association for Computational Linguistics, Portland, Oregon, USA, pp 752–762
14. Dzikovska M, Nielsen R, Brew C, Leacock C, Giampiccolo D, Bentivogli L, Clark P, Dagan I, Dang HT (2013) SemEval-2013 Task 7: The Joint Student Response Analysis and 8th Recognizing Textual Entailment Challenge. In: Second Joint Conference on Lexical and Computational Semantics (*SEM), Proceedings of the Seventh International Workshop on Semantic Evaluation (SemEval 2013), vol 2. Association for Computational Linguistics. Atlanta, Georgia, USA
15. Sawatzki J, Schlippe T, Benner-Wickner M (2021) Deep Learning Techniques for Automatic Short Answer Grading: Predicting Scores for English and German Answers. In: The 2nd International Conference on Artificial Intelligence in Education Technology (AIET 2021), Wuhan, China
16. Krishnamurthy S, Gayakwad E, Kailasanathan N (2019) Deep Learning for Short Answer Scoring. Int J Recent Technol Eng 7:1712–1715
17. Sung C, Dhamecha T, Mukhi N (2019) Improving Short Answer Grading Using Transformer-Based Pre-Training. In: Artificial Intelligence in Education, pp 469–481
18. Camus L, Filighera A (2020) Investigating Transformers for Automatic Short Answer Grading. Artif Intell Educ 12164:43–48
19. Liu Y, Ott M, Goyal N, Du J, Joshi M, Chen D, Levy O, Lewis M, Zettlemoyer L, Stoyanov V (2019) RoBERTa: A Robustly Optimized BERT Pretraining Approach
20. Devlin J (2019) BERT-Base, Multilingual Cased. https://github.com/googleresearch/bert/blob/master/multilingual.md
21. Wu Y, Schuster M, Chen Z, Le QV, Norouzi M, Macherey W, Krikun M, Cao Y, Gao Q, Macherey K, Klingner J, Shah A, Johnson M, Liu X, Kaiser L, Gouws S, Kato Y, Kudo T, Kazawa H, Stevens K, Kurian G, Patil N, Wang W, Young C, Smith J, Riesa J, Rudnick A, Vinyals O, Corrado G, Hughes M, Dean J (2016) Google's Neural Machine Translation System: Bridging the Gap Between Human and Machine Translation. CoRR. 1609.08144
22. Budur E, Özçelik R, Gungor T, Potts C (2020) Data and Representation for Turkish Natural Language Inference. In: Proceedings of the 2020 Conference on Empirical Methods in Natural Language Processing (EMNLP). Association for Computational Linguistics, pp 8253–8267
23. Stapleton P, Leung Ka Kin B (2019) Assessing the Accuracy and Teachers' Impressions of Google Translate: A Study of Primary L2 Writers in Hong Kong. In: English for Specific Purposes, vol 56, pp 18–34
24. Aiken M (2012) An Analysis of Google Translate Accuracy. Stud Linguist Lit 3:253
25. Aiken M (2019) An Updated Evaluation of Google Translate Accuracy. Stud Linguist Lit 3:253
26. Wikimedia: List of Wikipedias (2021). https://meta.wikimedia.org/wiki/List_of_Wikipedias#All_Wikipedias_ordered_by_number_of_articles
27. Rajapakse TC (2019) Simple Transformers. https://github.com/ThilinaRajapakse/simpletransformers

28. Wolf T, Debut L, Sanh V, Chaumond J, Delangue C, Moi A, Cistac P, Rault T, Louf R, Funtowicz M, Davison J, Shleifer S, von Platen P, Ma C, Jernite Y, Plu J, Xu C, Scao TL, Gugger S, Drame M, Lhoest Q, Rush AM (2020) Transformers: State-of-the-Art Natural Language Processing. In: Proceedings of the 2020 Conference on Empirical Methods in Natural Language Processing: System Demonstrations. Association for Computational Linguistics, pp 38–45

29. Kingma DP, Ba J (2015) Adam: A Method for Stochastic Optimization. In: Bengio Y, LeCun Y (eds) 3rd International Conference on Learning Representations, ICLR 2015, Conference Track Proceedings, San Diego, CA, USA

30. Schlippe T, Sawatzki J (2021) AI-Based Multilingual Interactive Exam Preparation. In: The Learning Ideas Conference 2021 (14th Annual Conference). ALICE—Special Conference Track on Adaptive Learning via Interactive, Collaborative and Emotional Approaches. New York, New York, USA

Application of Improved ISM in the Analysis of Undergraduate Textbooks

MeiLing Zhao, Qi Chen, Qi Xu, XiaoYa Yang, and Min Pan

Abstract In order to improve the quality and efficiency of the overall design and analysis of undergraduate professional courses, this paper analyzes and discusses Questionnaire Design through the improved ISM. As a result, 13 knowledge elements related to the content were successfully divided into 6 layers to form 5 effective teaching sequences. It fully verified the effectiveness of this method and provided data support for the analysis of the undergraduate textbook "Educational Technology Research Methods". Finally, some suggestions are put forward to related teachers to help them innovate and develop in curriculum teaching.

Keywords Interpretive Structural Model (ISM) · Hierarchical Analysis (AH) · Quantitative analysis

1 Introduction

As an important part of curriculum reform, textbooks play a very important role. The textbooks are formulated in accordance with certain curriculum standards. They are the communication link between curriculum designers, teachers and students. Whether the arrangement of the content structure of the textbook conforms to the curriculum standards and the law of the cognitive development of the students plays a very important role in the analysis of the textbook in the teaching design.

However, the analysis of textbooks usually has certain subjective intentions. Therefore, a scientific and reasonable analysis of textbooks helps teachers understand their educational goals and achieve effective education. Textbook analysis is an indispensable part of teaching system design. In actual teaching, due to the lack of scientific knowledge structure framework and systematic evaluation basis, some teachers cannot coordinate the connection between knowledge and students' learning enthusiasm is difficult to mobilize [1]. In September 2019, the Ministry of

M. Zhao · Q. Chen (✉) · Q. Xu · X. Yang · M. Pan
School of Computer and Information Engineering, Hubei Normal University, Huangshi, China
e-mail: 106062628@qq.com

Education issued the "Opinions on Deepening the Reform of Undergraduate Education and Teaching and Comprehensively Improving the Quality of Talent Cultivation" [2], which proposed "to comprehensively improve the quality of curriculum construction and strengthen the overall design of the curriculum system". The quality of undergraduate teaching still needs continuous attention. Reasonably digging out the relationship between the knowledge points in the undergraduate textbooks and determining the optimal organization of teaching content are the manifestations of the professional qualities and teaching abilities of university teachers, which help students form a complete subject system thinking and build a good knowledge system structure. "Research method" is the vitality of the continuous development of educational technology as a discipline [3]. Therefore, the research takes the undergraduate course of educational technology "Educational Technology Research Method" as an example, and conducts an in-depth analysis of the second section of Chap. "A Two-Stage NER Method for Outstanding Papers in MCM" "Questionnaire Design". The purpose is to grasp the hierarchical relationship between the knowledge elements in the professional curriculum, and provide reference and reference for teachers to analyze the textbook.

2   Research Methods

The research adopts the Interpretive Structural Model (ISM), which is widely used to solve the correlation between chaotic and disordered elements in complex systems, and can organize the scattered elements into a simple and orderly hierarchical structure through quantitative analysis [4]. Applying the ISM method to the content analysis of teaching materials can visualize complex knowledge and present a good logical relationship, which is in line with the cognitive development of students. At present, the ISM method has been successfully applied to junior high school [5], high school [6], higher vocational education and other stages [7], but there are few cases of applying this method to university textbooks, and its calculation process needs to be further optimized. In order to enhance the practicality of undergraduate textbook analysis, the research will optimize ISM. The steps to determine the improved ISM are: (1) Extract teaching elements and initially determine the high and low layer relationships between them; (2) Establish an adjacency matrix to visualize the relationship between knowledge elements; (3) Use Boolean algebra rules to connect adjacencies matrix is converted into a reachable matrix; (4) Determine the hierarchical relationship, optimize the reachable set, antecedent set and common set links of the statistical knowledge elements in the traditional ISM, and directly obtain the key knowledge elements of each level in a simple and easy-to-operate matrix calculation method; (5) Construct an Interpretive Structural Model of knowledge points, and analyze and discuss the content; (6) Form an effective teaching sequence, which is convenient for teachers to compare and make decisions. The flow chart is shown in Fig. 1.

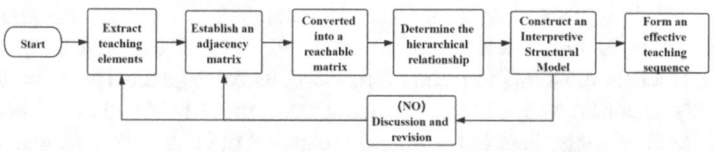

Fig. 1 Optimized ISM flow chart

The ISM model is a systematic conceptual model that can flexibly and simply handle diversified, hierarchical and complex structures, and the model can reveal the internal structure of the system by using the relationships between various elements. The characteristic of this method is to accurately and effectively incorporate practical experience into the teaching materials, and to express the content of the teaching materials in a visual form. Using the ISM model to analyze the content of textbooks can effectively overcome the disadvantages of traditional subjective textbook analysis and can make the analysis of textbooks more objective, scientific, concise and orderly.

3 Knowledge Element Analysis

3.1 Knowledge Element Extraction

According to the teacher's analysis of the textbook, the basic learning elements of the textbook can be extracted, which are called knowledge points. For example: Suppose the elements of the textbook are S1, S2, S3. The elements extracted here are often the important content contained in each chapter. For the extracted elements, determine the forming relationship between the elements according to the logical relationship between them and the various relationships between top and bottom.

In order to make the extracted knowledge more scientific, the research selected the "Twelfth Five-Year" planning textbook "Research Methods of Educational Technology" (Second Edition) for general higher education as the research object [8]. Through consulting the instructor of the course, interviewing the students who have taken the course about their understanding of the content of "Questionnaire Design", and consulting relevant course materials, the 13 sub-teaching elements of "Questionnaire Design" are determined. S1: The advantages and disadvantages of the questionnaire survey method; S2: The scope of application of the questionnaire survey method; S3: The type of questionnaire; S4: Questionnaire preface and name design; S5: Questionnaire main content design; S6: Questionnaire conclusion design; S7: Questionnaire category design; S8: Question structure; S9: Question design principles; S10: Question expression; S11: The transfer of relevant questions; S12: Answer design principles; S13: Questionnaire design skills.

Determine the core knowledge elements of textbooks are affected by many influencing factors, such as the teacher's teaching experience, student outlook, textbook outlook, etc. It is necessary to analyze the relationship between the knowledge elements of the teaching content in the textbook. In order to reach a higher level of goals, teachers must be premised on reaching its basic goals. This is logical and objectively scientific. When teachers begin to analyze textbooks, they not only need to consider the arrangement of the course content of the textbooks, but also need to understand the relationship between the content of each preparatory knowledge and the new knowledge. In addition, teachers should also consider the internal relationship between knowledge and determine teaching ideas, and then carry out teaching activities.

Taking the above 13 sub-teaching elements as the basic elements of the analysis of the interpretation structure model, if you want to complete the learning of knowledge element A, you must first complete the learning of knowledge element B, so A is the low-layer goal of B. For example, "S4's low-layer goals are expressed as {S2, S3}" and its actual teaching meaning is to master the design of the preface and name of the questionnaire, you must first understand the advantages and disadvantages of the questionnaire, the scope of application, and the type. Therefore, the initial direct target is determined: S1, S2, and S3 have no direct low-layer targets; the low-layer targets of S4 are {S2, S3}; the low-layer targets of S5 are {S1, S2, S3}; the low-layer targets of S6 are {S2, S3}; The low-layer goal of S7 is {S5}; the low-layer goal of S8 is {S5}; the low-layer goal of S9 is {S5, S7, S8}; the low-layer goal of S10 is {S7, S8, S9}; the low-layer goal of S11 is {S7, S10}; the low-layer goal of S12 is {S9}; the low-layer goal of S13 is {S7, S8, S12}.

3.2 Construction of Adjacency Matrix

We mark the 13 knowledge elements as S = {S1, S2, S3, S4, S5, S6, S7, S8, S9, S10, S11, S12, S13}, and build a two-dimensional matrix. The horizontal axis of the two-dimensional matrix is high-layer target knowledge elements and the vertical axis is low-layer target knowledge elements. If there is a direct high- and low-layer relationship between the knowledge elements Si and Sj, the position (Si, Sj) is recorded as Sij = 1 (true), otherwise Sij = 0 (false), the adjacency matrix A is shown in Fig. 2. To identify the direct connections between the elements of knowledge.

3.3 Reachability Matrix Construction

The reachable matrix is used to express the degree of reachability between the knowledge elements after a certain length of path. The reachable matrix M can be calculated by the rules of Boolean algebra, that is, when the two knowledge elements are added, they satisfy $0 + 0 = 0, 0 + 1 = 1, 1 + 0 = 1, 1 + 1 = 0$, and when they multiply, they satisfy $0 \times 0 = 0, 0 \times 1 = 0, 1 \times 0 = 0, 1 \times 1 = 1$. First construct a 13×13 identity matrix I. If the knowledge elements Si and Sj are equal, the position (Si, Sj)

$$A=$$

	S1	S2	S3	S4	S5	S6	S7	S8	S9	S10	S11	S12	S13
S1	0	0	0	0	1	0	0	0	0	0	0	0	0
S2	0	0	0	1	1	1	0	0	0	0	0	0	0
S3	0	0	0	1	1	1	0	0	0	0	0	0	0
S4	0	0	0	0	0	0	0	0	0	0	0	0	0
S5	0	0	0	0	0	0	1	1	1	0	0	0	0
S6	0	0	0	0	0	0	0	0	0	0	0	0	0
S7	0	0	0	0	0	0	0	0	1	1	1	0	1
S8	0	0	0	0	0	0	0	0	1	1	0	1	1
S9	0	0	0	0	0	0	0	0	0	1	1	0	0
S10	0	0	0	0	0	0	0	0	0	0	0	0	0
S11	0	0	0	0	0	0	0	0	0	0	0	0	0
S12	0	0	0	0	0	0	0	0	0	0	0	0	1
S13	0	0	0	0	0	0	0	0	0	0	0	0	0

Fig. 2 Adjacency matrix A

$$M=$$

	S1	S2	S3	S4	S5	S6	S7	S8	S9	S10	S11	S12	S13
S1	1	0	0	0	1	1	1	1	1	0	0	0	0
S2	0	1	0	1	1	1	1	1	1	0	0	0	0
S3	0	1	1	1	1	1	1	1	1	0	0	0	0
S4	0	0	0	1	0	0	0	0	0	0	0	0	0
S5	0	0	0	0	1	0	1	1	1	1	1	1	1
S6	0	0	0	0	0	1	0	0	0	0	0	0	0
S7	0	0	0	0	0	0	1	0	1	1	1	0	1
S8	0	0	0	0	0	0	0	1	1	1	1	1	1
S9	0	0	0	0	0	0	0	0	1	1	1	0	0
S10	0	0	0	0	0	0	0	0	0	1	0	0	0
S11	0	0	0	0	0	0	0	0	0	0	1	0	0
S12	0	0	0	0	0	0	0	0	0	0	0	1	1
S13	0	0	0	0	0	0	0	0	0	0	0	0	1

Fig. 3 Reachability matrix M

is recorded as $S_{ij} = 1$, otherwise $S_{ij} = 0$. Then calculate, where, if there is, then M can be expressed as, the above process can be completed by program design. The calculation result is shown in Fig. 3, which can reflect the connected relationship between each knowledge element.

3.4 Hierarchical Relationship Determination

The determination of the hierarchical relationship between the two elements can be determined with reference to the characteristics of the learner. To a large extent, it reflects the analyst's understanding, knowledge and experience of the problem. In the traditional ISM method [9], the reachable set $R(S_i)$, the antecedent set $A(S_i)$ and

Fig. 4 Hierarchical
structure model diagram

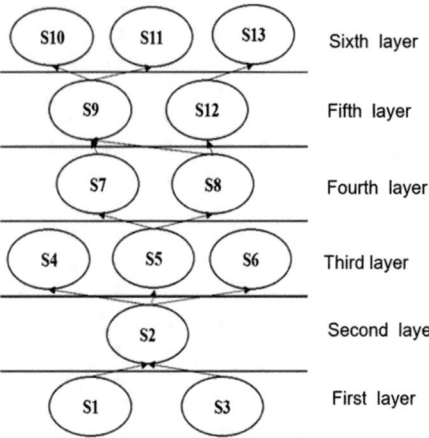

Sixth layer

Fifth layer

Fourth layer

Third layer

Second layer

First layer

the common set T(Si) of each element are mainly counted according to M. Among them, R(Si) is the set of all elements that can be reached from Si, A(Si) is the set of all elements that can reach Si, and T(Si) is the set of all elements that can be reached from Si and can reach itself. When Si satisfies the equationw (T(Si) ∩ A(Si) = T(Si)), first Si is extracted as the first-layer element, and then extracted again from the remaining knowledge elements to satisfy the equation was the second layer. The above process is iterated to obtain the first layer, the second layer and the last layer in turn, and finally form the knowledge element set.

Traditional calculation is too cumbersome and time-consuming, which is not conducive for teachers to quickly build a hierarchical structure model. Therefore, the study adopts DRM [10] to improve the above process and verify its feasibility. First of all, since the path of each element can reach its own path has little effect on the layer determination, the 13×13 identity matrix I is subtracted from M to obtain the matrix M* without its own path. We observe each column in M* in turn, and find that when each row element is 0, the knowledge element corresponding to this column is recorded as the first layer, and the row is removed. Repeat the above operation until M* equals 0. The operation is over, so that all layers can be solved quickly. After calculation and verification, the hierarchical relationships obtained by the above two methods are consistent, and the improved method is more efficient, and can directly transform M into a hierarchical structure model, as shown in Fig. 4.

4 Hierarchical Model Analysis

The 13 knowledge elements can be divided into 6 layers, and there are no more than 3 knowledge elements at each layer. The bottom-to-up layer teaching conforms to the law of students' cognitive development, and the teaching sequence within the same layer can be flexibly arranged by the teacher [11].

The first and second layers are the types of questionnaires, the advantages and disadvantages of the questionnaire survey method, and the scope of its application. These two layers of knowledge belong to the lower-layer goals. The third layer is the design of the foreword and name of the questionnaire, the design of the main content of the questionnaire and the design of the questionnaire conclusion. Among them, "the design of the main content of the questionnaire" is most closely related to the fourth layer, and involves many sub-knowledge points, which is the most difficult point in teaching. It requires teachers to design more abundant teaching activities to deepen students' understanding of the main framework of the questionnaire and grasp the overall design process of the questionnaire. The fourth layer is the type of problem and the structure of the problem. There is no absolute sequence between these two types of knowledge points. Teachers can choose one of them to explain and the teacher leaves the students to learn another knowledge point by themselves to strengthen the cultivation of students' initiative and initiative. The fifth layer is the design principle of the question and the design principle of the answer. This layer of knowledge is relatively abstract and not easy to control directly. Teachers may consider combining it with the expression of questions in the next layer, the transfer of related questions, and the skills of questionnaire design. In addition, teachers carry out group practice activities designed by questionnaires, so that students can check and fill vacancies in collaboration, and teachers can supplement and make suggestions to promote students' reflection.

In actual teaching, explaining the hierarchical structure model diagram is not the ultimate goal. Determining the order of teaching content can more effectively help teachers organize classroom teaching [12]. According to the principle of from the outside to the inside, from the easy to the difficult, the knowledge is connected in series, and Fig. 4 is reasonably split to form 5 effective teaching sequences, as shown in Fig. 5. The end points of the first two teaching sequences are the design of the preface and name of the questionnaire and the design of the conclusion of the questionnaire. Teachers can prioritize these shorter sequence content to improve teaching efficiency. The sequence ④ and ⑤ are more complicated. Teachers can flexibly arrange teaching time according to students' mastery, and add a variety of questionnaire design exploration activities to mobilize students' enthusiasm and achieve the best teaching effect.

The teaching sequence diagram obtained according to the research content should be deeply researched and discussed. In the process of analysis, attention should be paid to the symbols of the elements and the content they represent. Based on the

Fig. 5 Teaching sequence diagram

① S1、S3→S2→S4

② S1、S3→S2→S6

③ S1、S3→S2→S5→S8→S12→S13

④ S1、S3→S2→S5→S7、S8→S9→S10

⑤ S1、S3→S2→S5→S7、S8→S9→S11

content represented by the symbols, understand and analyze the obtained teaching sequence diagram, and modify the formation relationship diagram according to the actual situation. The formed sequence diagram can assist teachers in teaching, and is more conducive for students to master knowledge and understand the relationship between relevant knowledge elements.

5 Conclusion and Suggestion

Through the improved ISM method, the "questionnaire design" was analyzed in depth. The study found that the 13 knowledge elements in the content can be divided into 6 layers, forming 5 effective teaching sequences, which provide data support for the "Research Methods of Educational Technology", which can be used as a reference for teachers.

Teachers only need to pre-determine the direct high- and low-layer goals of each knowledge element, and then the originally messy set of knowledge elements can be summarized into a visual hierarchical structure. Compared with the traditional ISM method, the advantage of the improved ISM method is that it can avoid tedious and complicated mathematical calculations and reduce the error rate of calculations. Finally, this result can improve the efficiency of teacher's textbook analysis and improve the quality of the overall curriculum design.

The improved ISM can decompose the complex and messy element relationships in the textbook structure into a clear multi-layer hierarchical form. It can also provide an optimal way for teachers to establish their teaching goals, so that teachers can grasp the internal connections between different knowledge elements, so as to help students construct scientific and reasonable mind maps, and ultimately achieve the optimization of learning effects.

However, there are still some shortcomings in the improved ISM method, which is beyond doubt. When using the improved ISM to analyze textbooks, because of the lack of teaching experience, the extraction of various knowledge elements and the correlation analysis of the elements are only based on the order of curriculum standards and textbook layout, which results in insufficient consideration of element extraction. Moreover, in this study, there is no complete teaching demonstration on the content of the teaching materials after using the improved ISM method. Regarding the problems in this research, I hope that in subsequent research, we will actively conduct empirical research with experts in the field of teaching and teachers in the frontline of teaching, so as to prove the effectiveness and scientific of the improved ISM method.

Finally, some suggestions for teachers of related courses are as follows. First, improve information literacy and enhance the ability of information teaching. In the process of using the improved ISM method, although the calculation process is relatively simple, it puts forward certain requirements on the teacher's educational information processing ability. Especially when there are many knowledge elements in the textbook, manual calculations are likely to cause various errors. The teacher's calculation errors in a certain link may lead to deviations in the construction of the interpretation structure model. Therefore, teachers should continuously strengthen self-learning, improve the awareness of information application, actively participate in teaching and training activities supported by information technology, and try to use relevant software tools, such as MATLAB, SPSS, etc., to strive to promote innovation in teaching practice. Second, teachers should change their roles and establish advanced educational concepts. With the advent of the "Internet + " era, Education Information 2.0 has put forward higher requirements for modern teachers. While paying attention to the growth of students, teachers should also focus on their own professional development, try to understand new teaching concepts, learn new information technology, integrate them in teaching design, and sum up experience in teaching practice. In this way, we can set an example of informatization application so that we can keep up with the pace of education reform and development and not be eliminated by the times. Second, teachers should change their roles and establish advanced educational concepts. With the advent of the "Internet + " era, Education Information 2.0 has put forward higher requirements for modern teachers. While paying attention to the growth of students, teachers should also focus on their own professional development, try to understand new teaching concepts, learn new information technology, integrate them in teaching design, and sum up experience in teaching practice. In this way, we can set an example of informatization application so that we can keep up with the pace of education reform and development and not be eliminated by the times. Third, teachers should delve into teaching materials and innovate teaching. Textbooks are the main resources and important tools to start classroom teaching. Learning to extract and express scattered and abstract knowledge elements in the content of the textbook is an important skill necessary for teachers. Network teaching is becoming the mainstream teaching method, and it is particularly important to return to the essence of teaching materials. Therefore, teachers should use textbooks as the basis, combined with the background of the times, find other suitable auxiliary learning resources, design more abundant teaching activities, further diversify students' thinking, broaden their horizons, and realize meaningful teaching.

References

1. Suhomlinski (2005) Suggestions to teachers. Beijing: Educational Science Press
2. Ministry of Education of the People's Republic of China. Opinions of the ministry of education on deepening the reform of undergraduate education and teaching and comprehensively improving the quality of talent training [EB/OL] [2019–10–11]. http://www.moe.gov.cn/src site/A08/s7056/201910/t20191011_402759.html
3. Ziyun L (2015) The unique disenchantment and methodology interpretation of educational technology research methods. Audio-Vis Educ Res 36(03):17–21+28
4. Warfield JN (1976) Societal systems: planning, policy and complexity. John Wiley&Sons, New York
5. Tianyu X (2015) Research on the application of concept map supported by ISM method in junior middle school chemistry teaching. Northeast Normal Univ

6. Xi T (2020) Comparative analysis of new and old geography textbooks based on the ISM analysis method: Taking "Water on the Earth" as an example. Geograph Teach 03:42–45

7. Yan C (2013) Discussion on the design of the curriculum system of medical imaging technology in higher vocational colleges based on the ISM analysis method. China Adult Educ 11:154–156

8. Yi Z, Pinghong Z (2013) Research methods of educational technology (Second Edition). Peking University Press, Beijing

9. Takahiro S (1996) Introduction to the ISM structural learning method. Meiji Books, Tokyo

10. Yue Z (2015) Design of teaching content analysis system based on DRM. J Inner Mongolia Normal Univ (Natural Science Chinese Edition) 44(05):701–703+708

11. Jin Z (2014) Discussion on the practicability of comparative research on mathematics textbooks based on ism method. Central China Normal University

12. Qian L, Yujiao W, Zhengming T (2019) An improved ISM method suitable for textbook analysis. China Educ Technol Equip (18):53–54+57

Design of Student Management System Based on Smart Campus and Wearable Devices

Jun Li, Yanan Shen, Wenrui Dai, and Bin Fan

Abstract In view of the current military cadets' daily teaching management information level is not high, service function is weak, cannot effectively grasp the students' individual physiological and psychological state changes, the design of military cadets intelligent teaching accurate management system based on wearable devices. Relying on the smart campus wireless network platform, the system uses intelligent wearable devices, RFID technology and data analysis technology to collect the basic data of students' daily movement track, going out, exercise class, training, sleep and emotion, and master the comprehensive physiological and psychological state through big data analysis, so as to provide means and tools for accurate management of students under the condition of smart campus.

Keywords Wearable devices · Cadets · Precise management · System design

1 Introduction

Student management is an important daily work in the management of military academies. It is not only the need of normal work, but also directly affects the cultivation of students' comprehensive quality. With the continuous development of new management concept and management mode, the traditional "passive management" which takes command, restriction, control and supervision as the basic management mode has been unable to meet the needs of efficient and accurate management under the new situation, and there are the following prominent management problems [1, 2].

J. Li (✉) · Y. Shen · W. Dai · B. Fan
Army Academy of Artillery and Air Defense, Anhui Province, Hefei City 230031, China
e-mail: 18876203@qq.com

© The Author(s), under exclusive license to Springer Nature Singapore Pte Ltd. 2022   141
E. C. K. Cheng et al. (eds.), *Artificial Intelligence in Education: Emerging Technologies,*
Models and Applications, Lecture Notes on Data Engineering and Communications
Technologies 104, https://doi.org/10.1007/978-981-16-7527-0_11

1.1 The Low Level of Information Management Leads to Low Efficiency

At present, the informatization level of student management means is still at the general level, such as various inspection, supervision and attendance, and still stays in manual registration, on-site verification and other ways. This kind of management method is easy to cause students' psychological conflict, and it is also very cumbersome for managers. It does not reflect the people-oriented and information-oriented management idea, and the efficiency of management is often not satisfactory.

1.2 It is Difficult to Carry Out Differentiated Management According to the Characteristics of Individual Students

Differentiation management is an important purpose of precise management. Only by realizing the differentiation of management, abandoning the stereotyped management mode and adopting different management methods for different students and different goals, can we achieve the accuracy of management, implement the decision-making scheme, make the management instructions act on the students accurately and improve the pertinence of management. However, in the process of student management, there are many problems, such as incomplete information, inaccurate quantification, and too much fuzzy content, which lead to the failure of differentiated management.

1.3 Unable to Obtain the Psychological Changes of the Managed

Different from local colleges and universities, military colleges and universities are characterized by rigid constraints of management system. For example, the system of work and rest is strict, the system of asking for leave and selling leave is strict, and the requirements of orders, prohibitions, rules and discipline are very high. But some students can't adjust themselves well in this situation, and they are prone to depression. The main performance is: the personality is withdrawn, seldom exchanges and the communication with others, everything does not adapt and so on. Although there are some channels of psychological counseling, students often do not seek help because of face, privacy and other issues. If managers cannot find students' psychological problems in time, it may lead to serious consequences.

In view of the above problems, this paper studies through the secondary development of wearable devices, connecting it with the software and hardware facilities of smart campus, real-time acquisition of students' movement trajectory and various physiological and psychological data, through big data analysis technology, fully understand and predict the physiological and psychological development of students, to provide scientific and accurate management basis for managers, and also for students to understand Provide information service according to their own situation.

2 System Design

2.1 System Overall Design

The student precise management system based on wearable devices is to solve the problem of low degree of informatization and accuracy of traditional management mode. It can obtain the daily movement and physiological information of students through the motion sensor, heart rate sensor and blood pressure sensor in the smart bracelet, and locate the real-time position of students through the WIFI network of smart campus, and the data will be automatically collected Upload to the server, the server uses big data analysis technology to get students' physiological and psychological state information. Through the analysis of the data, managers can fully grasp the relevant situation of each student, and adopt differentiated management methods according to the characteristics of different students; at the same time, students can also fully understand their own situation through the relevant information, actively adjust their physical and mental state, and achieve the improvement of overall quality [3].

2.2 System Detailed Design

According to the key problems existing in the current student management, through the full investigation of the government, student brigade, student camp, teaching and research section and students, the requirements of various users for the system are determined, including the main functions of the system, the application of the system in various management activities, and the interaction between the system and various existing management information systems, Guide users to explore the potential application of the system in the framework of smart campus.

System composition. The whole system is composed of server (including database), wearable device, data acquisition terminal, student terminal and manager terminal (see Fig. 1). It is connected with wireless network system of smart campus, camp access control system, educational administration management system and quantitative evaluation system of administration to complete data sharing and interaction.

The system structure is divided into three layers, namely physical layer, data layer and application layer (see Fig. 2). The physical layer mainly collects all kinds of

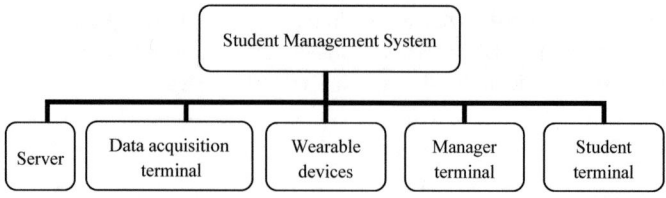

Fig. 1 System composition diagram

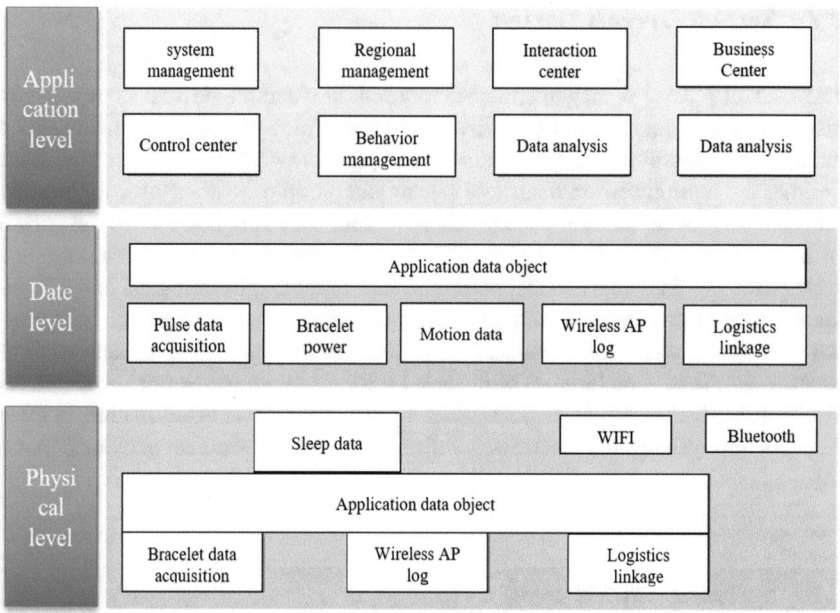

Fig. 2 System hierarchy

sensor data and provides initial data for the data layer; the data layer mainly processes all kinds of initial data and provides standard data objects for the application layer; the main function of the application layer is to provide all kinds of functions that users need.

System function. The main functions of the system include the following aspects: Students' real-time location information acquisition function; students' physiological information acquisition function; automatic data acquisition and upload function; students' psychological state analysis function; students' sports data statistical analysis function and students' behavior analysis function based on location.

System operation process. The operation process of the system is as follows: students wear wearable devices for daily learning, training and rest; wearable devices continuously obtain and save students' physiological information data and location data; wearable devices upload relevant data to the server through the data acquisition terminal; servers analyze students' class operation, movement, sleep, emotion and other information according to the data; managers communicate with each other Through the administrator terminal to obtain students' relevant information, it is convenient for accurate differential management; through the student terminal to obtain their own relevant information, it is convenient for self-regulation [4].

Design of data acquisition terminal. The traditional smart Bracelet usually adopts the way of manual pairing, which connects the mobile phone and the bracelet through Bluetooth, and then uploads the data stored in the bracelet. The disadvantage of this way is that every time the data is uploaded, it needs manual operation, and the data transmission is slow and inefficient, which cannot meet the needs of large-scale automatic data acquisition of the system. Therefore, this paper designs a data acquisition terminal, which is installed at the entrance and exit of each student brigade. When the student wears a bracelet through the entrance and exit, the system automatically reads the data in the bracelet and saves it in the server. The bracelet and data acquisition terminal are connected by WIFI, and the transmission rate is greatly improved compared with Bluetooth [5].

Design of data analysis and management system. Data analysis and management system is the core of the whole system software, which is composed of database, server, student client and manager client (see Fig. 3).

All the data of the system operation are stored in the database, including user basic information data, bracelet information data, WIFI base station information data, physiological and psychological analysis data, auxiliary management decision data, etc.

The server side is mainly composed of data analysis module and data management module. The server side reads the students' movement information data and physiological information data from the database. The data analysis module obtains each student's exercise class, self-study, entertainment, movement statistics, mental state information and so on according to the corresponding analysis algorithm. The data management module can add, query, delete and modify the relevant information in the database according to the needs.

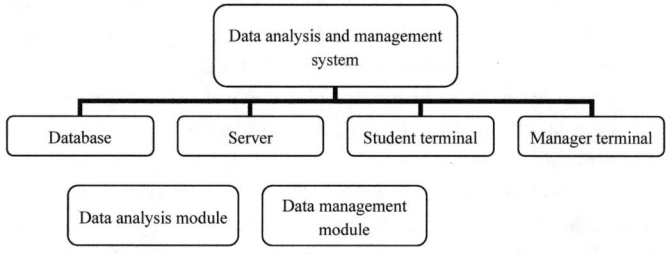

Fig. 3 Composition of data analysis and management system

Student client runs on the mobile terminal, each student can query their own relevant information through the student client, and can query part of other people's information according to the distribution of authority, and set the information of sports, exercise and so on into the form of ranking list, which is convenient for students to compare.

The manager client is for managers to use, can operate the student information, and can set some special rules to manage the students. For example, set the key person alarm: when the set key person's position, emotion, physiological indicators exceed a certain range, automatically remind the management personnel, so as to take timely measures.

3   Key Technology Analysis

3.1   *Location Algorithm Based on WIFI*

The existing indoor wireless positioning system mainly uses infrared, ultrasonic, Bluetooth, WIFI, RFID and other short-range wireless technology. Among them, wireless location technology based on WIFI network has attracted much attention due to its wide deployment and low cost. The radar system developed by Microsoft is the earliest positioning system based on WIFI network. It uses the RF fingerprint matching method to find the nearest K neighbors from the fingerprint database, and takes the average value of their coordinates as the coordinate estimation.

The location system is composed of server and mobile terminal. The server is mainly responsible for location calculation and response to the location request of the terminal. Based on the consideration of load balancing, the web server responding to the location request and the location server running the location calculation are separated, and the data exchange mode is the same as that of the client and web server. The client is mainly responsible for collecting the wireless signal strength of the peripheral AP and submitting the signal characteristics to the server. The server uses the signal characteristics collected by the client to calculate the location and obtain the location estimation of the mobile terminal.

Figure 4 is the information interaction flow chart of the positioning system. The mobile terminal submits the get request to the web server. The get request contains the signal strength eigenvector. After the web server receives the request, it transmits it to the location server in the same way. The location server queries the database and performs the relevant location operation, so as to obtain the location estimation of the mobile terminal.

Due to the complex indoor environment, the wireless signal propagation attenuation model is difficult to describe the mapping relationship between distance and signal strength. This paper proposes to use the radio frequency fingerprint matching localization method, which has good localization robustness.

Fig. 4 Flow chart of
information interaction of
positioning system

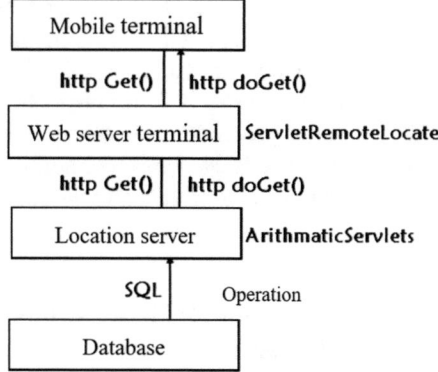

Fingerprint matching location algorithm is based on the experimental data, which mainly includes two stages: offline training and online location. The task of offline training stage is to establish a one-to-one correspondence between the RF signal strength vector and the location of the client to form a fingerprint database. In the localization phase, the real-time collected signal intensity vector is used to match the fingerprint database constructed in the training phase to obtain the target position estimation.

The existing methods based on RF fingerprint matching mainly include deterministic and probabilistic methods. In the deterministic localization algorithm, the centroids of the points with the minimum distance from the real-time RF fingerprint are selected in the fingerprint database as the position estimation of the target. The calculation efficiency of the deterministic location algorithm is high, but the accuracy is low. Probabilistic location algorithm generally uses Bayesian estimation theory, through different likelihood functions, such as likelihood function based on kernel function, the posterior probability of target location is calculated, and the location point with the largest posterior probability is taken as the final location estimation. Probabilistic location algorithm has high location accuracy and location robustness, but the amount of calculation is relatively large.

In the training stage, the RSSI mean value is used for fingerprint features, and in the positioning stage, the RSSI mean value accumulated step by step is used to match the fingerprint database, which greatly reduces the computational complexity.

The RSSI mean value of each AP is used for fingerprint feature.

$$F_L = \left(\overline{S}_{AP1}, \overline{S}_{AP2}, \ldots\right) \tag{1}$$

That is to say, in the training stage, the multiple data of each AP collected at the same location point is averaged, and so is in the positioning stage. The difference is that in the training stage, more data is collected in order to get as much information as possible, while in the positioning stage, less data is collected, which reduces the positioning delay and improves the real-time performance to a certain extent. For the RSSI value of each scanned AP, a selection interval is set $[RSSI - \sigma, RSSI + \sigma]$,

σ is the empirical value of many experiments, and the location points satisfying this interval range are found in the fingerprint database. If there are n location points falling in this interval range, the weight of these location points is 1/n, and the weight of other location points is 0; for all APS, the weight of each location point is 1/n After the above processing, the position with the largest weight is selected as the estimated position. If the weights of multiple position points are the same, the signal strength distance is compared, and the smallest one is selected.

3.2 Human Emotion Discrimination Algorithm Based on Wearable Devices

The key problem of emotion discrimination based on multi physiological information fusion is to use the appropriate classification and fusion algorithm to fuse and classify the feature parameters of various physiological signals collected to reflect the emotional state of human body, so as to diagnose the health status of human body. Figure 5 shows the framework of emotion discrimination model based on multi physiological information fusion [6, 7].

Support vector machine (SVM) is proposed from the optimal classification in the case of linear scoring, and its core is the principle of structural risk minimization. For the linear separable two class pattern classification problem, the best classification line is found to make the two classes of samples separated correctly, and make the classification interval of the two classes maximum. When extended to high dimensional space, the optimal classification line becomes the optimal classification surface.

Let the general form of the linear decision function in the d-dimensional space of the training sample set be as follows

$$w \cdot x + b = 0 \tag{2}$$

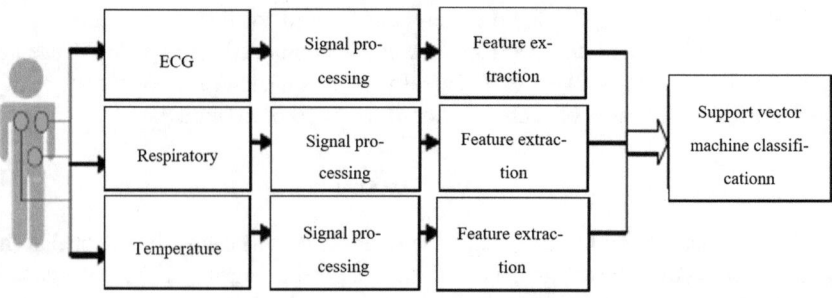

Fig. 5 Framework of emotion discrimination model based on multi physiological information fusion

The statistical learning theory points out that the optimal classification surface has a good generalization performance, so the problem of finding the optimal classification surface can be transformed into the problem of finding the optimization, that is, the minimum value under the constraint condition:

$$j(x) = \frac{1}{2}\|w\|^2 = \frac{1}{2}(w \circ w) \tag{3}$$

3.3 Location Algorithm Based on RFID

RFID tags are divided into active and passive, active tags can be regarded as a more generalized RFID, because the tag itself is active, signal processing can also be done more complex, positioning accuracy will be much higher. In the ideal situation, it can cover 100 m, and the positioning error is about 5 m. It is mainly completed by triangle positioning, but in this field, nodes such as UWB, ZigBee and so on can also be used to complete the positioning. Passive tag is referred to as RFID most of the time. Because the tag itself has no computing power, all signal processing is limited by the reflected signal received by the reader, so the choice of signal processing algorithm will be much smaller. And because the reader recognition range is basically within 10 m, it is generally very fine positioning to study the location of passive tags.

The indoor location method based on RFID is to locate the tag through the reader of known location, which can be divided into non ranging method and ranging method. The method based on ranging is to estimate the actual distance between the target device and each tag through various ranging technologies, and then estimate the location of the target device through geometric way. The common location methods based on ranging are: Based on the time of arrival (TOA) location, time difference of arrival (time of arrival) based on signal Difference of Based on arrival (TDOA), RSSI and angle of arrival. These technologies are consistent with the technical principles used in UWB and Wi Fi, but the propagation distance of RFID signal is very close due to the restriction of energy, generally only a few meters to tens of meters.

Non ranging method is to collect the scene information in the early stage, and then match the acquired target with the scene information, so as to locate the target. The typical implementation methods are reference tag method and fingerprint positioning method. The commonly used algorithm of reference tag method is centroid positioning method. Fingerprint positioning method is basically the same as Wi Fi positioning, beacon positioning and other technologies. When the target tag enters the scene, multiple readers can read the target tag information at the same time. The positions and lines of these readers form a polygon, and the centroid of the polygon can be regarded as the location coordinates of the target tag. Centroid positioning algorithm is simple and easy to operate, but the positioning accuracy is relatively low. It is often used in the scene where the positioning accuracy is not high and the hardware equipment is limited.

4 Conclusion

This paper studies the combination of wearable devices and cadet Management in military academies. In view of the current demand for accurate degree of informatization in cadet Management, relying on the construction conditions of smart barracks project, WIFI Positioning, big data acquisition and analysis technology and auxiliary decision-making theory and method are adopted to achieve accurate attendance, physiological data statistics, psychological state discrimination and differentiation of cadets. After the application of the system, it can provide accurate and intelligent management means for colleges and military barracks, greatly improve the management efficiency, reduce the contradictions and difficulties brought by the management mode from the perspective of managers, and broaden the channels for students and other managers to obtain intelligent information, which has good military management efficiency.

References

1. Guo X, Huang Q (2010) New exploration of the management of growing cadre students of the military academy. J Higher Educ Res 74–76
2. Guo Z, Ning X (2011) Sketchy research on administration innovation in current military academies operation. J Naval Univ Eng (Comprehensive Edition) 39–42
3. Xu S (2011) Precise management of the armed forces under conditions of informationization. Chinese Public Admin 85–87
4. Liu F, Jing W (2011) Application of precise management in cadet team management of military academies. Med Educ Res Practice 41–44
5. Liu Z (2013) Several key problems of mobile terminal teaching activity management. New Silk Road 80–81
6. Yin J (2012) The research and expectation on wearable health monitoring system. Chinese J Med Instrum 40–43
7. Wu X (2009) Multi-physiology information fusion for emotion distinction in smart clothing. Comput Eng Appl 218–221

Proving Geometric Problem by Adding Auxiliary Lines-Based on Hypothetical Test

Mingrui Zhou and Xinguo Yu

Abstract With the continuous development on computer science and technology, people have begun to use automatic reasoning systems to solve some logical mathematical proof problems. The traditional reasoning system can use the finite conditions to verify the conclusion of the problem, to achieve the effect of proving geometric problems. However, there is no unified standard for the proof of plane geometric problems, it is impossible to use formulas to prove directly. More importantly, some problems do not contain all the proof conditions, and have to add auxiliary lines to prove. However, there is still no unified standard for drawing auxiliary lines in the mathematical world. Therefore, there is no corresponding template for reference to add the function of automatic drawing auxiliary lines in the automatic reasoning system, which has a certain challenge and research value. Because the mathematical field can not summarize the standard implementation of auxiliary line construction, the research on automatic addition of auxiliary line in computer field is relatively less, and the common addition methods are mainly based on model classification or exhaustive. But these traditional algorithms have the problems of low efficiency or unlimited growth of problem scale. The main research goal of this paper is to optimize the auxiliary line adding algorithm based on exhaustive method. This paper proposes an auxiliary line adding algorithm for plane geometry problems based on hypothesis testing, which makes it more efficient in solving complex elementary mathematical and geometric problems, and avoids the problem of unlimited growth of traditional algorithm.

Keywords Plane geometry · Auxiliary line · Image understanding · Automatic inference

M. Zhou (✉) · X. Yu
Central China Normal University, Wuhan, Hubei, China
e-mail: cibaoxiaozi@vip.qq.com

© The Author(s), under exclusive license to Springer Nature Singapore Pte Ltd. 2022   151
E. C. K. Cheng et al. (eds.), *Artificial Intelligence in Education: Emerging Technologies,*
Models and Applications, Lecture Notes on Data Engineering and Communications
Technologies 104, https://doi.org/10.1007/978-981-16-7527-0_12

1 Introduction

Artificial intelligence is an important field of computer science. Among them, reasoning technology was one of the basic technologies for realizing artificial intelligence. Natural deductive reasoning in reasoning technology is based on common logical equivalence and common logical implication knowledge. Its core idea is to start from known facts and use reasoning rules to reason to get conclusions [1].

In 1965, Simon and Newll first proposed a machine reasoning system for propositional logic, thus opening up the field of machine proof [2]. In the same year, Robinson proposed a resolution method to deal with the formal formula set of clauses. This method is called the resolution principle [3]. For the proof of theorem of the form $(\forall x)\ (\exists y)P(x,\ y)$, the resolution principle can answer the question whether the conclusion is true, and common elementary mathematical geometry proof problems can be effectively transformed into this form, so this formula can solve most geometric proof problems.

After the automatic reasoning method, some scholars have made some progress in the research of automatically adding auxiliary lines to the elementary mathematical geometry problems. Matsuda N's paper proposes a proof of geometric prob-lems by adding geometric elements. But there are still many problems, for example, the addition of auxiliary lines lacks accuracy and has certain complexity.

Nowadays, almost all elementary mathematical and plane geometric problems can be classified and have corresponding typical models. However, this method encountered obstacles when it encountered the problem of adding auxiliary lines to prove, because its generalization is bad. With the variability of topic conditions and graphics, it is difficult to establish a set of auxiliary line addition standards applicable to most topics. Although some existing auxiliary line adding strategies and topics can be used for training through the idea of machine learning, but the generalization of this model is still not strong [4].

For computers, the most direct method is to verify the auxiliary line created by full connection of geometric vertices. But there are two problems.

1. If the problem graph is complicated, the number of auxiliary line combinations will increase exponentially.
2. When auxiliary lines need to be added to special points other than the endpoints in the question, the number of auxiliary line combinations will be further increased, and since the specific range of the special points cannot be determined, it is likely that the verification cost will increase greatly.

Based on the hypothesis testing method, this paper proposes a new auxiliary line adding algorithm, which combines the classical backtracking method and dichotomy method to improve the efficiency of verification of auxiliary line. The main work of this paper is to implement the new auxiliary line adding algorithm. This algorithm uses image understanding and resolution inversion functions, realizes the optimization of the traditional algorithm based on the exhaustive method, which makes it more efficient and solves the problem of unlimited growth of the problem scale.

2 Literature Review

2.1 Understanding of Plane Geometry Image

Image understanding technology will be used to obtain the new problem-solving conditions brought by the auxiliary line, which is an indispensable part of the implementation of the new auxiliary line adding algorithm in this paper. As one of the basic figures in plane geometry, straight lines have a wide range of applications. The recognition methods of straight lines can be divided into Hough transform method and Freeman chain code. Hough transform method was proposed by Hough in 1962, after that, many improved algorithms were derived [5, 6]. The basic idea is to use the dual characteristic of the point and the straight line to convert a point (x, y) in the image is mapped to a curve in the dual space, and then the straight line in the original image can be found through the intersection of the curves in the dual space [7, 8].

The original version of the hough transforms method is slow and inefficient. Therefore, the probabilistic Hough transform is proposed as an optimization scheme. Compare with the standard Hough transform, the algorithm will randomly extract a feature point in the image [9]; after the edge point is selected, the point will be Hough transformed and accumulated and calculated. Select the point with the largest value in the Hough space, if the value of the point is greater than the preset threshold, start from this point and move along the straight line until the two end points of the straight line are found. However, it sometimes causes the problem of dividing a straight line into multiple line segments because of the threshold [10].

The corner detection is to detect the vertices in the plane geometry, which is the key point in the understanding of the plane geometry. In 1988, Harris proposed the Harris corner detection algorithm, the algorithm uses the method of detecting the change in image gray level E_w in the sliding window to determine the gray level change of the image, thereby finding the inflection point [11]. The formula is as follows.

$$E_w(x, y) = \sum_{u,v} w_{u,v} |x \frac{\partial I}{\partial x} + y \frac{\partial I}{\partial y} + O(x^2, y^2)|^2 = Ax^2 + 2Cxy + By^2$$

In this formula, the $A = \frac{\partial I}{\partial x} * W$, $B = \frac{\partial I}{\partial x} * W$, $C = \frac{\partial I}{\partial x} \cdot \frac{\partial I}{\partial y} * W$, where $*$ represents the convolution operator.

And the sliding window adopts the Gaussian window mode

$$w_{u,v} = \frac{1}{\sigma} e^{-(u^2+v^2)/2}$$

Which (u, v) is the coordinate relative to the center of the window, and the E_w can be written as

$$E_w(x, y) = (x, y)M\begin{bmatrix} x \\ y \end{bmatrix}$$

Which $M = \begin{pmatrix} A & C \\ C & B \end{pmatrix}$

M is a symmetric matrix, and its eigenvalues α and β reflect the changes in the gray value of the image.

Define rotation-invariant functions here

$$R = \det(M) - k(\alpha + \beta)^2$$

Perform threshold processing on the function R, R is bigger than the threshold, that is, extract the local maximum of R. And Shi-Tomasi further optimized Harris algorithm. If the smaller one of the two eigenvalues is greater than the minimum threshold, the corner found is a strong corner, otherwise it is a weak corner.

2.2 Resolution Principle

Resolution inversion is a reasoning method based on predicate logic, it can be used to verify the correctness of adding auxiliary lines after adding new conditions of proof [12]. In the process of reasoning, quantified formulas need to be matched. Resolution is an important rule among inference rules, and its object is a type of formula called clause. The definition of a clause is a disjunctive formula composed of words, and the resolution method must be applied to the clause form. In this way, a clause set can be obtained, and it is easy to prove that if the formula I follows the formula set S, then I logically follows the clause set obtained by the transformation of S. Therefore, the clause can be used to express the formula for resolution inversion, The result is the same as the original formula [13]. Therefore, the reasoning process of the inversion is as follows, assuming that there is a formula set S, the target formula J needs to be proved.

It only needs to add the inverse formula of J to S, by transforming S into a set of clauses and applying the principle of resolution to derive an empty clause, you can prove that the target formula is established. The core idea is the contradiction method, that is, to prove that the inverse of adding the target formula J The formula set S after the form is unsatisfiable.

2.3 Common Guidelines for Adding Auxiliary Lines

Adding Principle Based on Classification

Although there is currently no set of auxiliary line addition principles that can be applied to all elementary mathematics and geometry problems, there are still scholars who have classified some topics and summarized some auxiliary line addition methods. For example, in the teaching and research materials published by Zhang Tiexia in 1986, the common elementary mathematics auxiliary line problems are divided into four categories, namely Complementing the manifestation type, Transform the stereotype, a bridging connection type, a centralized and flexible type [14].

This auxiliary line classification method is based on the basic mathematics exercise question bank, but the paper of Ulman found that the method is limited by question types, and it is likely to affect the efficiency of the system after a large change in the question type [15].

Adding Principle Based on Special Points.

It is impossible to try to draw all possible auxiliary lines, because there are an infinite number of corner points on a line segment. However, since the addition of auxiliary lines in this system is based on the plane geometry of elementary mathematics, there are certain rules to be found [16].

In elementary mathematics, the application of general geometric theorem properties can be described intuitively with a basic geometric figure, such as a triangle, parallel line, etc. Therefore, it is possible to obtain a correct auxiliary line only by completing these basic figures [17]. The addition of special point auxiliary lines is equivalent an improved algorithm based on exhaustive methods.

In the face of the geometry problem that auxiliary points need to be added to construct auxiliary lines, special auxiliary points can also be constructed, such as the midpoint of the line segment, but the effect is not as ideal as adding auxiliary lines directly [18]. For computers, after trying one by one, a suitable auxiliary line can always be found. The disadvantage is that the time efficiency is low, and if the problem requires more than one auxiliary line, the time for adding auxiliary lines will increase dramatically.

3 Algorithm Design and Experiment

3.1 Algorithm Design

The algorithm is based on the traditional special point auxiliary line adding algorithm. Because the traditional special point adding method can not determine the possible location of auxiliary line, the problem scale may grow infinitely. This algorithm makes full connection for all the common special points and endpoints, assuming that auxiliary line exists in these segments. Because the elementary mathematics plane geometry problem is not complex, in most cases, the auxiliary lines required by the problem can be included in the auxiliary line set. Therefore, this paper uses the idea of hypothesis testing to limit the scale of the problem, it is assumed that effective auxiliary lines exist in the initial auxiliary line set, and then the auxiliary line set is verified and deleted. The effectiveness of the auxiliary line is defined here. By adding the auxiliary line L, the clause set S that cannot be resolved by the inversion success can be changed into a new clause set S' after adding new conditions by L. If S' can be resolved by the resolution inversion, then L is called an effective auxiliary line. Finally, on the basis of the traditional algorithm, the dichotomy and backtracking method are used to reverse the calculation, so as to achieve the goal of improving the efficiency of the algorithm. The specific optimization algorithm flow is as follows.

1. Constructing possible auxiliary lines for all special points and line segments in the input graphics to obtain auxiliary line set A.
2. Using image understanding technology to obtain all the new conditions set W which is brought by auxiliary line set A, add W to the initial topic condition clause set S, and perform resolution inversion. If successful, go to step 3, otherwise back to step 1, and select more special points to construct Auxiliary line, if there are no more special points, the construction fails.
3. Dividing the auxiliary line set into two subsets AL and AR, both of which contain half of the auxiliary lines of the original set A (if the number of auxiliary lines is odd, the number of auxiliary lines in the left subset is 1 less than the number of auxiliary lines in the right subset). At the same time, W is also divided into two subsets WL and WR. If A can not be divided (there is only one auxiliary line), the current A is output, which is a valid auxiliary line.
4. Adding WL and WR to S respectively for resolution inversion. If successful, it can be proved that the auxiliary line subset corresponding to the subset contains valid auxiliary lines (only the conditions provided by the effective auxiliary line can prove the goal of the problem). Replace W and A with the current conditional subset and auxiliary line subset (For example, replace W and A with WL and AL), go back to step 3, continue to divide and verify. If the resolution inversion of the two subsets fails, go to step 5(This means that it is a multi auxiliary line problem).
5. If the current two auxiliary line subsets AL and AR both contain only one auxiliary line, then A is an optimal auxiliary line set (there may be multiple). If AL or AR contains more than one auxiliary line, the auxiliary line set A is subject to cyclic resolution inversion verification. Each round of verification will reduce one auxiliary line in set A until it meets the point where the verification cannot be successful. Skipping and adding this point to the output set, further reduce the auxiliary line in set A, until all the useful auxiliary lines are finding. The output is an optimal auxiliary line set.

3.2   Algorithm Complexity Proof

This algorithm is the core innovation of this paper. The space complexity of this algorithm is as same as the traditional algorithm, if there are n clauses, the space complexity of this algorithm is $O(n)$. Assuming there are x special points, and the time complexity of this algorithm can be proved:

The clause length converted from the elementary mathematics geometry problem condition is constant (the single condition given by the actual problem will not be too complicated, otherwise it is difficult for humans to calculate). Assuming there are n conditions, based on the principle of resolution inversion, at most $\frac{(n*(n-1))}{2}$ times resolutions are performed in each round (each clause is resolved with all other clauses). The time complexity of each time resolution is O (1), then the time complexity of each round of resolution is $O(n^2)$, for constant rounds of resolution, the overall time complexity is still $O(n^2)$.

Then drawing auxiliary line derived from a special point creates at most $\frac{x*(x-1)}{2}$ auxiliary lines, and for traditional algorithm, each auxiliary line added requires a resolution inversion verification, so the overall time complexity is $O(x^2) * O(n^2)$. Although the scale of general questions is relatively small, and a small number of questions are automatically answered on the computer without much impact, this method is not suitable for large-scale calculations. But if multiple auxiliary lines are included, it will greatly reduce the operating efficiency (It must find the correct answer by trying all combinations of the target auxiliary lines).

The algorithm in this paper first performs a resolution inversion to determine the set of auxiliary lines, and then finds the optimal auxiliary line by dichotomy. When the problem only needs to add one auxiliary line, the time complexity is $\log_2 x * O(n^2)$. When there is more than one auxiliary line, the actual application efficiency is also less than the pure special point addition method.

If the system needs to add more than one auxiliary line, in the worst case, it can find that the number of auxiliary lines required is greater than 1 by dividing the auxiliary line set for the first time (Inversion failed). At this time, system only needs to remove invalid auxiliary lines one by one, if the removed auxiliary line will cause inference failure, then it can be determined that this auxiliary line is one of the valid auxiliary lines. The overall time complexity is $O(x^2)*O(n^2)$, and the time complexity of the traditional special point algorithm is also $O(x^2) * O(n^2)$, but the number of resolutions in the worst case is about twice that of the new algorithm in actual work (after trying all the single auxiliary lines in the first round, the second round requires to try the combinations of two auxiliary lines, it needs a total of $C_x^2 = \frac{x*(x-1)}{2}$ times attempts in the worst case, so the total time complexity of the traditional special point algorithm will be $2 * O(x^2) * O(n^2)$).

3.3 Experiment

Unit Test

This experiment mainly tests two aspects. One is testing whether the algorithm can find auxiliary lines correctly, and the other is testing whether the new algorithm of adding auxiliary lines can be more efficient than traditional method. The unit test will show the problem-solving process in detail to verify the correctness of the automatic addition of auxiliary lines, while the batch test in the next section will compare the

Fig. 1 Input picture

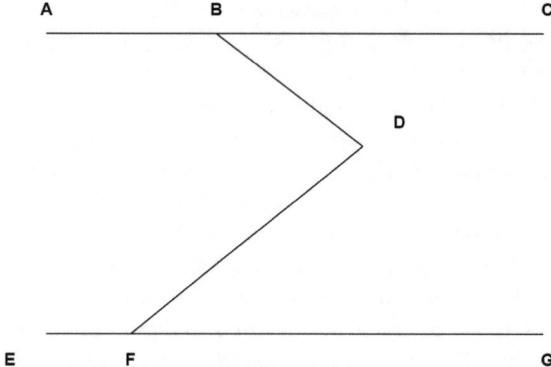

problem-solving time of new algorithm with the traditional algorithm. Because the image of the test questions on the website is not clear, this article will use drawing software to redraw the relevant test questions. In this paper, a simple geometric proof problem related to a parallel line theorem is selected for unit test.

Question: As shown in the Fig. 1, the straight line AC is parallel to the straight line EG, the point B and the point F are respectively located on the straight line AC and the straight line EG, and the point D is a point between the straight line AC and the straight line EG.

Prove: Angle BDF = Angle CBD + Angle DFG.

The available conditions given in the title can be transformed into a set of clauses: {Parallel (AC, EG), On (B, AC), On (F, EG)}.

Proof goal: Equal (BDF, Add (CBD, DFG)).

The whole process of the algorithm will be shown in Fig. 2.

In Fig. 2, the image on the top left is a preprocessed image, the letters are separated from the geometry, and the lines and corners are marked. The image in the upper right corner is a fully connected image, and the length of the parallel line extended by the special point is a random value. The image in the lower left corner is the last round of resolution inversion. Among them, parallel (DH, AC) and parallel (DH, EG) are added by auxiliary lines, ¬ consistent (DFG, HDF) ∨ equal (DFG, HDF) and ¬ consistent (DFG, HDF) ∨ equal (CBD, BDH) are derived from the inner stagger angle theorem, the auxiliary line DH is finally obtained. The last picture is the output picture.

Batch Test

The results of batch testing for 20 questions which screened from the Internet will be shown in the Table 1.

1. It can be drawn from the Table 1 that compared with the traditional algorithm, the optimized algorithm has a more stable running time. Other specific analysis is as follows.
2. For both algorithm, seven of the twenty questions cannot be resolved correctly, and the correct answer rate is 65%.

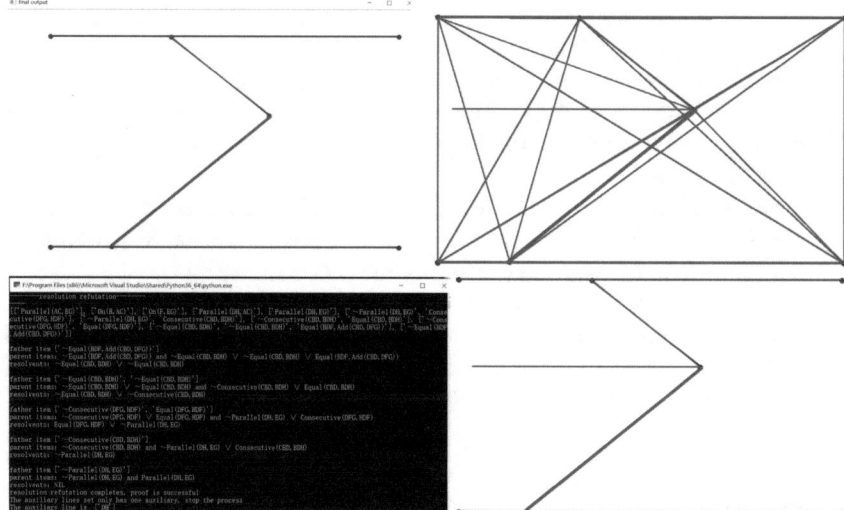

Fig. 2 Algorithm process

Table 1 Result of batch test

Sequence	Question type	Points	Run time (old)	Run Time(new)
1	Triangle	4	0.81 s	0.41 s
2	Triangle	4	0.32 s	0.42 s
3	Triangle	5	1.21 s	0.61 s
4	Triangle	5	0.66 s	0.69 s
5	Triangle	6	1.54 s	0.97 s
6	Triangle	6	Fail	Fail
7	Triangle	6	Fail	Fail
8	Triangle	7	3.89 s	2.45 s
9	Triangle	8	Fail	Fail
10	Triangle	9	Fail	Fail
11	Parallel lines	5	0.89 s	0.65 s
12	Parallel lines	5	0.68 s	0.73 s
13	Parallel lines	6	1.06 s	0.94 s
14	Parallel lines	6	Fail	Fail
15	Parallel lines	6	1.47 s	1.15 s
16	Parallel lines	6	0.93 s	1.04 s
17	Parallel lines	7	1.53 s	1.21 s
18	Parallel lines	8	Fail	Fail
19	Parallel lines	8	2.40 s	1.69 s
20	Parallel lines	10	Fail	Fail

3. Compared with the traditional special point auxiliary line addition algorithm, the optimized algorithm takes 12.96 s to solve the 13 problems successfully, and the traditional algorithm is 17.39 s, and the efficiency is increased by about 34.1% (the problem solving failure here is caused by the accuracy problems of image understanding). And the running time of the optimized algorithm is relatively stable.
4. If the image understanding accuracy is low, it will get some useless interference conditions, which reduces the efficiency of resolution or causes reasoning to fail. Therefore, failures often occur in complex questions.
5. In some questions, the traditional auxiliary line adding algorithm has faster problem-solving speed, such as question 2 in the unit test and questions 2, 4.12 and 16 in batch test. But in general, the new algorithm is more efficient and faster in the face of complex problems.

4 Conclusion

The main contribution of this paper is to optimize the traditional auxiliary line adding algorithm based on special points, improve the time efficiency of the algorithm, and make the scale of auxiliary line adding problem will not grow infinitely. Although the running time of the new algorithm is worse than that of the traditional algorithm in some cases, the overall running efficiency of the new algorithm is better than that of the traditional algorithm, and it has higher efficiency in the face of complex problems. Of course, there are still some problems in the algorithm. In the future research work, we will try to improve the efficiency of the algorithm in dealing with simple problems.

References

1. Guo XW, Xia QM (2019) Progress in the research of computability logic. J Guizhou Univ Eng Sci
2. Jiang J, Zhang J (2012) A review and prospect of readable machine proofs for geometry theorems. J Syst Sci Complex
3. Fu H, Zhong X, Li Q et al (2014) Geometry knowledge base learning from theorem proofs. Springer, Berlin Heidelberg
4. Ol´ak M (2020) GeoLogic-graphical interactive theorem prover for euclidean geometry
5. Vaca-Castano G, Lobo DV, Shah M (2019) Holistic object detection and image understanding. Comput Vis Image Understand
6. Yu W (2011) Research on automatic recognition of plane geometric figures. IEEE Transl. Guangzhou University, China
7. Seo MJ, Hajishirzi H, Farhadi A (2014) Diagram understanding in geometry questions. In: Twenty-Eighth aaai conference on artificial intelligence
8. Gan W, Yu X, Wang M, (2019) Automatic understanding and formalization of plane geometry proving problems in natural language: a supervised approach. Int J Artif Intell Tools 28(4):1940003

9. Zhang X, Hongguang Fu (2015) Graphic recognition and understanding of plane geometry problems. Comput Appl 35(s2):280–283

10. Marzougui M, Alasiry A, Kortli Y, et al. "A Lane Tracking Method Based on Progressive Probabilistic Hough Transform", IEEE Access, 2020, vol.99 pp.1–1.

11. Bansal M, Kumar M, et al. "An efficient technique for object recognition using Shi-Tomasi corner detection algorithm", 2020.

12. Mondal B, Raha S (2012) Approximate Reasoning in Fuzzy Resolution. International Journal of Intelligence 3(2):86–98

13. Buning HK (2020) Wojciechowski P, Subramani K, "NAE-resolution: A new resolution refutation technique to prove not-all-equal unsatisfiability." Mathematical Structures in Computer Science 30(7):736–751

14. Tiexia Zhang, Hui Zhang and Chuntian Li, "On the auxiliary line problem in the current geometry textbook", Middle School Teaching and Research (Mathematics), 1986, vol.7 pp.24–27.

15. Ulman S, "Model-Driven Geometry Theorem Prover", Computer Science and Artificial Intelligence Lab (CSAIL), 2005.

16. Hong Fan, "Discussion on the optimization of problem solving by adding auxiliary lines", Newsletter of Mathematics Teaching, 2019, vol.0(32) pp.75–77.

17. Alvin C, Gulwani S, Majumdar R et al (2015) Automatic Synthesis of Geometry Problems for an Intelligent Tutoring System. J Viral Hepatitis 15(6):434–441

18. Chen X, Song D, Wang D (2015) Automated Generation of Geometric Theorems from Images of Diagrams. Annals of Mathematics & Artificial Intelligence 74(3–4):333–358

A Virtual Grasping Method of Dexterous Virtual Hand Based on Leapmotion

Xizhong Yang, Xiaoxia Han, and Huagen Wan

Abstract Virtual hand interaction technology has been widely used in handheld information device evaluation, flight simulation, virtual assembly/maintenance as well as intelligent education, especially virtual experiments. In these applications, dexterous virtual hand is usually used as the embodiment of human hand, and virtual grasping is an important way of interaction. Grasping interaction usually involves whole hand motion tracking and finger tracking to drive virtual hand deformation and evaluate whether grasping conditions can be met. However, as far as finger tracking is concerned, accurate finger tracking is still a luxury for both economic and technical reasons. In view of this situation, the kinematics analysis results of all five fingers in the object grasping task is used, combined with the characteristics of LeapMotion and inverse kinematics, to develop a simple and effective virtual grasping method, which enables users to use a dexterous virtual hand as the avatar in the virtual environment without finger tracking to reach and grasp digital 3D objects naturally. In this paper, the specific description of our proposed method is given, and the preliminary verification results show that the proposed method has some advantages over the traditional virtual grasping method based on forward kinematics.

Keywords Virtual grasping · Grasp trajectory · Hand tracking · Finger tracking · Inverse kinematics

X. Yang
Science and Technology On Avionics Integration Laboratory, Shanghai 200233, China

X. Han
College of Information Science & Electronic Engineering, Zhejiang University, Hangzhou 310027, China

H. Wan (✉)
Center for Psychological Sciences, Zhejiang University, Hangzhou 310027, China
e-mail: hgwan@cad.zju.edu.cn

1 Introduction

In recent years, virtual hand interaction technology has developed rapidly in the fields of handheld information equipment evaluation, flight simulation, virtual assembly and virtual maintenance, virtual surgery, and digital games. In these contexts, a dexterous virtual hand is often used as an avatar of the user's hand.

Generally speaking, virtual hand interaction mainly includes two types: gesture interaction and grabbing interaction. Gesture interaction refers to online or offline recognition of user gestures based on hand motion data or static posture data. The recognition algorithms cover many strategies from experience-based heuristic algorithms to currently popular deep learning algorithms [1]. Grasping interaction refers to the real-time capture of human hand movement data (including the translation and rotation of the overall hand motion and finger bending) to drive the deformation of the virtual hand, and based on the relationship between the posture of the virtual hand and the object to be grasped in the scene, to determine whether the grasping conditions are met, and then to complete the digital simulation of the actual grasping action [2]. This paper mainly studies virtual grasping.

Grasping in the real world depends on many complex factors, such as the shape, material, and weight of the object, as well as the contact force between the finger and the object. However, in the virtual world, virtual capture is mainly achieved through heuristics and/or physical simulations.

The virtual grasping based on heuristics relies on the assumption that a stable grab can be performed with different finger combinations. For example, Ullmann et al. established a single-handed grasping condition, so that only when there is an angle greater than a given and experimentally determined angle β between the normal n_i ($i = 1, 2$) of the contact surface of a rigid virtual object, can an object be grasped. In this process, the thumb and at least one finger and/or palm are involved [3]. Holz et al. proposed a method of multi-contact grasping interaction, in which the concept of "grasp pairs" was proposed [4].

In contrast, simulation-based virtual grasping methods are more robust, but more computationally expensive [5–7]. For example, Jacobs et al. proposed a hand model based on the coupling of a flexible body and a rigid hand skeleton for physical manipulation of virtual objects. Their method allows precise and robust grasping and manipulation of virtual objects based on fingers, but the geometric structure of the virtual hand is relatively simple, and in order to avoid computational complexity, only a flexible component is used to represent the finger bones [6]. Moehring et al. proposed to integrate pseudo-physical simulation into a heuristic-based method to manipulate constrained objects [7].

Generally speaking, both heuristic-based methods and simulation-based methods need to track hand motion and finger bending for online calculation to drive the virtual hand model, and then support virtual grasping interaction. Various technologies including magnetic tracking, optical tracking, inertial tracking, etc. can realize real-time tracking of human hand movements with six degrees of freedom. They are mature enough to be used in various applications. However, as far as finger tracking is concerned, whether for economic or technical reasons, accurate finger tracking is still a luxury [8–12]. Although the data glove is considered a popular finger tracking device and has been used for more than 20 years, its inconvenience and high price greatly limit its application[8]. While the latest development of Kinect sensor and LeapMotion provides promising methods for free-hand finger tracking, they are not only limited by the tracking range, but also greatly affected by occlusion. Therefore, it would take more time to be mature enough [9–12].

In view of the contradiction between the demand for finger tracking for virtual hand grasping interaction and the maturity of finger tracking technology, we have formerly proposed a hybrid method combining heuristics and physical simulation to support fast, stable and visually controllable multi-finger grasping interaction [13, 14]. This method first detects the collision between the object and the predetermined finger trajectory, then derives the grasping structure according to the collision detection result, and finally calculates the feedback force according to the grasping condition. As far as we know, this is by far the first and only method to achieve natural grasping operations without finger tracking during online operation. However, this method still requires expensive data gloves in the offline construction process to build the basic database of finger bending, and the six-degree-of-freedom tracking scheme used in the online operation phase is still complicated. However, in practice, there is an urgent need to construct a simpler and more practical method to expand the application context of the method, and to promote the method.

Using the kinematic analysis results of all five finger movements in the reach-and-grasp tasks of various objects [2, 7, 15], combined with the characteristics of Leapmotion and Inverse Kinematics (IK), this paper proposes a simple and effective virtual grasping method, allowing users to use a dexterous virtual hand as an avatar for reach-and-grasp tasks in a virtual environment, without finger tracking. Compared with previous work, the contributions of this article are mainly manifested in the following four aspects:

1. A three-layer dexterous virtual hand model of "kinematics-appearance-grasping computation" is proposed, which effectively supports the separation of the visual feedback of the virtual hand and the grasping computation;
2. The whole implementation can be completed only by using LeapMotion, a cheap interactive device, which can greatly expand the applicability and facilitate the popularization of the method;
3. Since only the six-degree-of-freedom data of the entire hand is used and no finger bending data is required as input, the defect of inaccurate finger tracking data caused by LeapMotion due to occlusion is greatly avoided, and the robustness of the method is improved;
4. The effectiveness of the proposed method is preliminarily verified by comparing it with the traditional virtual grasping method based on forward kinematics.

2   Virtual Grasping Based on Forward Kinematics

In this section, we briefly introduce the traditional dexterous virtual hand grasping method based on forward kinematics [16]. This type of method mainly involves three parts, namely the modeling of the virtual hand, the kinematics of the virtual hand, and the judgment of the object grasped by the virtual hand.

2.1   Virtual Hand Modeling

In the traditional dexterous virtual hand grasping method based on forward kine-matics, virtual hand modeling usually includes appearance modeling and kinematics modeling.

The appearance model of the virtual hand usually adopts a mesh model (Fig. 1a) [13] or a stick-ball model (Fig. 1b) [12]. The mesh model can be either polygonal modeling or free-form surface modeling followed by polygonal discretization to generate a polygonal mesh; the stick-stick model is constructed based on the bone structure of the human hand.

For the virtual hand represented by the stick-ball model, its appearance model and hierarchical structure constitute its kinetics model; for the virtual hand represented by the mesh model, in order to realize the mesh deformation during the virtual hand movement, the human body animation technology is usually used to create the human hand. The "skin-muscle-skeleton" three-layer representation (Fig. 2a,b) [15], is widely used where the mesh model of the virtual hand is the skin layer. First, a hierarchical bone description (i.e., skeleton layer) needs to be established according to the anatomical structure of the human hand. The movement of the bone follows the principle of Forward Kinematics. Then, the bone is embedded in the mesh model through the vertex binding operation (That is, each mesh vertex of the skin layer is given the weight that it is affected by the bone). Finally, the muscle layer is established, and the movement of the bone layer is transmitted to the skin layer through the muscle layer.

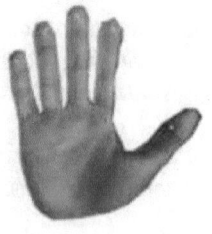   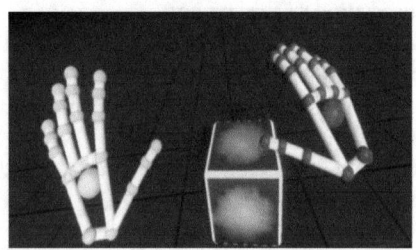

(a) Mesh model [13]   (b) Stick-ball model [12]

Fig. 1 Appearance model of the virtual hand

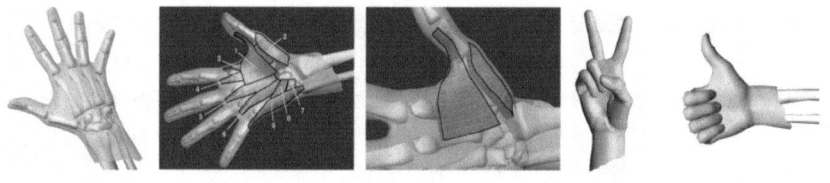

(a)   Skin layer   (b) Bone layer and muscle layer   (c) Kinematics deformation

Fig. 2   Kinematics model of Virtual Hand [15]

2.2   Virtual Hand Motion

By tracking the motion of the real human hand and mapping the motion data to the kinematics model of the virtual hand, the motion of the virtual hand can be driven.

For the virtual hand represented by the stick-ball model, the joint angle of the real human hand movement can be directly applied to the stick-ball model through rotation transformation.

For the virtual hand represented by the mesh model, first, similar to the driving ball and stick model, the joint angle from the real human hand movement is applied to the bone layer through rotation transformation, and then the change in the bone layer is used as the input of the muscle layer, which is calculated and passed through the muscle layer. The synthesis of the layers affects the vertex coordinates of the mesh model of the skin layer, causing the deformation of the virtual hand skin layer. Among them, the muscle layer model mainly controls the contraction/diastolic deformation of the local mesh of the skin layer during the bending process (Fig. 2c) [15].

2.3   Judgment of Objects Grasped by Virtual Hands

During the movement of the virtual hand, the collision between the virtual hand and the object to be grasped is calculated in real time, and the intersection point between the virtual fingertip and the object is calculated based on the real-time collision detection result. Once information such as the intersection point is obtained, we can determine whether the object to be grasped has been grasped according to a certain object grasping criterion.

Common criteria are divided into geometric criteria (non-force criterion) and force criterion [3]. If the virtual object can be grasped, the object is attached to the virtual hand as a child node, and the child node of the virtual hand moves with the movement of the virtual hand.

3   Dexterous Virtual Hand Grasping Based on Leapmotion

In the following, the Leapmotion-based dexterous virtual hand grasping method is proposed.

3.1   Framework of Dexterous Virtual Hand Grasping

In order to effectively support the LeapMotion-based dexterous virtual hand grasping without finger tracking, we propose a "kinematics-appearance-grasping computation" three-layer virtual hand model, and propose the framework of dexterous virtual hand grasping as shown in Fig. 3.

As shown in Fig. 3, the virtual hand model we propose consists of three layers, namely the kinematics layer (forward kinematics model), the appearance layer (shape model) and the grasping computation layer (grasping calculation model). These three levels are integrated into a relatively complex virtual hand grasping framework to effectively support virtual grasping tasks. In the interactive loop, with the traditional dexterous virtual hand grasping method, the six-degree-of-freedom movement data and the bending data of each finger need to be transmitted to the bone layer. The difference is that in the method proposed here, only the user's hand other than the fingers is needed. The 6-dof data (captured by LeapMotion), and the data driving the grasping computation layer, after the grasping computation and judgement, drive the skeleton structure of the virtual hand kinematics layer, and then drive the deformation of the appearance layer through the muscle model.

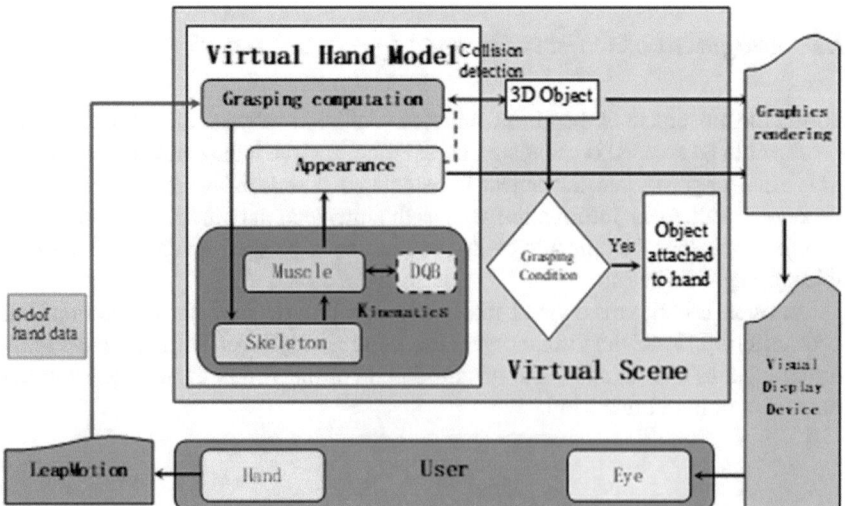

Fig. 3 The dexterous virtual hand grasping framework proposed in this paper

3.2 Virtual Model

In the appearance layer modeling, we apply the subdivision surface modeling method. In order to facilitate drawing and motion control, based on the knowledge of the shape and anatomical structure of the hand [16], the geometric structure of the triangle mesh is constructed. In addition, the balance between the number of triangles and the visual effect is also considered. Figure 4a shows the appearance of the virtual hand constructed.

In the modeling of the kinematics layer, we use the skinning algorithm to bind the geometric model to the internal skeleton, and set the weight for each vertex to describe the influence of the skeleton on them. Therefore, the skin layer will deform and move with the bone model.

The human hand is a complex structure with additional joints that enable us to grasp, hold and manipulate various objects. The human hand system is composed of soft tissues such as bones, ligaments, muscles, and tendons. There are 27 bones and 22 joints, of which 8 bones belong to the wrist, 5 bones belong to the palm, and 14 bones form 5 fingers [16]. Usually, when building a mathematical model, the common method is to replace the real human hand joints with simplified joints. In the simplified bone structure in this paper, 20 bones are retained among 27 bones, and every 3 bones make up a finger. The remaining 5 pieces belong to the palm, connecting the fingers and the wrist. This simplified bone layer is represented as a hierarchical structure and coupled with the skin layer. Each vertex is bound to its related bones to form a virtual hand kinematics model (Fig. 4b).

The construction of the grasping computation layer is the core of this paper. Neurophysiological research results show that when people make movements of natural reach-and-grasp, their fingertips often follow a specific curved path [2, 17]. Based on this result, this paper uses LeapMotion to record the motion trajectory of the fingertip when a person is grasping an object in real time, and uses it as a grasping computational geometric model of a virtual hand.

Fingertip grasping trajectory refers to a series of trajectory line segments when the fingertip grasps an object. The structure of the fingertip grasping trajectory is described as follows:

Fig. 4 **a** The appearance model for the virtual hand; **b** Virtual hand kinematics model

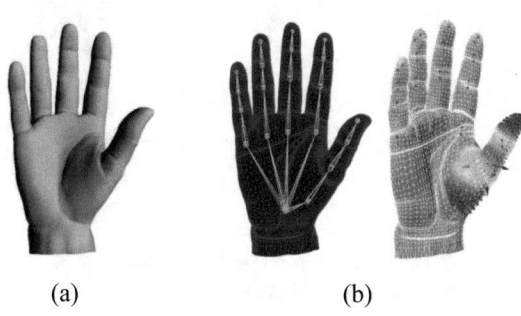

(a)   (b)

First, select a point for each fingertip surface of the virtual hand model as the seed point at the center of the fingertip surface.

Then, the user is required to complete the tasks of freehand reach-and-grasp, and use LeapMotion to record the angle of their finger joints in real time, and record the current sampling frame and the two-tuple of the joint angle for each joint, namely ($frame_k$, $angle_{ij\_k}$). Among them, $frame_k$ is the serial number of the current frame, and $angle_{ij\_k}$ is the bending angle of the jth joint of the ith finger in the current frame.

Next, simulate the user's grasping process according to the recorded finger joint angles, and according to the recorded joint angles, the seed point trajectory of each finger is generated based on the forward kinematics. The specific calculation formula is as follows:

$$P_i = l_{dp\_i} \times \sin(\theta_{1\_i} + \theta_{2\_i} + \theta_{3\_i}) + l_{mc\_i} \times \sin(\theta_{1\_i} + \theta_{2\_i}) + l_{pp\_i} \times \sin\theta_{1\_i} \tag{1}$$

where P_i represents the trajectory of the seed point of the ith finger, l_{dp_i}, l_{mc_i}, and l_{pp_i} represent the length of the far, middle, and near finger segments of the ith finger (Fig. 5a).

Finally, through uniform sampling, the grasping trajectory of each finger is discretized into straight line segments. Figure 5b shows the grasping trajectory of the fingertips connected to the virtual hand from two different perspectives. It is worth pointing out that in practical applications, the grasping model only participates in the calculation process and does not actually display it.

Once the grasping computation model is completed, it is attached to the fingertip seed point of the virtual hand appearance model, and moves with the appearance when the virtual hand does not meet the object grasping conditions; and when the virtual hand meets the object grasping conditions, its movement needs decoupling from the appearance model, because the virtual hand appearance model is driven by the muscle model at this time (see Sects. 3.3–3.4 for details).

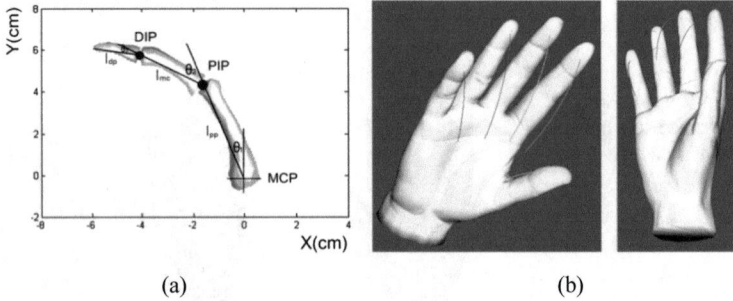

(a)   (b)

Fig. 5 **a** Calculation of fingers seed point trajectory based on forward kinematics; **b** Example of Virtual fingertip grasping trajectory

3.3 Judgment of Objects Grasping by Virtual Hand Based on Non-Force Grasping Criteria

In the real world, people will first reach out to grab the object before performing an object operation, and then perform the corresponding operation. Therefore, we track the movement of the human hand in the process of extending the hand, and predict in real time whether the virtual hand may grasp the object in the next process. The specific process is as follows:

1. Based on the 6-dof motion data of the human hand captured by LeapMotion, the translation and rotation of the virtual hand model (including the grasping computation layer) are driven.
2. Calculate the intersection of the fingertip gasping trajectory of each finger and the object to be grasped in the scene. Since the grasping trajectory of each fingertip is represented by a line segment, the problem of calculating the inter-section point between the trajectory and the virtual object is transformed into an intersection test between the line segment and the object. In order to speed up the calculation speed of the intersection of the straight line and the virtual object, we improved the RAPID collision detection open source package [18] to make it suitable for the calculation of the intersection between the straight line and the triangle.
3. Once the intersection point of all fingertip grasping trajectories and the object to be grasped is calculated, the pre-calculation can be performed. It is assumed that there is geometric relationship between the finger and the object to be grasped when the finger reaches the intersection point through bending motion, according to the literature [3], the one-handed non-force grasping criterion is used to evaluate whether the virtual object can be grasped.

3.4 Virtual Hand Motion Based on Finger Trajectory Resampling and Inverse Kinematics

Once it is determined that the virtual hand can grasp the virtual object, the finger bending motion simulation process is started, the fingertip is moved to the intersection point, and then the virtual object is attached to the virtual hand for further operations. The specific process is as follows:

1. As shown in Fig. 6, based on the intersection information between the fingertip trajectory and the object to be grasped, the time t (in frames) required for the fingertip to reach the intersection from the default position is calculated according to formula (2):

$$t = \begin{cases} t_k \|AC\| < \|BC\| \\ t_k + 1 \; else \end{cases} \tag{2}$$

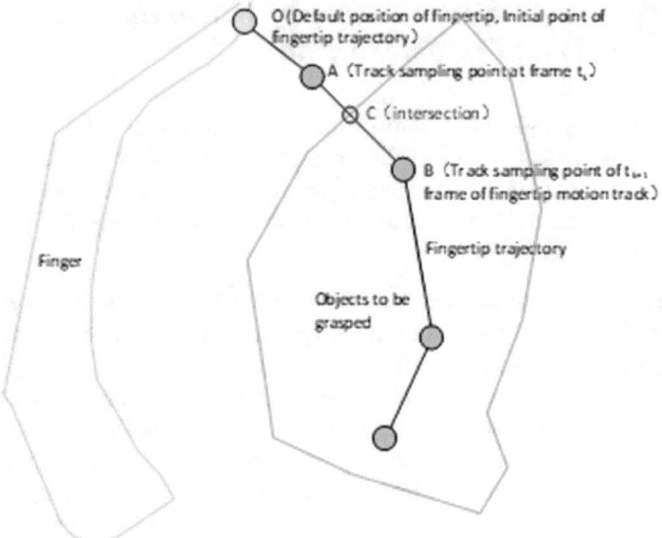

O(Default position of fingertip, Initial point of fingertip trajectory)

A (Track sampling point at frame t_i)

C (intersection)

B (Track sampling point of t_{i+1} frame of fingertip motion track)

Fingertip trajectory

Finger

Objects to be grasped

Fig. 6 Calculation of finger bending time

2. Resample the OC segment of fingertip trajectory to t sampling points, where O is the default position of the fingertip;
3. Based on Inverse Kinematics [19] at each resampling point in the previous step, calculate the angle that each finger joint needs to bend when the fingertip moves to the sample point;
4. Drive the bone model based on the bending angle of the finger joint calculated in the previous step;
5. Based on the skeletal model to drive the muscle model, this paper uses the double quaternion model [20] as the muscle model;
6. Drive the appearance model of the virtual hand based on the muscle model.

4   Implementation and Evaluation

The proposed LeapMotion-based dexterous virtual hand grasping method has been implemented on a computer equipped with Intel(R)Core (TM)i7-5820 K CPU@ 3.30 GHz/64 GB RAM/Microsoft Windows 10 64-bit operating system (Method A). To evaluate the proposed method, a traditional dexterous hand virtual grasping method (Method B) has been implemented based on the framework of Fig. 7, and made a preliminary comparison between the two methods (Method A vs. Method B) for grasping accuracy and grasping time.

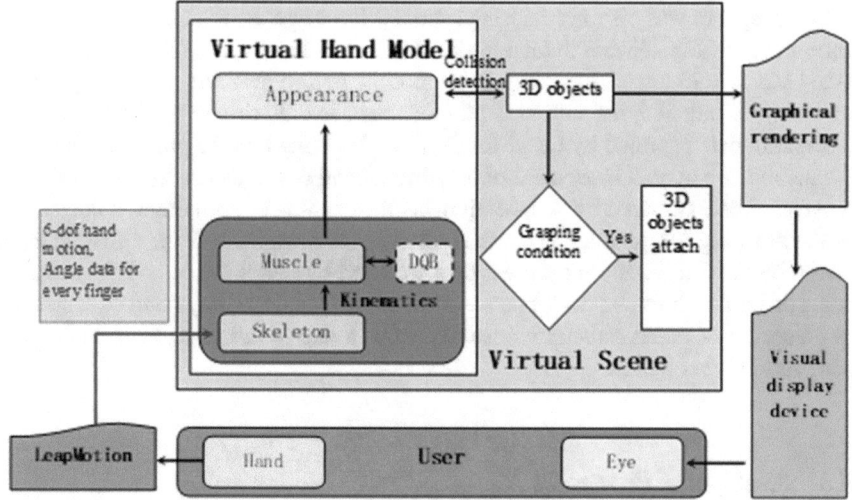

Fig. 7 The traditional dexterous virtual grasping framework

4.1 Participants

The evaluation consisted of 8 participants, 4 men and 4 women, aged between 18 and 25 years old. They are all right-handed, and none of them have experience in using six-degree-of-freedom hand tracking devices or finger tracking devices. Participants were divided into two groups: 2 men and 2 women in each group. In the experiment, one group is required to use method A first, and then use method B to complete the grasping task, the other group is the opposite. Before the start of the experiment, each participant has 5 min of warm-up practice time.

4.2 Experimental Setup

The experimental device includes a 24-inch LED screen, connected to a computer equipped with Intel(R)Core(TM)i7-5820 K CPU@3.30 GHz/64 GB RAM/Microsoft Windows 10 64-bit operating system, and one LeapMotion for capturing Participants' overall hand movement and finger bending angle. The LeapMotion is placed on the desktop. The user sits about 60 cm in front of the screen. The experimental application written in OpenGL shows a virtual sphere centered on the origin of the world coordinate system and a virtual hand determined by the position and direction of the user's hand. In order to enable users to easily perform reach-and-grasp operations, the coordinate system of the LeapMotion tracking system and the coordinate system of the virtual world have been calibrated carefully. In the experiment, three virtual spheres with diameters of 6, 8, and 10 cm were used, but only one sphere was displayed during each reach-and-grasp operation. The order of appearance of the spheres is randomly determined. The size of the virtual hand is 10.64 cm × 8.23 cm × 20.46 cm. In method A, before detecting that the grasping condition is satisfied, the virtual hand movement is only driven by the 6-dof hand movement data captured by LeapMotion; after detecting that the grasping condition is satisfied, the overall movement of the virtual hand is still driven by the6-dof hand movement data captured by LeapMotion, but the bending of each finger is controlled by the grasping computation layer. In method B, the movement of the virtual hand is driven by the 6-dof hand movement data captured by LeapMotion and the bending angle data of each finger, but when the collision between the finger and the object to be grasped is detected, an interference protection strategy will be set to avoid finger's penetration into objects.

4.3   Tasks and Processes

The participant's task is to reach and grab a virtual sphere. At the beginning of each task, the main experimenter issued an order to notify the participant to start the grasping operation, and press a key on the keyboard at the same time, instructing the computer to time. For both Method A and Method B, participants are required to successfully grab each of the three virtual spheres 6 times. Thus each participant should successfully complete the task of $2 \times 3 \times 6 = 36$ trials. For the first failed attempt, the grasping success rate will be counted as a failed task, but the attempt will be re-queued to the end of the task queue for grasping time statistics, but it will no longer be included in the grasping success rate statistics. The entire experiment lasts approximately 50 min.

4.4   Experimental Results and Discussion

Record whether the task was successful and when the task was completed. Table 1 shows the task completion time statistics of successful tasks, where MTA and MTB represent the average capture time of method A and method B (in milliseconds), and MAA and MAB represent the average success rate of method A and method B,

Table 1 Task completion time and accuracy rate statistics

Virtual sphere radius (cm)	MTA (ms)	MTB (ms)	MAA (%)	MAB (%)
6	3567.71	5058.35	68.89	55.81
8	2727.82	4637.97	78.72	65.22
10	2505.25	3225.24	87.50	77.08
All	2933.59	4307.19	78.57	66.42

respectively. In these two methods, if the participant can grasp a virtual sphere within 15 s, then the grasp operation is recorded as a success, otherwise it is recorded as a failure.

From the above preliminary experimental data analysis, it can be seen that in terms of grasping success rate and grasping, it has certain advantages in the sense of average value. However, the advantages of Method A over Method B are only a preliminary comparison on the simple shape of a sphere, although different sphere diameters have been evaluated. For complex shapes, it remains unclear whether these advantages will be retained. We believe that for objects with complex shapes, without loss of generality, using its convex hull representation to determine the grasping conditions may simplify the difficulty of grasping tasks. Therefore, the method proposed in this paper still has its applicability.

5 Conclusion

The traditional dexterous virtual hand grasping interaction method requires whole hand motion tracking and finger motion tracking, but the current precise finger tracking technology still has many shortcomings. Aiming at this, we propose a simple and effective dexterous hand virtual grasping interactive method based on the analysis of the movement of all five fingers in the object grasping task, the characteristics of LeapMotion and inverse kinematics, which enables users to use a dexterous virtual hand as an avatar in a virtual environment even without finger tracking.

The paper reviews the current research status and briefly introduces a typical dexterous virtual hand grasping interaction method based on forward kinematics. Based on this, a LeapMotion-enabled dexterous virtual hand grasping method is proposed. A three-layer dexterous virtual hand model (kinematics-appearance-grasping computation), and a dexterous virtual hand grasping interaction framework is given, which effectively supports the separation of the visual feedback of the virtual hand from the grasping computation. The LeapMotion, a cheap interactive device, is the only need to complete the virtual grasping interaction, through the construction of the grasping computation layer, the determination of the grasping condition and virtual hand deformation based on the grasping computation layer and inverse kinematics. This not only greatly expands the applicability of the method and facilitates the popularization of the method, but also avoids the defect of inaccurate finger tracking data caused by LeapMotion due to occlusion, and improves the robustness of the method. Compared with the traditional virtual grasping method based on forward kinematics, the effectiveness of the proposed method is preliminarily verified. Although for highly complex shapes, the advantages of the proposed method over traditional methods are not clear, we believe that by constructing a convex hull of complex shapes and performing computation based on the convex hull, the proposed method may still have its advantages.

References

1. Ahmed KHA, Abbas HHA (2019) Survey of hand gesture recognition systems. In: IOP Conf Ser: J Phys Conf Ser 1294:042003
2. Maitland ME, Epstein MB (2009) Analysis of finger position during two- and three-fingered grasp: possible implications for terminal device design. J Prosthets Orthot 21(2):102–105
3. Ullmann T, Sauer J (2000) Intuitive virtual grasping for non haptic environments. In: Proceedings of the 8th pacific conference on computer graphics and applications (PG '00). IEEE Computer Society, USA, pp 373–381
4. Daniel H, Sebastian U, Marc W, Torsten K (2008) Multi-contact grasp interaction for virtual environments. J Virtual Rity Broadcast 5(7):16–26
5. Borst CW, Indugula AP (2005) Realistic virtual grasping. In: Proceedings of the 2005 IEEE conference 2005 on virtual reality (VR '05). IEEE Computer Society, USA, Indugula, pp 91–98, 320
6. Jacobs J, Froehlich B (2011) A soft hand model for physically-based manipulation of virtual objects. In: Proceedings of the 2011 IEEE virtual reality conference (VR '11), IEEE Computer Society, USA, pp 11–18
7. Moehring M, Froehlich B (2005) Pseudo-physical interaction with a virtual car interior in immersive environments. In: Proceedings of the 11th Eurographics conference on virtual environments (EGVE'05), eurographics association, Goslar, DEU, pp 181–189
8. CyberGlove Systems. http://www.cyberglovesystems.com/. Accessed 21 Feb 2021
9. Vijitha T, Kumari JP (2014) Finger tracking in real time human computer interaction. Int J Comput Sci Netw Secur 14(1):83–93
10. Qiab C, Sun X, Wei Y, Tang X, Sun J (2014) Realtime and robust hand tracking from depth. In: Proceedings of the 2014 IEEE conference on computer vision and pattern recognition (CVPR '14), IEEE Computer Society, USA, pp 1106–1113
11. Togootogtokh E, Shih TK, Kumara WG, Wu SJ, Sun SW, Chang HH (2018) 3D finger tracking and recognition image processing for real-time music playing with depth sensors. Multimed Tools Appl 77(8):9233–9248
12. Leap Motioon. https://www.leapmotion.com. Accessed 21 Feb 2021
13. Zhu Z, Gao S, Wan H, Yang W (2006) Trajectory-based grasp interaction for virtual environments. In: Advances in computer graphics, computer graphics international conference. Springer, Berlin, pp 300–311
14. Wan H, Han X, Chen W, Ding Y, Ge L (2019) A comparative study for natural reach-and-grasp with/without finger tracking. In: Bagnara S, Tartaglia R, Albolino S, Alexander T, Fujita Y (eds) Proceedings of the 20th congress of the international ergonomics association (IEA 2018), advances in intelligent systems and computing, vol 822. Springer, Cham
15. Albrecht I, Haber J, Seidel HP, Breen D, Lin M (2003) Construction and animation of anatomically based human hand models. In: ACM SIGGRAPH/eurographics symposium on computer animation (SIGGRAPH-SCA-03), ACM, pp 98–109, 368
16. Kapit W, Elson L (1993) The anatomy coloring book. 2nd edn. Harper Collins
17. Kamper D, Cruz E, Siegel M (2003) Stereotypical fingertip trajectories during grasp. J Neurophysiol 90(1):3702–3710
18. RAPID - Robust and Accurate Polygon Interference Detection. http://gamma.cs.unc.edu/OBB/. Accessed 21 Feb 2021
19. Inverse Kinematics. https://www.rosroboticslearning.com/inverse-kinematics. Accessed 21 Feb 2021
20. Chen W, Zhu S, Wan H, Feng J (2013) Dual quaternion based virtual hand interaction modeling. Sci China Inf Sci 56:1–11

Performance Evaluation of Azure Kinect and Kinect 2.0 and Their Applications in 3D Key-Points Detection of Students in Classroom Environment

Wenkai Huang, Jia Chen, Xiaoxiong Zhao, and Qingtang Liu

Abstract Kinect, which has the functions of depth data acquisition and human skeleton 3D key-points detection, is one of the most popular consumer-grade depth data acquisition devices in the past decade. In 2019, Microsoft released the latest generation of Kinect (Azure Kinect). The contribution of this paper is to design experiments to evaluate the performance of this sensor in depth data acquisition and human skeleton 3D key-points detection. In particular, we explored the effect of Azure Kinect and Kinect 2.0 in 3D key-points detection of students in the classroom environment. In order to qualitatively compare their performance in the detection of human skeleton 3D key-points in classroom environment, we conducted a multi-person experiment in the classroom environment. The experimental results show that: (1) Azure Kinect is better than Kinect 2.0 in depth data acquisition accuracy; (2) The performance of Azure Kinect is better than Kinect 2.0 in acquiring depth data at different distances; (3) In terms of human 3D key-points detection accuracy, the difference between Kinect 2.0 and Azure Kinect is not big; Under the condition that the detected people are occluded, the detection effect of Azure Kinect is better than Kinect 2.0; (4) In the experiment of multi-person 3D key-points detection, Azure Kinect has better performance than Kinect 2.0; Azure Kinect is more accurate than Kinect 2.0 in assigning ID to the tested person.

Keywords Kinect · Depth data · 3D key-points detection · Classroom environment

W. Huang
CCNU Wollongong Joint Institute, Central China Normal University, Wuhan 430079, China

W. Huang · J. Chen (✉) · X. Zhao · Q. Liu
Faculty of Artificial Intelligence in Education, Central China Normal University, Wuhan 430079, China
e-mail: jiachenccnu@163.com

© The Author(s), under exclusive license to Springer Nature Singapore Pte Ltd. 2022   177
E. C. K. Cheng et al. (eds.), *Artificial Intelligence in Education: Emerging Technologies, Models and Applications*, Lecture Notes on Data Engineering and Communications Technologies 104, https://doi.org/10.1007/978-981-16-7527-0_14

1 Introduce

1.1 Research Background

Kinect has attracted much attention because of its low cost and broad application prospects [1]. Different from ordinary cameras, Kinect can obtain the color and depth data (i.e. RGB-D data) at the same time [2]. The first generation Kinect is equipped with ps1080 and it can get depth data through Light Coding technology [3]. Kinect 2.0 adopts × 871,141–001 SOC processor. Kinect 2.0 chooses to use time-of-flight (TOF) technology instead of Light Coding technology [4, 5]. Kinect 2.0 gets the phase difference between the transmitted signal and the received signal by transmitting the sinusoidal signal whose intensity changes with time [6]. Kinect 2.0 uses this to calculate depth data. While acquiring depth data, Kinect's color camera will take color images of the scene in real time. In order to facilitate the subsequent texture mapping operation, the system corrects the parallax generated by the two cameras to align the two target images at the same position. Kinect for windows SDK is a software development package designed by Microsoft for developing Kinect sensors [7]. This software development package enables the computer to call its own "eyes" and "ears" (i.e. infrared projector, RGB camera, microphone array, etc.) to achieve audio-visual functions [8]. Microsoft released the Kinect. NET website interface to facilitate developers to use. These two versions of Kinect have been widely used in various fields [9, 10], such as target detection and recognition [11], simultaneous localization and mapping (SLAM) [12], gesture recognition [13], human–computer interaction [14], telepresence [15], virtual reality [16], hybrid reality [17, 18], medicine and rehabilitation [19], etc. In 2015, the research of Zennaro et al. compared the first-generation Kinect and the second-generation Kinect in many aspects and provided experimental data [20]. These works have important reference value for the subsequent development of the second-generation Kinect. This research is also of great significance to promote the development of Kinect in the field of education. However, above two versions of Kinect have been discontinued and are no longer officially distributed and sold.

In 2019, Microsoft released a new generation of Kinect (newly named Azure Kinect) development kit. The volume of Azure Kinect is $103 \times 39 \times 126$ mm, and the volume of Kinect 2.0 is $249 \times 66 \times 67$ mm (see Fig. 1).

Depth image is also called range image. In the image frame provided by the depth data stream, each pixel represents the distance (in mm) from the object to the plane at the specific (x, y) coordinates in the field of vision of the depth sensor [21].

Fig. 1 Two generations of Kinect (Kinect 2.0 on the left and Azure Kinect on the right)

In the field of intelligent education, Kinect makes the concept of human–computer interaction more thorough because of its powerful interactive function [22]. At the same time, Kinect's powerful ability of depth data acquisition makes it a powerful means to collect students' information in real time in the classroom environment. Researchers from all over the world have introduced Kinect into the field of education and made many achievements. Kinect has great potential in the field of education, including parent–child educational game, children's physical interaction, specific subject education and special education for disabled children.

1.2 Research Challenges

Kinect has broad application prospects in many fields because of its depth data acquisition function and 3D human skeleton key-points detection function [23]. In the field of intelligent education, Kinect makes the concept of human–computer interaction more thorough because of its powerful 3D human skeleton acquisition ability. At the same time, in the daily teaching environment, Kinect's powerful depth data collection ability can help teachers effectively and comprehensively grasp the students' state and learning effect. The latest generation Kinect (Azure Kinect) was released in 2019. At present, there is a lack of comprehensive performance test for Azure Kinect, which affects the application of Kinect in many fields including education. Because the related test often has the cost of the instrument, site and other constraints, so we need to find a suitable experimental program to test all aspects of the performance. This work is to prepare for developers to understand and use Azure Kinect. At the same time, there is no specific data about the actual detection effect of Azure Kinect in the classroom environment, so we need to design experiments to test in the real classroom environment. How to collect various key-points of skeleton information in the classroom environment is also the problem faced by this research.

1.3 Goals and Contributions

The goal of this study is to test the performance of Azure Kinect in many aspects. These works enable us to obtain detailed and reliable experimental data, including human detection performance data in various scenes in the classroom. These data can provide a reference for later researchers, and help researchers to carry out further research in many fields including education. We need to explore the design of reliable performance test experiment for similar depth data and human detection equipment in the classroom environment. In order to achieve this goal, this study will collect depth data of two generations of Kinect and quantitatively compare their accuracy and information collection performance at different distances. We need to collect the human skeleton 3D key-points detection data of the two generations of Kinect and make a quantitative comparison, and carry out multi-person detection experiments in various scenes in the classroom. These works help us to qualitatively compare the performance of two generations of Kinect.

This study is helpful for later developers to better understand the specific performance of the two-generation Kinect and the detection performance of the two-generation Kinect for objects and people in various scenes in the classroom. These works further help later researchers to use Kinect 2.0 and Azure Kinect in the development of education and related fields.

1.4 Paper Structure

The Sect. 1 of this paper introduces the research background and progress of Kinect, and describes the research objectives. From Sects. 2–4, we introduce the experiments designed by us to evaluate the performance of the sensor in depth data acquisition and 3D human skeleton key-points detection. Based on these experiments, we explored and evaluated the performance differences between the latest Kinect and the previous Kinect. The Sect. 5 summarizes the research content.

1.5 Overview of Related Research

After Microsoft released Kinect 2.0, Sarbolandiv et al. [24] released a study. The study made a detailed comparison between Kinect 2.0 and the first-generation Kinect. For the sake of comparison, the author proposes a scheme for evaluating this depth camera equipment. The scheme is a research framework composed of seven parts, which provides ideas for subsequent related research.

Smisek et al. [25] planned to evaluate Kinect's 3D data collection capabilities. The resolution and error characteristics of depth measurement are studied by experiments. In this paper, the stereo reconstruction and motion accuracy of SLR camera and 3D time-of-flight camera are quantitatively compared. The performance of depth data acquisition of the sensor is studied through experiments. The author makes a detailed study of the new technology (TOF) used in Kinect 2.0, and discusses the technical improvement of the new generation products compared with the first generation Kinect.

In the research of Choo et al. [7], The random error characteristics of Kinect sensor in each axis direction are given. A new random error model based on measurement and statistics is proposed. The model is based on the position and depth values of the data measured by each pixel in the whole field of view. Depth (z) error, horizontal (x) error and vertical (y) error are measured by plane method and a new three-dimensional method respectively. The results show that the influence of the depth of the measured object on the randomness of dynamic measurement error can not be ignored.

Pagliari et al. [3] proposed a Kinect calibration method based on depth camera for Xbox one imaging sensor. The mathematical model of sensor error is established and described as a function of the distance between the sensor itself and the object. All the analyses described in this paper are aimed at the second generation Kinect and the first generation Kinect, and the purpose is to quantify the improvement of imaging sensors.

Zennaro et al. [20] compared the data provided by the first generation of Kinect and Kinect 2.0 to explain the achievements of technology transformation. After analyzing the accuracy of the two kinds of sensors under different conditions, the application examples of 3D reconstruction and human tracking were given, and the performance of the two kinds of sensors were compared.

2 Evaluate the Accuracy of Depth Data Collected at Close Range from Azure Kinect and Kinect 2.0

2.1 Experimental Design

The first part of the experiment is to evaluating the accuracy difference of depth data collected in close range between Azure Kinect and Kinect 2.0. We set a group of object scenes 2–6 m away from the sensing device as the objects to be detected. We collected the depth data collected by two generations of Kinect for this group of scenes, and transformed each pixel information of the depth data into a 3D coordinate point in the point cloud data through transformation. We could get the accuracy difference of depth data collected by different versions of Kinect by comparing the obtained point cloud data with the ground truth point cloud data obtained by professional 3D scanner.

In this study, we first obtained the detection data of Kinect 2.0 in close range environment. The data were depth data. We obtained Kinect 2.0 depth data under the close range scene (2–4 m) through the official interface.

To get Kinect depth, we first needed to create two mat type variables to store the captured data.

```
colorImage.create(480, 640, CV_8UC3);
depthImage.create(480, 640, CV_8UC1);
```

Then we initialized Kinect. Image acquisition is realized in the while (1) loop. The images we captured were stored in the project directory for further use.

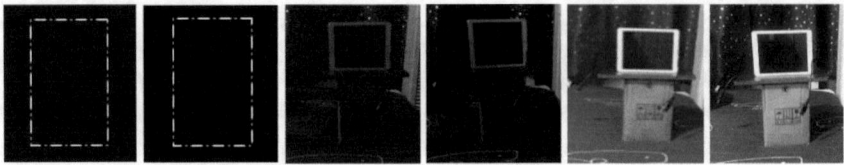

Fig. 2 In the figure above, from left to right are data from Azure Kinect in dark environment, data from Kinect 2.0 in dark environment, data from Azure Kinect under 20 W lamp, data from Kinect 2.0 under 20 W lamp, data from Azure Kinect under 60 W lamp, data from Kinect 2.0 under 60 W lamp. The dashed box indicates the test object

```
pColorIageFrame   hr = NuiImageStreamGetNextFrame(color-
StreamHandle, 0, &pColorImageFrame);
INuiFrameTexture * pTexture = pColorImageFrame-
>pFrameTexture;
NUI_LOCKED_RECT LockedRect;
pTexture->LockRect(0, &LockedRect, NULL, 0);
```

Then we extracted and stored the point cloud from the image. After that, we integrated them into a directory file.

```
cloud_a.points[num].b = color.at<cv::Vec3b>(u, v)[0];
cloud_a.points[num].g = color.at<cv::Vec3b>(u, v)[1];
cloud_a.points[num].r = color.at<cv::Vec3b>(u, v)[2];
```

The depth data were obtained in bright light (60 W), dark light (20 W) and no light environment (see Fig. 2). Through calculation, the depth data is transformed into point cloud under 3D coordinate axis.

After that, we get the depth data obtained by Azure Kinect under the same environment. The depth data were obtained in bright light (60 W), dark light (20 W) and no light environment respectively. We needed to turn on the default device to obtain the depth image.

```
k4a::device device = k4a::de-
vice::open(K4A_DEVICE_DEFAULT);
std::cout << "Done: open device. " << std::endl;
```

Next, we retrieved and saved the Azure Kinect image data. After that, we configured and started the device. We used NFOV (narrow field-of-view depth mode) unbinned mode.

```
k4a_device_configuration_t config =
K4A_DEVICE_CONFIG_INIT_DISABLE_ALL;
config.camera_fps = K4A_FRAMES_PER_SECOND_30;
config.color_format = K4A_IMAGE_FORMAT_COLOR_BGRA32;
config.color_resolution = K4A_COLOR_RESOLUTION_1080P;
config.depth_mode = K4A_DEPTH_MODE_NFOV_UNBINNED;
config.synchronized_images_only = true;
```

These works ensured that depth images are available.

```
device.start_cameras(&config);
```

Firstly, we got the value of (m, n) in the depth map.

```
ushort d = depth.ptr<ushort>(m)[n];
```

We added a point to the point cloud. Then we calculated the spatial coordinates of this point.

```
p.z = double(d) / camera.scale;
p.x = (n - camera.cx) * p.z / camera.fx;
p.y = (m - camera.cy) * p.z / camera.fy;
```

We added the point to the point cloud and saved it.

```
cloud->width = cloud->points.size();
```

After that, we used a high-precision 3D scanner to obtain the 3D scanning data of the scene under the condition of sufficient light. We took the 3D scanning data as the reference data. The 3D scanner we use was Cubify Sense 3D. Then we compared Kinect 2.0 with the scan data obtained by the 3D scanner using the software Cloud-Compare. Through point cloud registration and point cloud distance calculation, we got the error between Kinect 2.0 scanning data and the ground truth point cloud data obtained by the 3D scanner.

After that, we compared the Azure Kinect scanning data with the ground truth point cloud data obtained by the 3D scanner using CloudCompare. After point cloud registration and point cloud distance calculation, we got the error between the Azure Kinect scanning data and the ground truth point cloud data obtained by the 3D scanner. After that, we compared the error between Kinect 2.0 scanning data and 3D scanner's ground truth point cloud data, and the error between Azure Kinect scanning data and 3D scanner's ground truth point cloud data. We used these data to compare the accuracy of depth data obtained by Kinect 2.0 and Azure Kinect in close range environment.

2.2 Experimental Result

We converted the depth data into point clouds. Then we used CloudCompare to process the data. Since depth data and point cloud data could be transformed into each other and there is a one-to-one correspondence between the two, by comparing these point cloud data with the ground truth point cloud data (see Fig. 3), we could evaluate the accuracy difference of depth data collected by different versions of Kinect. The basic unit of these data is centimeter (see Table 1).

There is almost no difference in depth data accuracy of Kinect 2.0 and Azure Kinect under different light conditions. There is little difference in the depth data accuracy of Azure Kinect under different light conditions. In this experiment, the depth data accuracy of Azure Kinect is slightly better in the range of 2–6 m.

Fig. 3 Point cloud data transformed from depth data collected by two generations of Kinect (data on the left is from Kinect 2.0, and data on the right is from Azure Kinect)

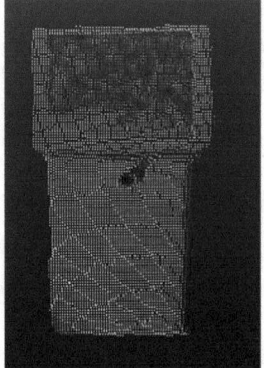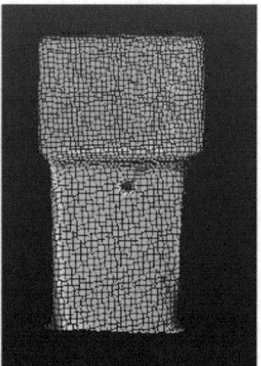

Table 1 The error between the ground truth point cloud data and the point cloud data converted from depth data obtained by Kinect 2.0 and Azure Kinect in a close environment (cm)

	60 W lamp	20 W lamp	No light	Average
Azure Kinect	1.606 6	1.608 9	1.602 4	1.605 9
Kinect 2.0	1.827 1	1.810 9	1.820 1	1.819 4

3 Evaluate the Performance of Depth Data Collected at Different Distances from Azure Kinect and Kinect 2.0

3.1 Experimental Design

The second part is to evaluate the performance difference of depth data collection between Azure Kinect and Kinect 2.0 at different distances. In this part of the study, we planned to evaluate the difference of depth data performance between Kinect 2.0 and Azure Kinect for the same object at the same distance. And we planned to evaluate the difference of effective depth data distance between Kinect 2.0 and Azure Kinect. We collected the depth data collected by two generations of Kinect for the same object at different distances, and transformed each pixel information of the depth data into a 3D coordinate point in the point cloud data through transformation. By counting the number of valid point clouds in the point cloud data, we could evaluate the performance difference of depth data collection between Azure Kinect and Kinect 2.0 at different distances.

In this part of the study, we first fixed Kinect 2.0 position. After that, we used an accurate infrared range finder to place a vertical plane every 0.5 m within the range of 0.5 to 10 m from Kinect 2.0 as the detection target. The vertical plane is 185 mm × 260 mm white paper (see Fig. 4). This was to ensure infrared reflectance. Then we turned on the depth image capture function of Kinect 2.0. The depth image acquisition was similar to the first part of the experiment, and the image was obtained through the official interface. Then we took the depth image and recorded it into the computer.

Then we fixed the Azure Kinect location. We used an accurate infrared range finder to place a vertical plane every 0.5 m from Azure Kinect from 0.5 m to 10 m as the detection target. Then we turned on the depth image capture function of Azure Kinect. We took depth images and recorded them on the computer. We use NFOV unbinned mode. The depth image acquisition was similar to the first part of the

Fig. 4 Experimental scenario

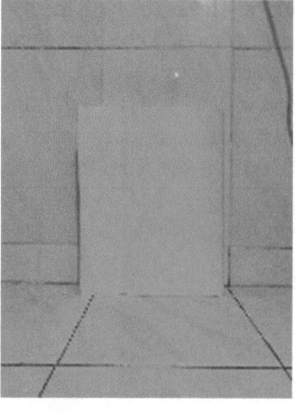

Fig. 5 Data obtained during
the experiment

experiment, which was obtained through the official interface. We would get the depth image through the camera parameters into the point cloud import professional software. The transformation process was similar to the first part. We compared the point clouds obtained by two generations of Kinect (see Fig. 5). Through this work, we could evaluate the performance difference of depth data collection between two generations of Kinect at different distances.

3.2 Experimental Result

In this experiment, the number of effective point clouds collected by Azure Kinect is significantly higher than Kinect 2.0 for the same area. Kinect 2.0 has a depth camera resolution of 512 × 424. The depth camera resolution of Azure Kinect in NFOV unbinned mode is 640 × 576. Under this resolution comparison, the number of effective point clouds collected by Azure Kinect is about 1.7 times that collected by Kinect 2.0 (see Table 2), that is, the gap between the number of effective point clouds collected by Kinect 2.0 and Azure Kinect at 1.5 m to 2.5 m is in line with the expectation. The experimental results beyond 2.5 m show the performance advantages of Azure Kinect. Azure Kinect can collect twice as many effective point clouds as Kinect 2.0. That is to say, Azure Kinect has better performance in collecting depth data than Kinect 2.0. Kinect 2.0 can not get the depth data of the target when it is 850 cm away from the sensor device (see Fig. 6).

Table 2 The number of effective point clouds converted from depth data collected by Kinect 2.0 and Azure Kinect at different distances (pieces)

Distance (cm)	Kinect 2.0	Azure Kinect
100	5879	11,838
150	3232	5554
200	1969	3444
250	1240	2139
300	784	1628
350	565	1251
400	451	806
450	328	665
500	259	515
550	221	421
600	170	387
650	148	292
700	132	280
750	103	202
800	95	203
850	–	123
900	–	131
950	–	89
1000	–	69

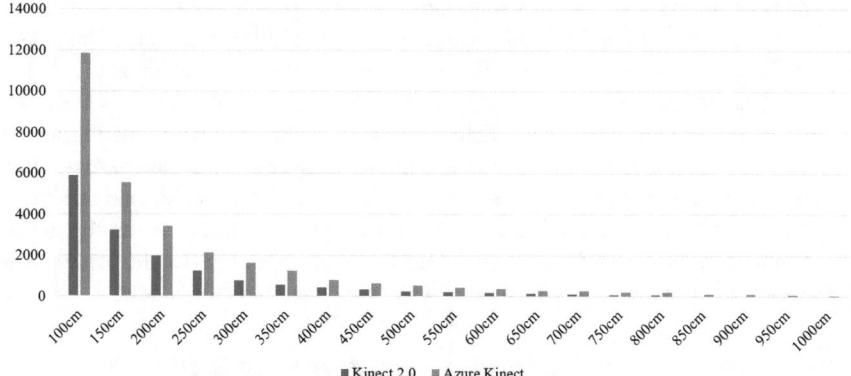

Fig. 6 The number of effective point clouds converted from depth data collected by Kinect 2.0 and Azure Kinect at different distances (pieces)

4 Evaluate the Effect of 3D Key-Points Detection and the Effect of ID Assignment for Multiple Persons in Various Scenes in the Classroom of Azure Kinect and Kinect 2.0

4.1 Experimental Design

Human body 3D key-points information (also named as human skeleton information) in classroom environment can be used in the teaching and learning behavior recognition and analysis [26]. The fourth part is to evaluate the difference between Azure Kinect and Kinect 2.0 in the effect of 3D key-points detection and the difference in the effect of ID assignment for multiple persons in various scenes in the classroom. An open environment with fewer desks and chairs was the first test scenario. In this part of the study, we first let multiple people walk back and forth in the open environment. We used Kinect 2.0 and Azure Kinect to detect these people at the same time and record the detection records. Azure Kinect and Kinect 2.0 would simultaneously detect the 3D skeleton key-points of all people. After that, we qualitatively analyzed the skeleton key-points detected by two generations of Kinect to test the performance difference of Kinect 2.0 and Azure Kinect in detecting skeleton key-points of multiple people in an open environment. At the same time, we evaluated the difference of person ID recognition effect according to the person ID assigned by two generations of Kinect.

The second test scenario was to simulate the classroom environment of many desks and chairs (see Fig. 7). We let the people sit behind the desk in the classroom and walk around in the classroom. We used Kinect 2.0 and Azure Kinect to detect these people at the same time and record the detection records. Azure Kinect and Kinect 2.0 would simultaneously detect all the skeleton key-points of the tested people (see Fig. 8). After that, we qualitatively analyzed the skeleton key-points detected by two generations of Kinect to test the performance difference of Kinect 2.0 and Azure Kinect in detecting multi-person skeleton key-points in the classroom environment. At the same time, we evaluated the difference of person ID recognition effect according to the person ID assigned by two generations of Kinect.

Fig. 7 RGB data collected by two generations of Kinect (data on the left is from Kinect 2.0, and data on the right is from Azure Kinect)

Fig. 8 Skeleton key-points information collected by two generations of Kinect (data on the left is from Kinect 2.0, and data on the right is from Azure Kinect)

4.2 Experimental Result

The following conclusions were obtained in this experiment:

1. Azure Kinect is easy to detect skeleton key-points for the detected people whose hands and heads are obvious, which means that it is not easy for Azure Kinect to detect skeleton key-points when the two hands and heads of the detected people are together in the specific detection environment.
2. When the detected person was sitting behind the desk, the skeleton structure established by Azure Kinect (the connection of the human key-points) tended to be that of the detected people standing behind the desk. The skeletal structure established by Kinect 2.0 tended to be that of the detected people curled up on the desk, because all the human skeleton structure established by Kinect 2.0 for the detected people was built in the detected part of the people.
3. In the process of detecting the people behind the desk, the skeleton structure established by Azure Kinect was stable. The skeleton structure established by Azure Kinect basically followed the movement of the detected people. Kinect 2.0 was not stable. Kinect 2.0's position recognition of detected people had been jumping. Even if the hand of the detected people does not move, the detection result of Kinect 2.0 always jumps.
4. Kinect 2.0 sometimes even created abnormal skeleton structures (the connection of key-points does not conform to the shape of normal human body). If the detected people was occluded, the skeletal structure established by Kinect 2.0 device was abnormal in most cases. However, there was no deformity in the skeleton structure established by Azure Kinect.
5. It is speculated that due to Azure Kinect's efforts to construct a complete and reasonable skeleton structure, once the known part of the detected people does not meet the requirements of establishing an appropriate skeleton architecture, its skeleton key-points will not be detected. So when the detected people were occluded, Kinect 2.0 had better detection and recognition effect than Azure Kinect, that is, when the detected people were occluded, Azure Kinect's missed detection rate was higher than Kinect 2.0.
6. When the detected people suddenly appeared in the field of vision, the recognition speed of Azure Kinect was significantly better than Kinect 2.0. When

the detected people were completely occluded by other people and reappears, Azure Kinect would immediately detect its skeleton key-points, while Kinect 2.0 would react very slowly (when the detected people does not move obviously, Kinect 2.0 can hardly detect its skeleton key-points). Kinect 2.0 sometimes assigned a new person ID (with new skeleton color) to the temporarily occluded detected people standing in the same place, while Azure Kinect did not assign a new person ID to the temporarily occluded detected people standing in the same place. However, when it was occluded for more than 2 s, Azure Kinect would probably assign a new person ID to the detected people that is temporarily occluded and standing in the same place.

7. There is no significant difference in recognition distance between the two generations of Kinect. The effective recognition distance of the two versions of Kinect is one meter to four meters.

8. If the purpose of key-points detection is to determine the position of each key-point, Azure Kinect has great advantages. If only the detected people are identified, Azure Kinect has better stability and reaction. However, most of the time, the missed detection rate of Azure Kinect is higher than Kinect 2.0, because Kinect 2.0 can recognize the people and detect its key-points based on only a small number of features. In a few cases, Kinect 2.0 will detect skeleton key-points for the detected people that do not exist in the field of vision, that is, the false detection rate of Kinect 2.0 is higher than that of Azure Kinect.

9. Azure Kinect is better than Kinect 2.0 in person ID assignment of the detected people. When a person leaves the field of view of the lens, another person immediately enters the field of view of the lens, and the two detected persons appears and disappears at similar positions. There is a high probability that Azure Kinect will not assign the same ID to the two persons (but when three or more persons interact, this recognition will be problematic). For Kinect 2.0, even if two detected persons intersect at the edge of the recognizable range of the lens, they will be recognized as one person with a high probability (pulling the skeleton temporarily to accommodate the interleaving of the two persons).

10. In this study, when we ran on GPU (gtx1050mq) on notebook, the frame number and processing speed of Azure Kinect were obviously inferior to Kinect 2.0, about 0.5 s slower.

11. In terms of the number of identifiable parts, Azure Kinect has increased from 25 to 32 in Kinect 2.0. Azure Kinect adds two eyes, two ears, two clavicles, chest and abdomen as new recognizable key-points, and removes the key-points of the middle spine.

12. We quantitatively analyzed the performance difference of two generation Kinect in person ID assignment for multiple persons in complex environment (see Table 3). FP is the number of false detections, that is, the number of person IDS assigned to the non-existent people. FN is the number of missed detections, that is, the number of times that the detected people does not assign person ID. ID SW. Is the number of wrong assignment of person ID, that is, the number of different person ID assignment for the same detected people before and after the frame. ACIA is the accuracy of person ID assignment (see

Table 3 The performance of Kinect 2.0 and Azure Kinect's multi-person ID assignment		Azure Kinect	Kinect 2.0
	ACIA	90.8626%	69.5847%
	FP	40	120
	FN	369	1296
	ID Sw	20	12

Eq. 1). This value is calculated by the following formula, where GT is the actual number of detected people. It can be seen that the accuracy of Azure Kinect's person ID assignment is significantly higher than that of Kinect 2.0.

$$ACIA = 1 - \frac{\sum FN + FP + IDSw}{\sum GT} \tag{1}$$

5 Conclusion

In this study, we designed experiments to evaluate the performance of the sensor in depth data acquisition and 3D human skeleton key-points detection. Based on these experiments, we made innovative exploration and evaluation on the performance differences between the latest generation Kinect and the previous generation products. In particular, we explored the effect of Azure Kinect and Kinect 2.0 in the detection of student skeleton key-points in the classroom environment. According to the particularity of the classroom environment, we designed a large number of partial occlusion and mutual occlusion between human bodies in our experimental design. These designs are different from the previous comparative experiments of similar sensors. Through these special designs, we have collected the data of the new generation Kinect in a variety of complex environments such as classroom, which provides a reference for the future research of intelligent education. We collected depth data of two generations of Kinect, and quantitatively compared the accuracy and performance of depth data acquisition of two generations of Kinect; In order to qualitatively compare the performance of two generations of Kinect in the detection of human skeleton key-points in the classroom environment, we conducted a multi-person experiment in the classroom environment. The experimental results show that: (1) Azure Kinect is better than Kinect 2.0 in depth data acquisition accuracy; (2) The performance of Azure Kinect is better than Kinect 2.0 in acquiring depth data at different distances; (3) In terms of human skeleton 3D key-points detection accuracy, the difference between Kinect 2.0 and Azure Kinect is very small; Under the condition that the detected people is occluded, the detection accuracy of Azure Kinect is better than Kinect 2.0; (4) In the experiment of multi-person 3D key-points detection, Azure Kinect has better performance than Kinect 2.0; Azure

Kinect is more accurate than Kinect 2.0 in assigning ID to the tested person. The above conclusion proves that the performance of new generation Kinect is better than Kinect 2.0 in depth data acquisition and 3D human skeleton key-points detection. Since the launch of Kinect 2.0, researchers from all over the world have introduced Kinect 2.0 into education and other fields. We believe that with the promotion of Azure Kinect, researchers from all over the world will make better achievements in education by virtue of this new generation sensor with more powerful comprehensive performance. Azure Kinect has just been released, and its potential capabilities still need to be explored. We plan to make more comprehensive testing for this device in the future.

Acknowledgements This work was supported by the Hubei Provincial Natural Science Foundation (2021CFB539), National Natural Science Foundation of China (61803391 and 62173158), Open Fund of Hubei Research Center for Educational Informationization, Central China Normal University (HRCEI2020F0205), and the Fundamental Research Funds for the Central Universities (CCNU19TD007 and CCNU20TS032).

References

1. Pop N, Ulinici I, Pisla D, Carskeleton G, Nysibalieva A (2020) Kinect based user-friendly operation of LAWEX for upper limb exercising task. In: 2020 IEEE international conference on automation, quality and testing, robotics, Cluj-Napoca, Romania, pp 1–5
2. Tan T, Gochoo M, Chen H, Liu S, Huang Y (2020) Activity recognition based on DCNN and Kinect RGB images. In: International conference on fuzzy theory and its applications, Hsinchu, Taiwan, pp 1–4
3. Pagliari D, Pinto L (2015) Calibration of Kinect for Xbox one and comparison be-tween the two generations of microsoft sensors. Sensors 2015(15):27569–27589
4. Sarbolandi H, Lefloch D, Kolb A (2015) Kinect range sensing: structured-light versus Time-of-Flight Kinect. Comput Vis Image Underst 139:1–20
5. Corti A, Giancola S, Mainetti G, Sala R (2016) A metrological characterization of the Kinect V2 time-of-flight camera. Robot Auton Syst 2016(75):584–594
6. Huang W (2021) Performance comparison between Azure Kinect and Kinect V2 and its application in classroom environment. Master Thesis, CCNU Wollongong Joint Institute, University of Wollongong
7. Choo B, Landau M, Devore M, Beling P (2014) Statistical analysis-based error models for the microsoft KinectTM depth sensor. Sensors 2014(14):17430–17450
8. Ma Y, Sheng B, Hart R, Zhang Y (2020) The validity of a dual Azure Kinect-based motion capture system for gait analysis: a preliminary study. In: Asia-Pacific signal and information processing association annual summit and conference (APSIPA ASC), Auckland, New Zealand, pp 1201–1206
9. Rehouma H, Noumeir R, Bouachir W, Jouvet P, Essouri S (2018) 3D imaging system for respiratory monitoring in pediatric intensive care environment. Comput Med Imaging Graph 70:17–28
10. Byrom B, Breedon P, Siena F, Muehlhausen W (2016) Enhancing the measurement of clinical outcomes using microsoft Kinect. In: International conference on interactive technologies and games. IEEE
11. Asilah N, Saidin S, Shukor A (2020) An analysis of Kinect-based human fall detection system. In: 8th international conference on systems, process and control, Melaka, Malaysia, pp 220–224

12. Muallimi H, Dewantara B, Pramadihanto D, Marta B (2020) Human partner and robot guide coordination system under social force model framework using Kinect sensor. In: International electronics symposium, Surabaya, Indonesia, pp 260–264
13. Ma T, Guo M (2019) Research on Kinect-based gesture recognition. In: IEEE international conference on signal processing, communications and computing, Dalian, China, pp 1–5
14. Fankhauser P, Bloesch M, Rodriguez D, Kaestner R, Hutter M, Siegwart R (2015) Kinect v2 for mobile robot navigation: evaluation and modeling. In: Proceedings of the 2015 international conference on advanced robotics, Istanbul, Turkey, 2015, pp 388–394
15. He X, Zhang J (2020) Design and implementation of number gesture recognition system based on Kinect. In: 39th Chinese control conference (CCC), Shen-yang, China, pp 6329–6333
16. Xu Z et al (2020) Back shape measurement and three-dimensional reconstruction of spinal shape using one Kinect sensor. In: 17th international symposium on biomedical imaging, Iowa City, IA, USA, pp 1–5
17. Hebbur M, Sheshadri OM (2020) Kinect based frontal gait recognition using skeleton and depth derived features. In: National conference on communications, Kharagpur, India, pp 1–5
18. Ding X (2020) Research on kinect calibration and depth error compensation based on BP neural network. In: International conference on computer vision, image and deep learning, Chongqing, China, pp 596–600
19. Salamea H, Auquilla A, Alvarado-Cando O, Onate C (2020) Control of a telerobotic system using wifi and kinect sensor for disabled people with an industrial process. In: IEEE ANDESCON, Quito, Ecuador, pp 1–6
20. Zennaro S, Munaro M, Milani S, Zanuttigh P, Bernardi A, Ghidoni S, Menegatti E (2015) Performance evaluation of the 1st and 2nd generation Kinect for multimedia applications. In: Proceedings of the 2015 IEEE international conference on multimedia and expo (ICME), Turin, Italy, pp 1–6
21. Wang L, Huynh Q, Koniusz P (2019) A comparative review of recent Kinect-based action recognition algorithms. IEEE Trans Image Process 29:15–28
22. Zhou Y, Yu Z, Xu X, Zhai J, Wang H (2020) Practice research of classroom teaching system based on Kinect. In: International conference on computer science & education, delft, Netherlands, pp 572–576
23. Pan W, Zhang X, Ye Z (2020) Attention-based sign language recognition network utilizing keyframe sampling and skeletal features. IEEE Access 8:215592–215602
24. Sarbolandi H, Lefloch D, Kolb A (2015) Kinect range sensing: structured-light versus time-of-flight Kinect. Comput Vis Image Underst 2015(139):1–20
25. Smisek J, Jancosek M, Pajdla T (2013) 3D with Kinect. Consumer depth cameras for computer vision. Springer, Lomdom, UK, pp 3–25
26. Wu D, Chen J, Deng W, Wei Y, Luo H, Wei Y (2020) The recognition of teacher behavior based on multimodal information fusion. Math Probl Eng 2020:8269683

Research on Remote Sensing Object Parallel Detection Technology Based on Deep Learning

Chengguang Zhang, Xuebo Zhang, and Min Jiang

Abstract In recent years, the method based on deep learning has become a hot spot and trend in the field of remote sensing object detection, and has achieved a series of encouraging results. This method uses deep convolution neural network to extract object features adaptively and hierarchically, which has stronger and richer feature expression ability, better invariance of illumination, geometry and position, and stronger generalization ability. This paper first introduces the development of remote sensing object detection technology, then describes the working mechanism of remote sensing target detection using convolution neural network, and focuses on the YOLT algorithm which is suitable for large-scale remote sensing image target detection, On this basis, a parallel remote sensing target detection algorithm based on yolt model is proposed. The parallel processing of multi CPU + GPU is realized by mpi4py. The experimental results show that the parallel algorithm has high efficiency of remote sensing object detection.

Keywords Remote sensing object · Object detection · Parallel processing

1 Introduction

Remote sensing object detection technology is of great significance in both military and civil fields. In military aspect, high-precision remote sensing object detection is helpful to accurately locate enemy targets and evaluate strike effect; In the civil aspect, high-precision remote sensing target detection is the basis of resource exploration, urban planning, environmental monitoring and other applications.

There are several problems to be solved in detecting small target in large-scale remote sensing image:

1. Small object detection: the objects to be detected in satellite images are usually very small and dense. For example, in high-resolution satellite images, the car is usually only about 15 pixels in size.

C. Zhang · X. Zhang (✉) · M. Jiang
Space Engineering University, Huairou District of Beijing, Beijing, China

© The Author(s), under exclusive license to Springer Nature Singapore Pte Ltd. 2022   195
E. C. K. Cheng et al. (eds.), *Artificial Intelligence in Education: Emerging Technologies,*
Models and Applications, Lecture Notes on Data Engineering and Communications
Technologies 104, https://doi.org/10.1007/978-981-16-7527-0_15

2. Complete rotation without deformation: because CNN does not have the property of rotation without deformation, but the object in the satellite image will not change its properties according to the rotation of the object, so it is necessary to train a rotation invariant network (through data enhancement).
3. Small number of data samples: small number of data samples may cause over fitting or poor training results.
4. Super high resolution, large size: the size of the input image is very large, simple underground sampling is not feasible, after down sampling, it will cause data loss.

Remote sensing object detection algorithms can be roughly divided into three categories: detection methods based on classical pattern recognition, detection methods based on traditional machine learning and detection methods based on deep learning.

The classical pattern recognition methods mainly include template matching method, knowledge-based method, object-based image analysis method [1] and so on. The method based on template matching uses the template of the object to be detected to find the object and its position in the graph. The knowledge-based method transforms the object detection problem into a hypothesis testing problem by establishing various knowledge and rules [2]. Object based analysis method realizes object detection through two steps of image segmentation and object classification. The detection effect of this kind of method depends on the quality of image segmentation to a large extent.

The traditional method of remote sensing object detection based on machine learning firstly selects the candidate regions that may contain the object to be detected, then extracts the features of the candidate regions, and then discriminates the specific categories through the classifier. Remote sensing object detection methods based on traditional machine learning have many problems, such as large amount of computation, low efficiency, limited expression ability of feature model, unable to extract implicit and abstract semantic information, weak generalization ability and so on.

In recent years, the method based on deep learning has become a hot spot and trend in the field of remote sensing object detection, and has achieved a series of encouraging results. This method uses deep convolution neural network to extract object features adaptively and hierarchically, which has stronger and richer feature expression ability, better invariance of illumination, geometry and position, and stronger generalization ability. With the development of remote sensing technology and Internet technology, a large number of training samples have been gradually alleviated. At present, R-CNN [3], YOLO [4], SSD [5] and other object detection algorithms based on deep learning are directly applied to high-resolution optical remote sensing images, the effect needs to be improved, and the detection accuracy and efficiency need to be improved.

2 YOLT Object Detection Model (You Only Look Twice)

2.1 Summary

In 2018, Adam van Eten proposed a object detection algorithm: YOLT (you only look twice) [6] to detect small targets in large-scale remote sensing images. The algorithm is based on the network of yolov2 [7], and some improvements are made to make it possible to detect objects about 10 pixels in 250 million pixel images.

2.2 Mechanism of Convolutional Neural Network

Convolution neural network in deep learning consists of input layer, convolution layer, pooling layer, full connection layer and output layer. The input layer is mainly used for multi-dimensional data input and standardized processing. The convolution layer is used to extract the features of the input data, which contains multiple convolution kernels. When the convolution kernel works, it scans the input features according to certain rules, and convolutes the input features to get the output features. Pooling layer uses a location combined with the overall statistical characteristics of adjacent output to replace the output of the network at that location, which is a method of image compression (downsampling), including maximum pooling, average pooling and so on. When the input makes a small amount of translation, most of the output will not change after pooling layer, that is translation invariance. Convolution layer and pooling layer are usually set with multiple and alternate connections. Convolution layer has the characteristics of local connection and weighted summation, which is similar to convolution process, so it is called convolution neural network. In the problem of object detection, the output layer includes the center coordinates, size and category information of the object.

2.3 Brief Introduction of Yolov2 Model

The author of yolov2 proposes a new network as the foundation of yolov2—darknet-19, the model structure is shown in Fig. 1. Similar to vgg-16, the network uses a large number of 3 * 3 convolution layers. After each pooling, the channel is doubled. The global average pooling is used for prediction and the 1×1 convolution kernel is used to compress the feature representation between 3×3 convolutions. Batch normalization is used to stabilize training, accelerate convergence, and regularize the model.

First, we train the network structure of darknet-19 on imagenet1000 classification data set. After the initial training of 224 * 224 images, we fine tune the network structure to make it train on 448 * 448 images.

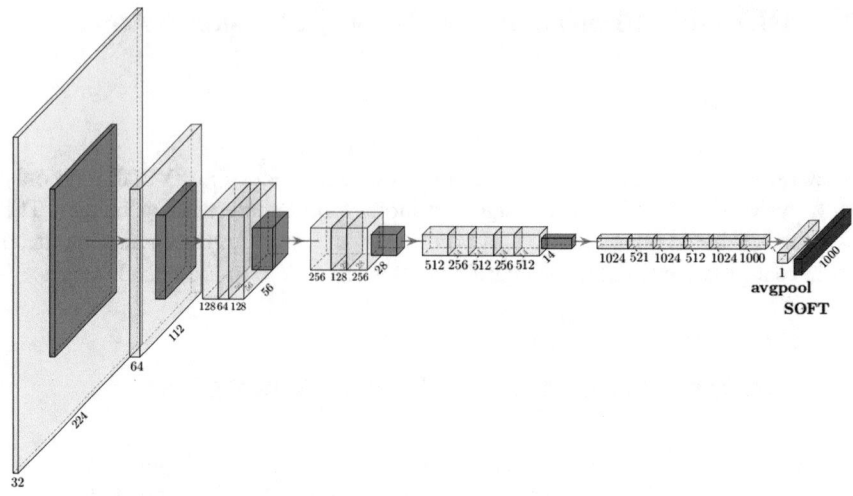

Fig. 1 Darknet-19, the main structure of yolov2

Then, in the process of detection training, the network structure is modified to delete the last convolution layer, add three convolution layers with 1024 3 * 3 convolution cores, and then the last 1×1 convolution layer and the number of outputs required for detection. For VOC data set, five bounding boxes are predicted, each bounding box has five coordinates and 20 categories, so there are 125 convolution cores. A pass through layer from the last $3 \times 3 \times 512$ layer to the penultimate convolution layer is also added so that fine-grained features can be used in the model.

2.4 Network Structure of YOLT Model

The structure of yolt network is shown in Fig. 2, by using 22 layer neural network and 16 times down sampling, the roughness of the model is reduced, the detection accuracy of dense object (such as cars or buildings) is improved, and a 416×416 pixel input image generates a 26×26 prediction grid. Inspired by 30 layer Yolo network, the network architecture is optimized for small and dense objects. In order to improve the accuracy of small targets, the network also includes a transfer layer, which connects 52×52 layers to the last convolution layer, and allows the detector to access the fine-grained features of the extended feature map.

Input Data Processing of Yolt Model

Yolt model divides any large-scale image into manageable slices, and forecasts the segmented slices on the trained yolov2 model. The segmented image is obtained by sliding window, which has a custom size and 15% overlap.

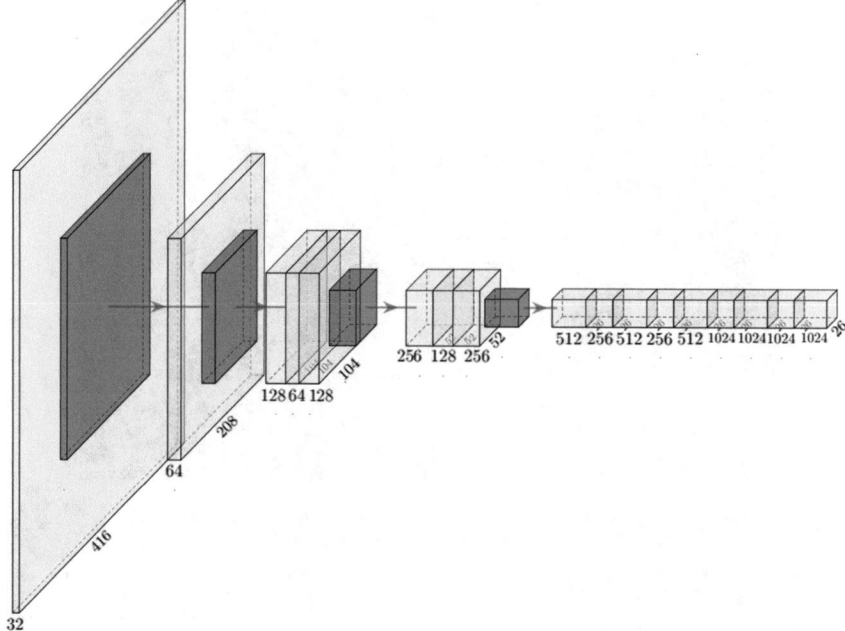

Fig. 2 Yolt network structure

Goal Integration of Yolt Mode

For each segmented image, the predicted value of the bounding box returned from the model is adjusted according to the cut row and column values, and the global position of the predicted value of each bounding box in the original input image is obtained. The 15% overlapping area ensures that the whole input image can be predicted, as shown in Fig. 3. NMS non maximum suppression is applied to the overlapping area to eliminate the influence of repeated detection.

3 Research on Parallel Target Detection Based on Yolt

3.1 Parallel Algorithm Description

1. The image is segmented according to the number of CPUs, and the main process segments the image according to the total number of processes and image size, the adjacent area will have 100 pixels overlap, and the slave process will read the corresponding data on the shared memory according to the location information passed by the master process.

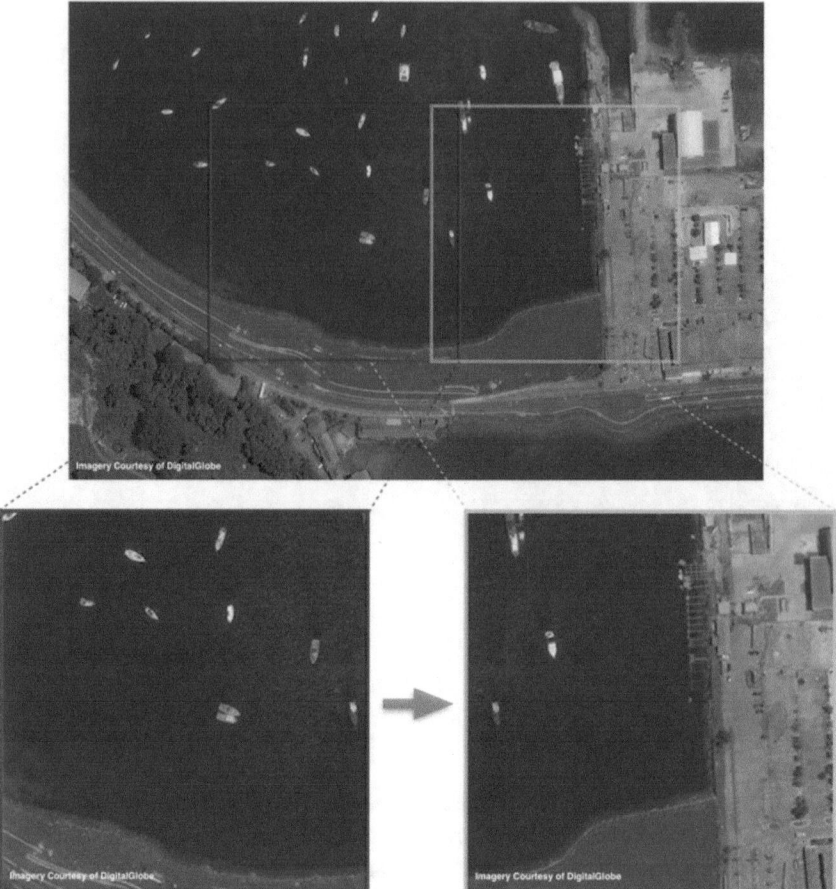

Fig. 3 Input data processing

2. The Image Data Divided from the Slave Process is Predicted by Using the Yolt Model, and Finally the Prediction Result is Transferred from the Slave Process to the Main Process.
3. The main process summarizes the prediction information, removes the prediction results of overlapping areas by NMS algorithm, and finally draws the bounding box, score and category of all targets in the original image.

3.2 Image Segmentation Algorithm Description Method

Assuming that the number of CPUs is K and the image size is m * n, the data block size is $\sqrt{\frac{mn}{k}} * \sqrt{\frac{mn}{k}}$ after the average allocation to a single GPU, the grid number of m dimension is $a = m // \sqrt{\frac{mn}{k}}$, the grid number of n-dimension is $b = n // \sqrt{\frac{mn}{k}}$, use floor division to round down, If the edge of the image is not allocated, change the grid size evenly, so that the image is divided into a * b grids, a * b < = k. Because of the uncertainty of the image format and the number of GPUs, the GPUs may not be fully utilized, but this algorithm can increase the general adaptability of the program. Just input the number of images and GPUs to run the program.

3.3 Parallel Goal Integration

According to the number of CPUs, the same number of image slices with the same scale are obtained for large-scale image through image segmentation algorithm. Each image slice is predicted by using the trained yolt model, and the boundary box, category and score of the target are obtained, then, the parallel algorithm is used to summarize the bounding boxes, categories and scores, and NMS algorithm is used to remove the prediction results of overlapping areas. Finally, the bounding boxes, scores and categories of all detected objects are drawn in the initial image.

3.4 Result Analysis

Based on the parallel target detection algorithm proposed in this paper, mpi4py program is used for multi process programming, and experiments are carried out in a cluster environment composed of eight computers. Each computer is equipped with a graphics cards.

SP stands for speedup, which is a measure of speedup performance to show the benefits of solving problems in parallel. It is defined as the time TS taken to solve a problem on a single processing unit divided by the time TP taken to solve the same problem on P identical processing units.

We use $SP = \frac{T_S}{T_P}$ for acceleration. If SP = P, then the acceleration is linear, which means that if the number of processors increases, then the execution speed will also increase, which is ideal. When TS is the best execution time of serial algorithm, acceleration is absolute; when TS is the execution time of parallel algorithm on a single processor, acceleration is relative.

When the program is executed, all processes run the same set of MPI code, and process 0 is the main process, which is responsible for basic control tasks. When the number of execution processes is equal to 1, the serial calculation is executed, and at this time, one process is responsible for the entire data processing process. When the number of processes is greater than 1, parallel calculations are performed, and the main processing node is responsible for the control of the entire parallel program and part of the calculation tasks, distributes data and calculation tasks to all processes, collects the calculation results from the processing nodes, and obtains the predicted results. Therefore, when two processes are used to perform parallel calculations, the communication time between the processes takes up a relatively large amount, and the execution time of the two processes is not much different from the execution time of one process.

Table 1 shows the experimental results, showing the predicted execution time, speed-up ratio, and parallel efficiency.

It can be seen from the parallel execution time list in Fig. 4 that the execution time gradually decreases as the number of processes increases, and the parallel processing speed is significantly improved.

It can be seen from the speedup change curve in Fig. 5 that the speedup of the parallel detection algorithm is basically linear with the number of processes. Because the main factor that affects the performance of parallel algorithms is the synchronization between processes and the communication between processes, as the number of processing processes increases, the proportion of communication time to computing time in parallel algorithms increases, and the processing process is idle and communication increases. The acceleration ratio is slightly reduced. It can be seen from the parallel efficiency graph in Fig. 6 that the efficiency of parallel detection of remote sensing targets is higher, which shows that the resource usage of multiple processors is better.

The experimental results show that: with the increase of the number of processors, the speedup ratio increases gradually, which plays an accelerating effect.

Table 1 Experimental result

Number of processes	Time(s)	SP	Efficiency(%)
1	114.86	1	100
2	105.91	1.08	54
3	79.14	1.45	48.33
4	62.42	1.84	46

Fig. 4 Parallel execution time change graph

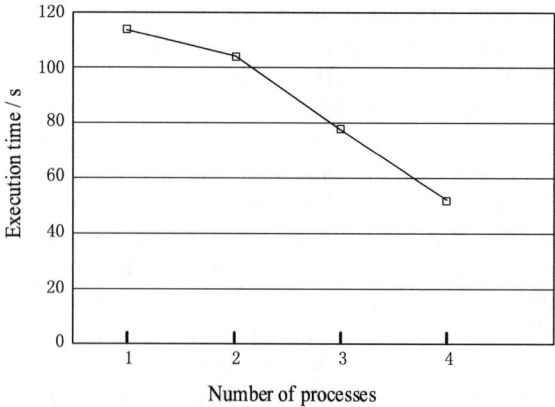

Number of processes

Fig. 5 Parallel execution speedup

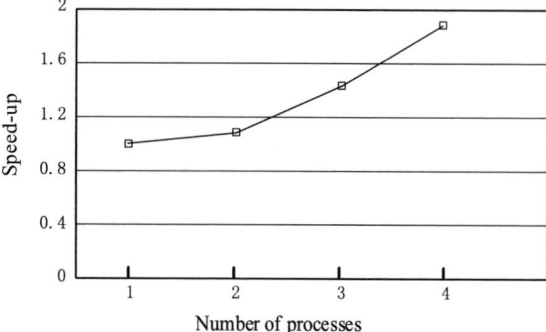

Fig. 6 Parallel efficiency

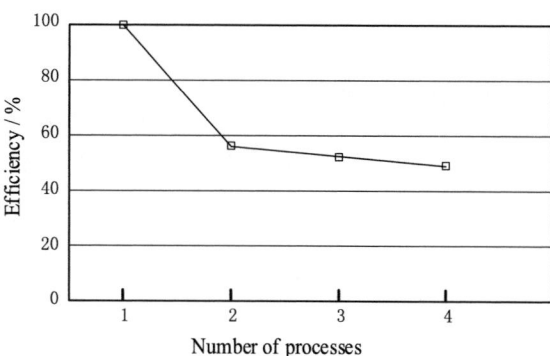

References

1. Cheng G, Han J (2016) A survey on object detection in optical remote sensing images. ISPRS J Photogramm Remote Sens 117:11–28
2. Liu G, Sun X, Fu K et al (2013) Aircraft recognition in high-resolution satellite images using coarse-to-fine shape prior[J]. IEEE Geosci Remote Sens Lett 10(3):573–577
3. Girshick R, Donahue J, Darrell T et al (2014) Rich feature hierarchies for accurate object detection and semantic segmentation. In: Proceedings of the IEEE conference on computer vision and pattern recognition, pp 580–587
4. Redmon J, Divvala S, Girshick R et al (2016) You only look once: Unified, real-time object detection. In: Proceedings of the IEEE conference on computer vision and pattern recognition, pp 779–788
5. Liu W, Anguelov D, Erhan D et al (2016) SSD: single shot MultiBox detector. In: European conference on computer vision. Springer, Cham, pp 21–37
6. Van Etten A (2018) You only look twice: rapid multi-scale object detection in satellite imagery
7. Redmon J, Farhadi A (2017) YOLO9000: better, faster, stronger[C]. In: IEEE conference on computer vision & pattern recognition. IEEE, pp 6517–6525

Research on Expression Processing Methods of Children with Autism in Different Facial Feature Types

Yishuang Yuan, Kun Zhang, Jingying Chen, Lili Liu, Qian Chen, and Meijuan Luo

Abstract Facial expression recognition disorders were common in patients with Autism Spectrum Disorder (ASD), and were considered to be the main cause of their social disorders. In order to further understand the facial expression recognition disorders of children with ASD, this research adopted a 2×3 multi-factor experimental design and used eye-tracking technology to explore the expression processing of 3–7 years old ASD children and Typical Developing (TD) children with matching psychological development levels, under three different facial feature presentation types, including Low Spatial Frequency (LSF) facial expression stimulation, LSF and eye feature mixed facial expression stimulation, and LSF and mouth feature mixed facial expression stimulation. The results showed that ASD children's visual scanning strategies for facial expressions were more disorganized than TD children. They were also weaker than TD children in processing facial expression structure information, but still had a "happiness advantage" similar to TD children, that was, the highest recognition rate of happiness expression. They were more accustomed to using local processing methods based on mouth features to process facial expression information, and their visual attention was more likely to be affected by the mouth rather than the eyes.

Keywords Autism · Expression processing · Facial feature · Eye-tracking · Spatial frequency

1 Introduction

Facial expressions have a wealth of biological and sociological information, and are of great significance to individuals' social perceptions. The "preference" of typical developing individuals for facial stimulation appears as early as infancy and is fully developed in early childhood [1]. However, facial expression recognition disorders

Y. Yuan · K. Zhang (✉) · J. Chen · L. Liu · Q. Chen · M. Luo
Faculty of Artificial Intelligence in Education, National Engineering Research Center for E-Learning, Central China Normal University, Wuhan, China
e-mail: zhk@mail.ccnu.edu.cn

are common in patients with Autism Spectrum Disorder (ASD), and are believed to be the main cause of their social disorders [2].

Artificial intelligence aims to allow intelligent technology to expand human capabilities in different fields [3]. Aiming at the difficulties in expression recognition of ASD children, artificial intelligence is expected to provide a new intervention strategy for their expression recognition capabilities. Relevant researches had been carried out both at home and abroad. For example, Charlop et al. combined 3D animation with artificial intelligence technology to allow autistic children to perform role play to observe the facial expressions of characters [4]; Professor Chen's team had integrated avatar technology to design an AR system, aiming at the intervention of expression recognition for ASD children [5]. However, these studies had not considered the choice of facial expression stimulation, the priority of presentation and the optimization and improvement of intervention methods, which were closely related to the processing methods and characteristics of facial expression of ASD children. Therefore, more in-depth research on the processing methods of facial expression of ASD children was needed, so as to provide theoretical support for the intervention training of ASD children.

It was generally believed that ASD children used atypical facial processing methods to perceive facial information, that was, they tended to use feature-based local processing methods to process facial information, which was mainly manifested as relying on the mouth and ignoring eye features. And Typical Developing (TD) children usually used configural processing methods based on the overall structural features to process facial information, and they were more sensitive to the eyes. However, there were also inconsistent conclusions. For example, Donck et al. believed that boys with autism used similar eye-gazing strategies to process facial information like ordinary boys [6], Cañigueral et al. also found that autistic children could use facial expressions as social signals [7].

Some researchers believed that analysis from the spatial frequency domain could further explain the local-total processing method. Low Spatial Frequency (LSF) mainly made facial features more ambiguous and was used to represent the overall structural information of the object, which was related to the configural processing method based on the overall structural feature. High Spatial Frequency (HSF) mainly made facial details more prominent and was used to represent the local fine information of the object, which was related to the processing method based on local features. When the stimulus was presented as the Broad Spatial Frequencies (BSF), that was the original image, the face recognition effect was the best [8]. Deruelle et al. used the means of spatial frequency separation to prove that ASD children adopted facial processing methods based on local features, and found that the expression matching performance of ASD children under HSF conditions was better than that under LSF conditions [9]. Kikuchi et al. combined the inverted paradigm with spatial frequency technology in the experiment, which also confirmed this view [10]. Yan used eye-tracking technology to explore the characteristics of 5–13 years old ASD children's processing of fuzzy expressions in different scenes [11]. None of the above studies directly proved whether ASD children relied on mouth features for expression processing, and most of the research participants were ASD teenagers. However, it was particularly important to explore the facial processing methods of ASD children, because the earlier the intervention, the better the effect.

In view of this, this study intended to explore the processing methods and characteristics of different facial expression information of ASD children through the combination of spatial frequency technology and mixed expression paradigm. By presenting LSF expression stimulation, LSF and eye mixed expression stimulation, and LSF and mouth mixed expression stimulation in sequence, this study used eye-tracking technology to explore the difference in expression processing between the ASD children and TD children. This experiment proposed the following hypotheses: (1) According to the results of Pelphrey's research [12], it was assumed that the overall performance of the ASD children was weaker than that of the TD children; (2) According to the results of Qiu's research [13], it was assumed that ASD children relied on mouth features to visually process human faces, while TD children relied on the overall facial structure to visually process human faces, and paid more attention to the eyes. Therefore, it could be assumed that the performance of ASD children under the condition of "LSF and mouth mixed expression stimulation" was better than that under the other two conditions. In the performance of eye-tracking features, ASD children's fixation duration on the mouth area was longer than that on the eye area. Under the condition of "LSF and eye mixed expression stimulation", TD children performed better than ASD children.

2   Experimental Method

2.1   Participants

16 ASD children diagnosed by medical institutions were selected as the experimental group (ASD group). These participants were all from a special education school in Wuhan, with normal or corrected vision, and the ability to express simple words, such as: Laugh/cry etc. In the experiment, 2 children could not follow the instructions of the experiment operator, and the other 2 children failed to complete all the experimental items. Therefore, after excluding these 4 children, there were 12 effective participants in the end. Their physiological age was 3–7 years old; the average monthly age was 67.42; and the male to female ratio was 9:3.

11 TD children were randomly selected from a kindergarten in Xinyang City as the control group (TD group). These participants had normal or corrected vision. Their physiological age was 3–5 years old; the average monthly age was 47.89; and the male to female ratio was 7:4.

Through talking with their teachers, the possibility of mental illness, developmental disorders and other mental illnesses was ruled out. Before the experiment, the "Parent Informed Consent" were signed with their parents, and the Peabody Picture Vocabulary Test-Revised (PPVT-R) was used to assess the psychological development of the two groups of children. The T-test was performed on the results,

Table 1 Basic information of participants

Item	ASD (n = 12)		TD (n = 11)	
	M	SD	M	SD
Physiological age (month)	67.42	5.99	49.08	3.32
PPVT-R results	51.08	14.66	53.54	4.23

and it was found that there was no significant difference in the level of verbal IQ between the two groups of children (P > 0.05), which met the requirements of the experiment. The following Table 1 contained basic information of all participants.

2.2 Experimental Design

Low Spatial Frequency (LSF) could blur facial expression features and weaken local features, so participants could only rely on global features for facial expression processing. Eyes and mouth were the two most important local features in facial expression processing. Therefore, this experiment used the mixed expression matching paradigm based on the image-image matching LSF face and the original image facial features. The experiment used a 2 × 3 multi-factor design. The inter-subject variables were ASD children and TD children, with a total of 2 levels. The intra-subject variables were facial expression presentation types, with a total of 3 levels: LSF face, "LSF face + eyes in the original image" and "LSF face + mouth in the original image". The dependent variables were the participant's eye-tracking characteristics and the average recognition rate of facial expressions under three conditions.

2.3 Experimental Materials

The experimental materials used the BU-3DFE facial expression data set of the State University of New York, from which 4 Asian youths (2 males and 2 females, with an average age of 22 years) were selected as the objects. Each object contained 4 basic emotions: happiness, sadness, anger, and fear. After normalization and grayscale processing using Photoshop software, the MATLAB2018b software was used to make the stimulus materials. The Fourier transform was carried out first, and then a Gaussian filter was used for LSF processing. The filtering standard was: LSF parameter < 2cycle/face [9]. After the LSF face was obtained, the eyes of the original image were mixed with the LSF face, and the mouth of the original image was mixed with the LSF face. Finally, a total of 48 images in three categories were formed, including LSF facial expression images, LSF facial expression images mixed with the eyes of the original image, and LSF facial expression images mixed with the

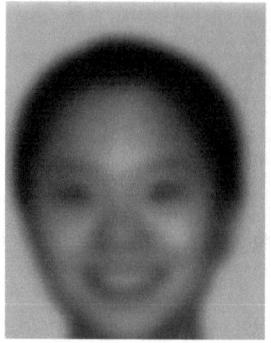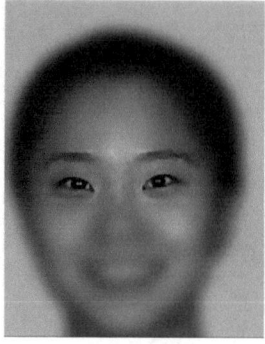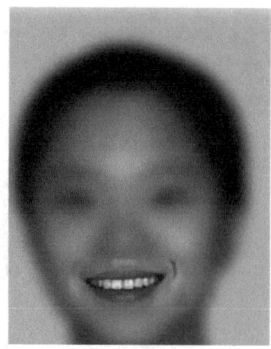

Fig. 1 Experimental materials (LSF face, LSF face + Original eyes, LSF face + Original mouth)

mouth of the original image. Figure 1 showed a sample of the experimental stimulus materials.

Figure 1 was an example of experimental materials. The size of each image was 1024 × 800 pixels, and the background gray scale was 20%. The facial expression matching item used the same expression of the target person as the true probe, and different expressions of the target person as the fake probe. The matching stimulus had the same size as the target stimulus and was located at the bottom left and bottom right of the screen. In addition, 12 images with different expressions were randomly selected from the Internet as practice materials.

2.4 Experimental Equipment

The Tobii Eye Tracker 4C with a frequency of 90 Hz was used to record the eye-tracking characteristics of each participant during the task. The eye tracker was fixed directly below the screen of a 23-inch all-in-one computer with a screen resolution of 1920 × 1080 pixels.

2.5 Experimental Procedures

The experiment was carried out in a quiet and comfortable room. The individual measurement method was adopted. First, participants needed to sit on a chair 60–65 cm away from the all-in-one computer screen, and complete the seven-point eye-tracking calibration. After that, experiment operator A prompted the participants to watch the images, and experiment operator B was responsible for controlling the experiment program and eye tracker, and giving instructions when presenting each target image: "Kid, look! Which of the following expressions do you think is the same as this one?" The participant needed to click on the matching expression option within

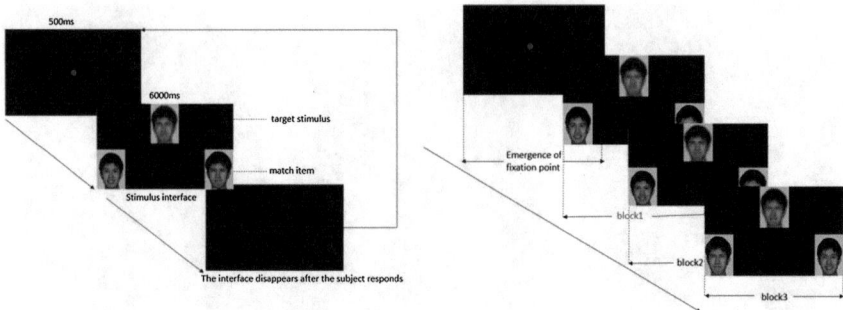

Fig. 2 Example of experimental procedures

the specified time, or verbally inform the correct option. The experiment operator would assist the participant to match facial expressions and record the participant's reaction until the end of the experiment. The score for facial expression recognition used the 0/1 scoring method. Correct recognition was counted as 1, and incorrect recognition or no response was counted as 0. When the participant clicked on the correct expression option, a cartoon character would appear on the feedback interface as a reward. The experimental procedure was shown in Fig. 2.

The practice materials before the experiment were images of different expressions randomly selected from the Internet. The main purpose was to help participant understand the different basic expressions and know the process of the experiment. Experiment operator A was responsible for explaining to participants the different types of facial expressions that appear on the screen. Before the formal experiment, participants had several opportunities for practice, and only those who had mastered the experimental process could enter the formal experimental process.

This experiment included 3 blocks: LSF face expression, LSF and eye mixed expression, LSF and mouth mixed expression. Each block included 24 trails. Each trail contained 4 basic emotions (happiness, sadness, anger, and fear) of 4 characters (2 males and 2 females). The same expression of each character appeared randomly once in each of the 3 blocks. The presentation order of facial expression stimuli in the 3 blocks was balanced with the matching order of facial expressions. Between two blocks, the participant could take a break of 10–15 min. The sequence of stimulus presentation was as follows: first, a red fixed point was presented on the black screen for 500 ms, and then the stimulus material was presented for 6000 ms, and participants were required to quickly match the target expression with the different expressions presented at the bottom of the screen. There was an interval of 2000 ms between each trail, showing a black screen. The experiment cycle was repeated until the end of the experiment program. Excluding the participants' rest time, the entire experiment took about 5 min.

2.6 Data Analysis Indicators

The supporting tool of the Tobii Studio 2010 software was used to divide the three core processing Areas of Interest (AOI) of the target image: face, eyes and mouth. The indicators of eye-tracking were: (1) Fixation Count (FC): the number of fixation points in each AOI during the stimulus presentation process, reflecting the spatial feature of facial expression processing; (2) Fixation Duration (FD): the total time of fixation in each AOI during the stimulus presentation process, reflecting the time of facial expression processing.

3 Results

3.1 Expression Recognition Rate Analysis

For three different facial expression feature types, the average correct rate of facial expression recognition of ASD children and TD children was shown in Table 2.

Although under these three expression presentation conditions, the average correct rate of facial expression recognition of ASD children was lower than that of TD children, it could be seen that ASD children had the highest expression recognition rate under the condition of LSF and mouth mixed expression, while TD children had the highest expression recognition rate under the condition of LSF and eye mixed expression.

The results of facial expression recognition rate were analyzed by repeated measures variance analysis of 2 (group) × 3 (facial feature presentation type), and the results showed that the main effect of facial feature presentation type was significant, $F(2,44) = 5.376$, $p < 0.05$. The interaction effect between the group and facial feature presentation type was significant, $F(2,44) = 5.615$, $p < 0.05$. A simple effect analysis of the interaction was carried out, and it was found that different facial feature presentation types had different effects on the expression recognition rate of ASD children and TD children. The expression recognition rate of ASD children increased significantly ($p < 0.05$) under the condition of LSF and mouth mixed expression. However, that of TD children increased significantly ($p < 0.05$) under the condition of LSF and eye mixed expression.

Table 2 Average correct rate of facial expression recognition for different facial feature types

Item	ASD children		TD children	
	M	SD	M	SD
LSF face expression	24%	14%	77%	12%
LSF and eye mixed expression	26%	11%	91%	19%
LSF and mouth mixed expression	33%	12%	82%	15%

3.2 Analysis of Eye-Tracking Data

Average Fixation Count

(1) Average fixation count on the facial area

As shown in Fig. 3, the average fixation count of ASD children for these three types of facial features was significantly lower than that of TD children, especially in the facial feature type with prominent eye features.

Taking facial fixation count as the dependent variable, a repeated measures variance analysis of two factors (group × facial feature presentation type) was carried out. The results showed that the main effect of facial feature presentation type was not significant, $F(2,44) = 2.78$, $p > 0.05$. The main effect of the group was significant, $F(1,22) = 23.62$, $p < 0.05$. The interaction effect between the group and facial feature presentation type was significant, $F(2,44) = 4.07$, $p < 0.05$. A simple effect analysis of the interaction was carried out, and it was found that the average facial fixation count of ASD children increased significantly ($p < 0.05$) under the condition of LSF and mouth mixed expression. However, that of TD children increased significantly ($p < 0.05$) under the condition of LSF and eye mixed expression.

(2) Average fixation count on the AOI

The average fixation count of the two groups of children on the AOI (eyes, mouth) was shown in Table 3. It could be seen that the average fixation count of TD children on the eye area increased significantly under the condition of LSF and eye mixed expression, while the average fixation count of TD children on the mouth area increased significantly under the condition of LSF and mouth mixed expression. It showed that

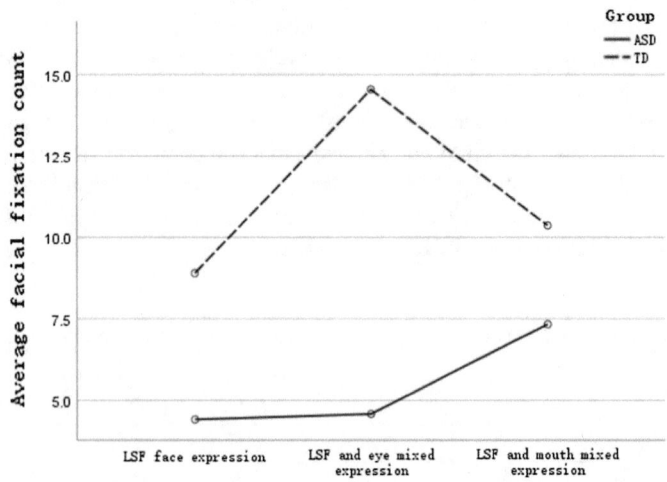

Fig. 3 Comparison of average facial fixation count between the two groups of children

Table 3 The average fixation count of the two groups of children on the AOI (eyes, mouth)

Item		ASD children		TD children	
		M	SD	M	SD
LSF face expression	Eyes	1.00	0.85	4.36	1.86
	Mouth	1.83	1.74	1.82	1.66
LSF and eye mixed expression	Eyes	1.17	1.27	8.36	2.46
	Mouth	1.25	1.13	3.09	1.92
LSF and mouth mixed expression	Eyes	1.67	1.97	1.82	1.47
	Mouth	3.17	2.12	3.73	2.37

TD children could effectively use the prominent local feature information for further facial expression processing when the overall structural features were blurred. ASD children could also be affected by the prompting effect of mouth features in the case of LSF and mouth mixed expression, which was manifested by a significant increase in their fixation count on the mouth area. However, it was difficult for them to obtain the local feature information of the eyes, which was manifested in that their fixation count on the eye area did not increase significantly.

Taking the fixation count on the facial AOI (eyes, mouth) as the dependent variable, a repeated measures variance analysis of three factors (group × facial feature presentation type × AOI) was carried out. The results showed that the main effect of facial feature presentation type was significant, $F(2,44) = 5.78$, $p < 0.05$. The main effect of the group was significant, $F(1,22) = 54.91$, $p < 0.05$. The main effect of the AOI was not significant, $F(2,44) = 3.98$, $p > 0.05$. The interaction effect between the group and the AOI was significant, $F(1,22) = 22.09$, $p < 0.05$. The interaction effect among the three factors of the group, facial feature presentation type, and the AOI was significant, $F(2,44) = 6.98$, $p < 0.05$.

Average Fixation Duration

(1) Average fixation duration on the facial area

As shown in Table 4, the fixation duration of TD children was significantly longer than that of ASD children. In the facial feature type with prominent eye features, the fixation duration of TD children increased significantly, while the fixation duration of ASD children increased under the condition of prominent mouth features.

Table 4 The average facial fixation duration of the two groups of children (unit: s)

Item	ASD children		TD children	
	M	SD	M	SD
LSF face expression	0.80	0.23	1.77	0.43
LSF and eye mixed expression	0.81	0.29	3.19	1.39
LSF and mouth mixed expression	1.24	0.64	2.38	0.51

Table 5 The average fixation duration of the two groups of children on the AOI (eyes, mouth) (unit: s)

Item		ASD children		TD children	
		M	SD	M	SD
LSF face expression	Eyes	0.16	0.11	0.75	0.42
	Mouth	0.23	0.25	0.38	0.25
LSF and eye mixed expression	Eyes	0.20	0.16	1.59	0.73
	Mouth	0.21	0.20	0.60	0.42
LSF and mouth mixed expression	Eyes	0.25	0.23	0.35	0.29
	Mouth	0.76	0.51	0.93	0.59

A repeated measures variance analysis of two factors (group $\times$ facial feature presentation type) was carried out. The results showed that the main effect of facial feature presentation type was significant, $F(2,44) = 6.78$, $p < 0.05$. The main effect of the group was significant, $F(1,22) = 75.04$, $p < 0.05$. The interaction effect between the group and facial feature presentation type was significant, $F(2,44) = 7.57$, $p < 0.05$.

(2)   Average fixation duration on the AOI

The fixation duration of the two groups of children on the AOI (eyes, mouth) under the three types of facial expression features was shown in Table 5. It could be seen that when different AOI was prominently presented, TD children would have different performances. When the eye features were prominent, the fixation duration of TD children on the eyes was significantly increased, and when the mouth features were prominent, the fixation duration of TD children on the mouth was also significantly increased. But relatively speaking, TD Children were more affected by eye features, and they had the longest fixation duration on the eye area under the condition of LSF and eye mixed expressions. In contrast, when the eye features were prominent, the fixation duration of ASD children on the eyes did not increased significantly, but when the mouth features were prominent, the fixation duration of ASD children on the mouth was significantly increased.

A repeated measures variance analysis of two factors (group $\times$ AOI) was carried out. The results showed that the main effect of AOI was not significant, $F(1,22) = 0.21$, $p > 0.05$. The interaction effect between the group and the AOI was significant, $F(1,22) = 9.31$, $p < 0.05$. A simple effect analysis of the interaction was carried out. It was found that under the two conditions of LSF face, LSF and eye mixed expression, the fixation duration of TD children on the eye area was significant ($p < 0.05$), but the fixation duration of ASD children on the eye area was not significant ($p > 0.05$). The fixation duration of the two groups of children on the mouth area was significant ($p < 0.05$) under the condition of LSF and mouth mixed expression.

Heat Map Analysis

The Ogama5.2 software was used to visually analyze the eye tracking data of ASD children and TD children.

Among the three types of facial expression features, the visual scans of ASD children were scattered and unorganized, and they paid less attention to the facial area of the target person. When the mouth features were prominent, ASD children would perceive and consciously pay attention to the target person's mouth area. When the eye features were prominent, ASD children still tended to focus on the lower half of the face. In contrast, TD children showed a high degree of attention to the facial area of the target person. Especially when the eye features were prominent, TD children's attention to the target person's facial area increased significantly. The results were shown in Fig. 4.

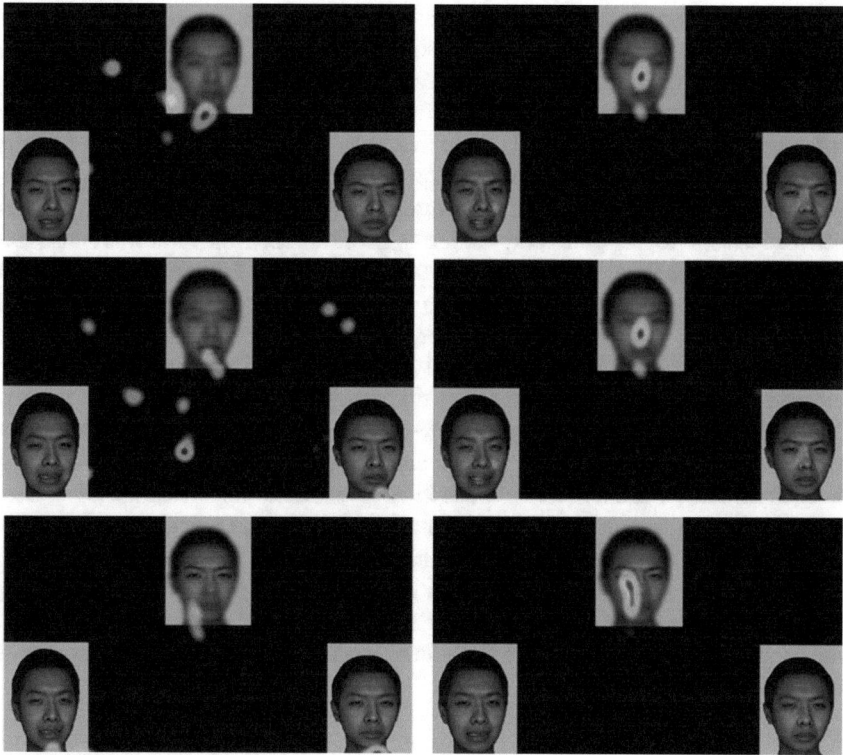

Fig. 4 Comparison of heat maps between the two groups of children (Left: ASD children; Right: TD children; The red area indicated the concentration area)

4 Discussion

4.1 Expression Processing Characteristics of ASD Children Under LSF Conditions

The purpose of this research was to study the facial expression processing methods of ASD children under the conditions that the structural information was completely preserved, the structural information existed while the local features were prominent. The experimental results showed that under the LSF facial expression type, the overall performance of ASD children was weaker than that of TD children. This result was similar to Pelphrey's research on the recognition characteristics of clear facial expressions in children with ASD [12]. It showed that ASD children had a certain degree of facial expression recognition obstacles regardless of the condition of clear expression or fuzzy expression. This result was also consistent with the results of Kikuchi's research [10], which to a certain extent showed that ASD children were weaker than TD children in processing facial expression structure information.

Further analysis found that ASD children had different recognition rates for different expressions under LSF conditions. Among them, ASD children had the highest recognition rate for happiness expression, followed by sadness expression, and the lowest recognition rate for fear expression. ASD children had the same recognition advantage for happiness expression as TD children. This result was consistent with the results of Pino et al., suggesting that there was no difference between ASD children and TD children in recognizing the happiness expression of virtual and real faces [14]. But the result was different from the results of Qiu's research [15], who believed that there was no significant difference in ASD children's cognition of happiness and sadness expressions. This difference might be related to two factors. One was the age of the participants. The experimental participants in this study were 3–7 years old, while Qiu's participants were 12–16 years old. With age, the recognition rate of sadness expressions in ASD children might naturally increase. The second was the experimental material. The experimental materials of this study were the LSF facial expression images, while Qiu used the original facial expression images. The LSF retained the structural information but weakened the local features, which might affect the recognition of sadness expressions, making the recognition rate of sadness expressions of ASD children in this experiment lower than that of happiness expressions.

4.2 Expression Processing Characteristics of ASD Children Under the Condition of Prominent Local Features

The results of this experiment showed that the expression recognition of ASD children under LSF and mouth mixed expression condition was significantly higher than that under the other two conditions, and the fixation count and the fixation duration on the face area were also significantly increased. When the eye features were prominent, the visual attention of ASD children was not attracted by the eye features, but still paid attention to the lower half of the face. However, TD children could not only perceive the mouth features, but also discover and pay attention to the eye features. When the eye features were prominent, the fixation duration of TD children would increase significantly. It showed that ASD children used atypical facial processing methods to process expression information, that was, they were more inclined to perform local processing on the mouth area. This was consistent with the findings of Thomas F Gross [16]. However, Gross used mixed expression paradigm to explore the expression processing of ASD children about 9 years old. In this research, the mixed expression paradigm and spatial frequency technology were combined to study the facial processing characteristics of ASD children aged 3–7. This atypical facial processing might be related to age. Dunsworth believed that there was a developmental delay in the holistic facial processing of ASD children, and with the increase of age, the holistic facial processing of ASD children would be gradually revealed and processed [17]. It had also been suggested that ASD children regarded the real eyes as threatening and therefore adopted the eye-avoiding visual scanning strategies [18]. However, the specific reasons still need further research.

5 Conclusion

Through the analysis of the experimental results, the conclusions of this study were as follows: (1) Compared with matched TD children, ASD children aged 3–7 had a more disorganized visual scanning strategy for facial expressions. They paid less attention to faces and AOI, and had a certain degree of facial recognition obstacles. (2) ASD children were also weaker than TD children in processing facial expression structure information, but still had the "happiness advantage" similar to TD children, that was, the highest recognition rate of happiness expression. (3) When the mouth features were prominent, ASD children could perceive and consciously paid attention to the mouth area, but when the eye features were prominent, they were still accustomed to paying attention to the lower half of the face, while TD children paid more attention to the eye area. It showed that ASD children mainly relied on facial expression processing methods based on mouth feature information, and their visual attention was more likely to be affected by the mouth rather than the eyes. It had a certain guiding effect on how to design expression intervention training materials for ASD children.

This research mainly used eye-tracking technology to study the facial processing methods of ASD children. In future research, intelligent bracelets could be used to monitor the heart rate changes and emotional states of ASD children during the experiment, so as to better analyze the real reasons behind the processing methods of ASD children. In addition, relevant EEG techniques could be used to analyze the close relationship between facial processing strategies and brain tissue abnormalities in ASD children. With the continuous development of artificial intelligence technology, multimodal behavior perception based on artificial intelligence technology, such as gestures, facial expressions, speech, head posture, etc., could also be combined with relevant physiological characteristics, such as EEG and heart rate, so as to comprehensively analyze the behavioral state of ASD children's processing strategies.

Acknowledgements The Project supported by National Natural Science Foundation of China (No. 61807014) and the Fundamental Research Funds for the Central Universities (No. CCNU20QN026, CCNU19QN039).

References

1. Johnson MH, Senju A, Tomalski P (2015) The two-process theory of face processing: modifications based on two decades of data from infants and adults. Neurosci Biobehav Rev 50(3):169–179
2. Dawson G, Webb SJ, McPartland J (2005) Understanding the nature of face processing impairment in autism: insights from behavioral and electrophysiological studies. Dev Neuropsychol 27(3):403–424
3. Deng L, Lei J (2021) The application of artificial intelligence in special education: an analysis of knowledge graph. Chin J Spec Educ 3:18–25 (in Chinese)
4. Charlop MH, Lang R, Rispoli M (2018) Play and social skills for children with autism spectrum disorder. Springer International Publishing, Switzerland
5. Liu L, Zhang M, Chen J, Li D (2017) An avatar-enhanced intervention method for autistic children's facial expression recognition. Chin J Spec Educ 9:35–42 (in Chinese)
6. Donck SV, Vettori S, Dzhelyova M et al (2021) Investigating automatic emotion processing in boys with autism via eye tracking and facial mimicry recordings. Autism Res 3
7. Cañigueral R, Ward JA, Hamilton AF (2021) Effects of being watched on eye gaze and facial displays of typical and autistic individuals during conversation. Autism Int J Res Pract 25(1):210–226
8. Goffaux V, Rossion B (2006) Faces are spatial—holistic face perception is supported by low spatial frequencies. J Exp Psychol Hum Percept Perform 32(4):1023–1039
9. Deruelle C, Rondan C, Gepner B, Tardif C (2004) Spatial frequency and face processing in children with autism and asperger syndrome. J Autism Dev Disord 34:199–210
10. Kikuchi Y, Senju A, Hasegawa T, Tojo Y, Osanai H (2013) The effect of spatial frequency and face inversion on facial expression processing in children with autism spectrum disorder. Jpn Psychol Res 55(2):118–130
11. Yan Z (2020) Study of eye movement in children with autism spectrum disorder for recognizing ambiguous expressions. Zhejiang Normal University (in Chinese)
12. Pelphrey KA, Sasson NJ, Reznick JS, Paul G et al (2002) Visual scanning of faces in autism. J Autism Dev Disord 32(4):249–261
13. Qiu T, Du, Xiaoxin, Zhang W et al (2013) The effect of the weakening information from the eyes and the mouth on the recognition of facial expressions of children with autism. Chin J Spec Educ 5:37–41 (in Chinese)
14. Pino MC, Vagnetti R, Valenti M, Mazza M (2021) Comparing virtual vs real faces expressing emotions in children with autism: an eye-tracking study. Educ Inf Technol 5
15. Qiu T, Du, Xiaoxin, Cai Y et al (2014) Influence of changing expression information characteristics on facial visual scanning of autistic children. J Psychol Sci 37(3):756–761 (in Chinese)

16. Gross TF (2008) Recognition of immaturity and emotional expressions in blended faces by children with autism and other developmental disabilities. J Autism Dev Disord 38:297–311
17. Dunsworth S (2016) Holistic face processing in children and adults along the autism spectrum as measured by the complete composite face test. Holist Process ASD 1–64
18. Tanaka JW, Sung A (2016) The "eye avoidance" hypothesis of autism face processing. J Autism Dev Disord 46(5):1538–1552

Teaching and Assessment Across Curricula in the Age of AI

Improving Java Learning Outcome with Interactive Visual Tools in Higher Education

Yongbin Zhang, Ronghua Liang, Ye Li, and Guowei Zhao

Abstract The object-oriented theory has been widely used in software design and development and Java programming language is a popular object-oriented programming language for computer science related majors in higher education. With the huge demand for intelligent manufacturing employees in the work market, more and more universities have listed Java programming language in their curriculum for non-CS undergraduates. However, learning object-oriented programming languages, such as Java, is tough even for CS students. This paper presents how to improve non-CS undergraduate students' learning outcomes of java language by adopting industry standard professional integrated development tool Eclipse and interactive visual educational tool BlueJ. Our Java programming language course is an alternative 8-week course including 16 h lecture and 16 h lab. Forty-seven junior students from mechanical engineering major were divided into two groups, 23 in the control group (CG) and 24 in the experimental group (EG). Students in both groups used Eclipse to write and run programs during the course while BlueJ was adopted to demonstrate the object-oriented concepts during the fifth and sixth weeks only for students in the EG. A pre-test, which was taken by students from both groups, shows no statistical differences in programming knowledge between the two groups. Results from the post-test at the end of the sixth week show students in EG achieved better performance than students in CG. Results of the survey for the EG reveal that students thought BlueJ was helpful for mastering the concepts of object-oriented programming.

Keywords Java programming · Learning output · Higher education · Interactive · Visual · BlueJ · Eclipse

Y. Zhang (✉) · R. Liang · Y. Li · G. Zhao
Beijing Institute of Graphic Communication, Beijing, China
e-mail: Zhangyongbin@bigc.edu.cn

R. Liang
e-mail: liangronghua@bigc.edu.cn

Y. Li
e-mail: liye@bigc.edu.cn

1 Introduction

The applications of artificial intelligence techniques can be found in everyday life and different industries, such as social networks, education, and manufacturing [1]. Many countries have issued policies to encourage the development of artificial intelligent technologies [2]. At the same time, the workforce market has a huge demand for new generation employees who are not only skillful in their majors but also proficient in applying artificial intelligence in their majors.

Programming abilities are prerequisite skills for the application of artificial intelligence. To meet the workforce market demand, universities have been providing basic programming courses for non-computer majors and advanced ones for computer majors. There are many reasons to provide students with programming courses as programming is a creative and developing activity [3].

However, learning to program is not an easy thing for beginners and object-oriented programming languages, such as Java, can be even tougher. Studies have shown that many students do not know how to program when finishing their introductory programming course. And for undergraduate students majoring in computer science, programming languages taught at university are obstacles that account for students' disengagement with the subjects [4]. Students majoring in information science could transfer into other majors or choose other courses to avoid programming demands [5].

To learn object-oriented programming, it is not enough to learn about just lines of code [6], essential concepts such as class, object, and method should be understood. Program visualization has a positive effect on student motivation [7]. Studies showed that students are more engaged with the concepts and methods used in the teaching of programming when animation and visual tools were used [4].

However, studies have also suggested that even though some visual interactive tools are helpful for students understanding programing concepts, these tools should not be used after the courses and students should be familiar with more professional tools. At the same time, most of those visual tools have been designed and developed by their researchers and are not well known by the public.

This leads us to think if well-known open-source visual tools could be adopted in the Java programming course. On the other hand, at present most studies have been based on either pre-university students or university undergraduate students who study for information technology or computer communications degrees. How to improve learning outcomes of non-computer science students in java introductory programming courses is still a puzzle for most teachers in universities.

In this paper, we will focus on how to improve Java programming learning outcomes of undergraduate students in university who study for a mechanical engineering degree with open source, visual, and interactive Java programming tools, including BlueJ and Eclipse. Participants were third-year students majoring in mechanical engineering at Beijing Institute of Graphic Communication, a university in Beijing. With the same Java programming concepts and skills covered for both the experimental group and the control group, two groups were evaluated with the same test. The results show that with visual and interactive Java programming tools, students achieved better learning outcomes than students did with traditional tools.

2 Literature Review

Java is a class-based, object-oriented programming language that is well known as write once, run anywhere. With the characteristics of simplicity, object orientation, robustness, and high performance, Java has grown in popularity with the increasing demand for platform-independent software. Therefore many curricula add Java programming as alternative courses for non-computer science majors.

Many researchers have presented how to improve students' learning output in programming language courses. In higher education, lecturing is still the most prevalent teaching method but students regard many lectures as ineffective. The software was developed to improve undergraduate students' Java programming learning outcomes by providing instant feedback on students' understanding [8].

A framework was used to learn algorithms, which is in advanced Java courses, for students who understand object-oriented programming and have Java programming skills provided in fundamental courses [9].

The visual tool was developed to help students understand the Java virtual machine by showing the internal state of the JVM during the execution of a Java program in java advanced courses[10]. Students' programming patterns were analyzed with logger data to help educators to understand students' behavior [11]. In traditional programming courses, programming, especially coding skills are taught first before advanced ones such as abstract data type, object-oriented concepts. Recently, another approach to teaching object-oriented skills has been proposed, which starts from abstract concepts before coding skills. This abstraction-first way can be regarded as building lower-level cognitive skills first followed by higher-level ones by using Bloom's Taxonomy [12]. Also, one kind of software was developed to improve undergraduate students' Java programming learning outcomes by providing instant feedback on students' understanding [8].

But studies have shown that undergraduate Students in universities did not live up to educators' expectations in programming courses, which is not related to what programming languages are taught but how the course is taught matters [4]. Learning to program is a key objective for most introductory computer programming language courses. Therefore it is appropriate to teach students to code for solving a specific problem in beginning programming classes because students need to learn syntactic rules of a new programming language. But it is more important for students to understand a higher level of abstraction about programming theory [9].

Therefore when selecting development tools for the Java programming course, we keep it in mind that aligning programming languages and environments is helpful for programming subjects [3].

3   Interactive Visual Tools for Learning Java

Visual tools for learning Java language are kinds of software that users see the dynamic changes or each step during program execution. There are various types of visual tools for Java programming. One kind of such tool provides the ability of visual programming with which users can describe the process using illustration instead of writing codes while non-visual tools usually are viewed as black boxes.

Interactive mode means users can send a request to and get a response from the tools. Almost every integrated development tool for Java language provides such a function. For example, with Eclipse IDE, users can input a variable in debug mode and will be able to watch the state or value(s) of the variable.

Graphics provide a more intuitive perception for a human being than texts do. Researchers have been working on visual tools for a long time. Visual programming tools have been designed and developed for observing the dynamic behavior of parallel Fortran programs [13], tuning nested parallel loops [14], programming graphs [15], robots programming [16], and so on.

An interactive tool was developed for beginners of object-oriented programming. This tool could generate relationships between parents and children classes automatically after a programmer draws a class diagram. In this way, this visual tool prevents mistakes from overriding methods [17]. However, this visual tool just provides a small part of functions for object-oriented programming. At the same time, its target programming language is Smalltalk, which is not suited for Java programming language.

A visual programming environment for Java was introduced which was adopted for producing Java software [18]. This IDE for Java was powerful and had been wildly used by programmers for a while. But now it has been out of date and been replaced by other IDEs, such as Netbeans and Eclipse [19]. Eclipse was developed by IBM and turned over to Eclipse Foundation as an open-source platform [20]. Eclipse platform has been used for different goals. Researchers design a fault prediction tool for Java programs based on Eclipse by integrating machine learning algorithms [21]. An editor for a library management system was also developed based on Eclipse [22].

Besides Eclipse, BlueJ is another visual tool for learning Java. As an integrated development environment, BlueJ has been designed to teach and learn Java programming. One advantage of BlueJ is that novice students can see classes and interact with objects without writing code, which is helpful for students to understand object-oriented concepts, such as class, object, instance variable, class variable, method, and so on. BlueJ was used in teaching and learning computer science courses by many researchers [23–25].

Another visual interactive tool for learning Java is Alice. As a block-based programming environment, Alice is designed to teach object-oriented programming by dragging and doping blocks without writing code. Middle school students used Alice to learn basic programming successfully and found Alice is both fun and easy to use [26].

Besides being used in introductory programming courses, Alice was even used in higher-level object-oriented programming courses along with industry-standard development tools [27].

4 Research Design and Experiment

4.1 Java Course

Our Java programming course, which is an elective 32 h course with 2 h lectures and 2 h tutors in each week lasting for 8 weeks, has been added to the curriculum of mechanic engineering, logistics engineering, and other non-computer science majors since 2016 at the Beijing Institute of Graphic Communication (BIGC), a university located in Beijing of China. The C programming language is a prerequisite for enrolling in the Java programming course. For mechanical engineering students, the C programming language is a compulsory course that includes 28 h lectures and 20 h labs in their first year.

4.2 Interactive Visual Tools

In our Java programming course, Eclipse- an interactive and integrated development environment, was chosen as a development tool for two main reasons: one is that Eclipse is an industry-standard environment [20] that has been widely used in software development projects. Commercial-related tools enhance students' engagement in programming learning. Mastering the skills of developing with Eclipse will benefit students and provide students advantages of proficiency in Eclipse; the other reason is that students experienced the process of developing software with an integrated development environment during their C programming course. Therefore students have no difficulties with the IDE during the java course. Students can transform from the integrated development environment for c to Eclipse smoothly and start their first Java programming without much difficulty as shown in Fig. 1.

But after students had finished their Java programming course, we found that students had not understood object-oriented key concepts such as class, object and inherence clearly, and correctly. Researchers have shown the same problems for first-year undergraduate students who are introduced to object-oriented programming for the first time [28, 29].

A study has pointed out that not enough time is one of the factors that account for unsatisfying results. In most of the available studies, their Java introductory course has a longer duration than ours. But on the other hand, our students were familiar with programming instead of students who were novices to program as stated in other studies.

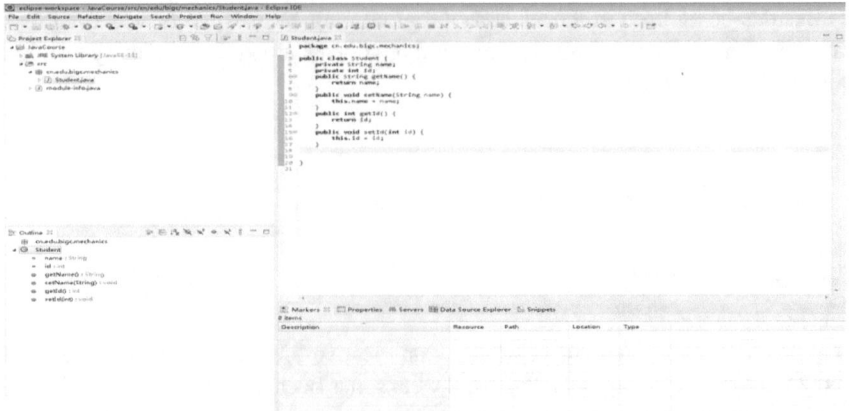

Fig. 1 Eclipse development environment

Therefore, our question is how to improve students' Java learning outcomes in a short-duration course in which students are familiar with writing code. We hypothesize that visual interactive tools will increase students' Java learning outcomes.

Besides Eclipse IDE, another teaching tool- BlueJ is also adopted in our study. Both tools are visual and interactive ones with different features. Eclipse has been widely used by software engineers all around the world. We chose Eclipse so that our students will be able to design and develop Java applications in their future academic or engineer career as stated before. BlueJ was selected in our research because studies have shown that BlueJ is helpful for java learning although researchers suggested other professional development environments should be learned and BlueJ should not be used after first-year study. With BlueJ, several important concepts of object-oriented can be demonstrated easily while it is difficult or even impossible to be realized with Eclipse. For instance, students can view one class and objects created from the class in the BlueJ environment as shown in Fig. 2.

4.3 Participants

In the 2020–2021 autumn term, 47 third-year undergraduate students, whose major is mechanical engineer, selected the Java introductory course which lasted for 8 weeks with 2 h lectures and two-hour labs for each week. All of the 47 students completed the C programming language course in the first year of their study and students are familiar with IDE to create projects and write code. Students were randomly divided into two categories: one consisted of 23 students as the control group (CG), and the other had 24 students as the experimental group (EG).

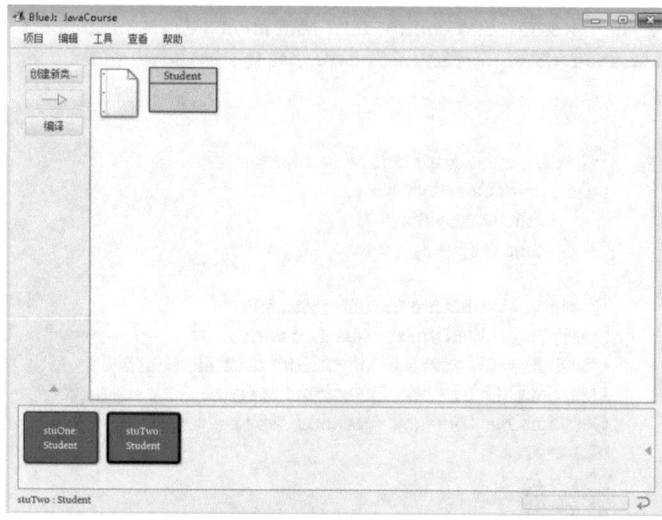

Fig. 2 Interactive and visual tool BlueJ

4.4 Experiment

In the Java programming course, both control group students and experimental group students used the Eclipse development environment for the first four weeks. Then a test was used to check whether there were differences between students in the two groups. In the following two weeks, both groups continued as usual besides that BlueJ was adopted as a visual tool in the two 2 h lectures for the experimental group students. After that, a survey and another test were assigned to all participants. Then the results were analyzed. Figure 3 depicts the whole process.

Eclipse was introduced to both groups in the two-hour lab of the first week and was used during the whole course. For the first four weeks, 47 students were taught by the same teacher in the same room at the same time. The content covered in the first four weeks includes class, object, method, and java language basics such

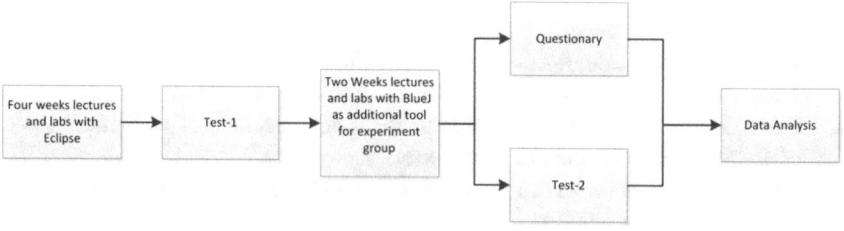

Fig. 3 Experiment process

as operators, expressions, and control flow statements. Then all students took a test (Test-1) to evaluate their programming skills. Test-1 is shown as follows:

```
What is Consider the following class:
public class QeustionOne {
        public static int x = 7;
        public int y = 3;
}
Question 1: What are the class variables?
Question 2: What are the instance variables?
Question 3: What is the output from the following code:
QeustionOne  q1= new QeustionOne ();
QeustionOne  q2 = new QeustionOne ();
q1.y = 5;
q2.y = 6;
q1.x = 1;
q2.x = 2;
System.out.println("q1.y = " + q1.y);
System.out.println("q2.y = " + q2.y);
System.out.println("q1.x = " + q1.x);
System.out.println("q2.x = " + q2.x);
System.out.println("QeustionOne.x = " + QeustionOne.x);
```

In the following two weeks, the fifth and sixth weeks, two groups were taught in different classrooms at different times by the same teacher. Students in CG were still taught as usual while BlueJ was employed by the instructor in EG during the lectures to demonstrate the main object-oriented concepts, such as class, objects, instance variables, and class variables. Students in EG were not required to install and to use BlueJ as a development tool therefore Eclipse IDE was used by students in both CG and EG.

At the end of the sixth week, a survey (3 questions for CG and 4 questions for EG) and a programming test (Test-2) were given to both groups. 5-point Likert Scale (Strongly agree, agree, neutral, disagree, strongly disagree) was adopted for the options of a questionnaire, which consists of the following questions:

1. I understand the concept of Java class
2. I know how to create objects from a Java class
3. I know the difference between instance variable and class variable
4. I found BlueJ to be very helpful to my understanding of class and object concepts. (this is for the EG group only)
5. I would like to use BlueJ as a Java development tool in my future project. (this is for the EG group only).

The test (Test-2) was designed to evaluate the understanding of basic java language and programming ability and is shown as following:

1. Code below has class QuestionB which inherits another class QuestionA, study following code carefully and choose the correct option about the output of the program.

```
QuestionA.java is shown as:
package cn.edu.bigc;
public class QuestionA {
protected int x;
    public QuestionA(int x) {
        this.x = x;
    }
}
QuestionB.java is shown as:
package cn.edu.bigc;
public class QuestionB extends QuestionA {
    public QuestionB(int x, int x2, int y) {
        super(x);
        x = x2;
        this.y = y;
    }
    private QuestionB(int x, int y) {
        super(x);
        this.x = x;
        this.y = y;
    }
    private int x;
    private int y;
    public static void main(String[]args) {
        QuestionA qa1 = new QuestionA(10);
        QuestionB qb1 = new QuestionB(20, 10);
        qa1 = qb1;
        System.out.println(qa1.x + " " + qb1.y);
    }
}
```
A. 10 10
B. 10 20
C. 20 20
D. 20 10
E. None

2. To design a Java program to check whether a number is prime or not. Notice: A prime number is a number that is divisible by only two numbers: 1 and itself. So, if any number is divisible by any other number, it is not a prime number.

5 Results

The results for Test-1 are shown as follows. For Question 1: What are the class variables? Results show that 21 students in CG answered correctly and 2 did wrongly while 22 in EG chose the right answer and 2 did not in EG; as for Question 2: What are the instance variables? The result was the same as Question 1.

For Question 3: What is the output from the following code, 5 students and 4 students provided wrong answers in CG and EG, respectively.

As for the survey, results were shown as follows:

1. I understand the concept of the Java class. And the results are shown in Fig. 4.
2. I know how to create objects from a java class. And the results are shown in Fig. 5.

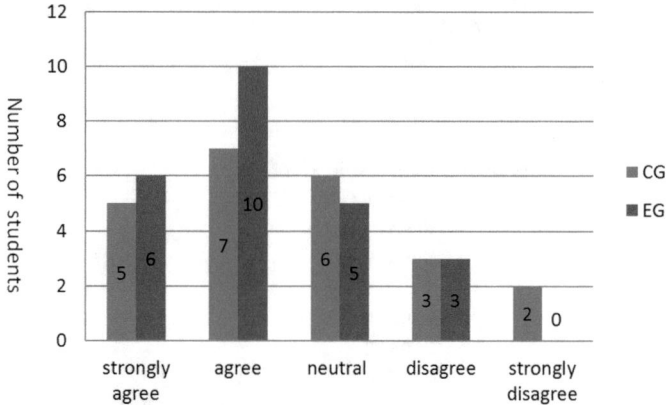

Fig. 4 Survey for I understand the concept of java class

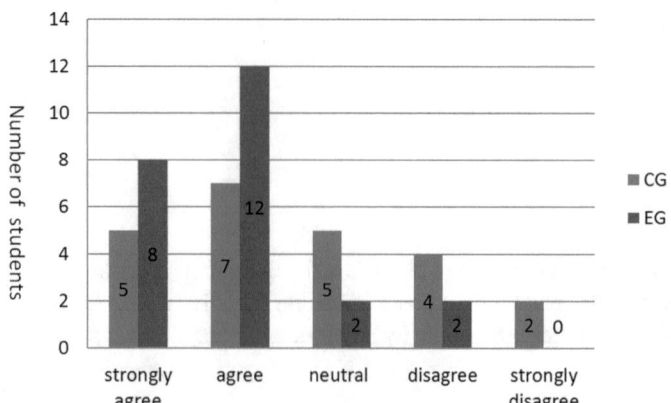

Fig. 5 Survey for I know how to create objects from a java class

3. I know the difference between an instance variable and a class variable. And the results are shown in Fig. 6.
4. I found BlueJ to be very helpful to my understanding of class and object concepts. (This is for the EG group only and the results are shown in Fig. 7).
5. I would like to use BlueJ as a Java development tool in my future project. (This is for the EG group only and the results are shown in Fig. 8).

With regards to the first selection question in Test-2, 18 students in CG answered correctly and 5 did wrongly while 22 in EG chose the right answer and 2 did not, so more students in the EG made the correct choice.

For the second problem in Test-2, students wrote code, compiled and ran their program individually; two instructors checked students' code and validated the running results. 15 students in CG answered correctly and 8 did wrongly while 22 chose the right answer and 2 did not in EG.

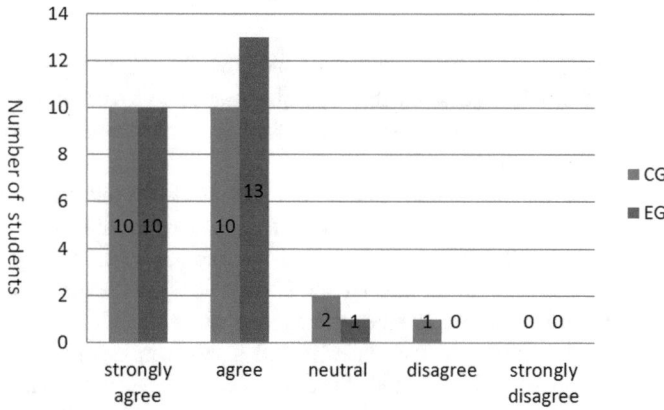

Fig. 6 Options for I know the difference between instance variable and class variable

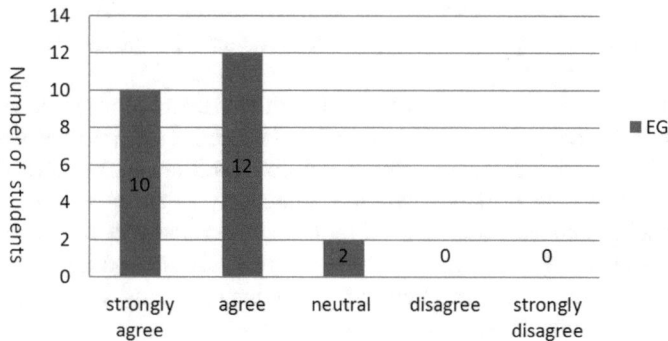

Fig. 7 I found BlueJ to be very helpful to my understanding of class and object concepts

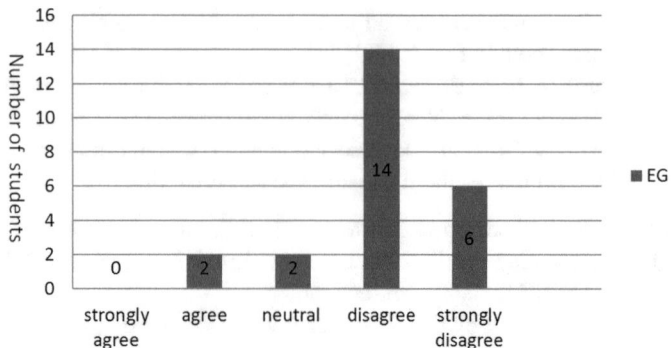

Fig. 8 I would like to use BlueJ as java development tool in my future project

6   Discussion

With results from Test-1, we can draw a conclusion that there is no significant difference ($p = 0.9655 > 0.05$) between the control group and experimental group for the first four weeks of lectures and labs. Therefore, it is reasonable to carry out the experiment with two groups.

After the interactive visual tool- BlueJ was introduced in the two 2 h lectures in the experimental group, there is no significant difference in understanding of the Java class concept ($p = 0.2771 > 0.05$) for the first question of the survey.

But there is a significant difference ($p = 0.0346 < 0.05$) in creating objects from a Java class regarding the results from the second question of the survey, which indicates that more students in the experimental group can distinct objects from classes. Concerning the ability to tell the difference between class variables and instance variables, there is no statistical difference ($p = 0.5792 > 0.05$) between the two groups which is consistent with the results of the first question from Test-1.

It is safe to conclude that BlueJ did help students understand the concepts of class and object according to the results of the fourth question which is only for students in EG because 22 students in EG thought it helped them and only two students were not sure. This finding complies with published results [23, 25].

The options about whether students will use BlueJ in their future projects show that students regard BlueJ only as an educational tool instead of a development tool. This find is consistent with previous research conclusion that BlueJ can be used to illustrate several important object-oriented concepts effectively as an introductory learning environment but students should leave BlueJ and be familiar with more professional tools [23]. This is also the reason why we adopted Eclipse as a development tool while BlueJ was used only for interpreting and demonstrating key object-oriented concepts to help students learn.

Results from the first question of Test-2 show that there is no significant difference ($p = 0.2051 > 0.05$) in understanding provided Java codes between the two groups although more students (22 out of 24) chose the answer correctly in the experimental group than students (18 out of 23) in the control group.

With the second part of Test-2, writing code and running programming, 15 students in CG and 22 students in EG run their code and returned correct answer for any random integer number provided while 8 students and 2 students in CG and EG did not, which means that the difference between the two groups is statistically significant ($p = 0.0268 < 0.05$).

However, we also noticed that several students from both groups wrote statements instead of invoking methods in the main method which breaks the object-oriented design principles [30]. This phenomenon reminds us that other instructional strategies should be studied and applied to solve object-oriented design problems instead of adopting interactive and visual tools alone.

According to the experimental results, we can conclude that by combing the professional development environment, Eclipse, with the interactive visual teaching environment, BlueJ, we could improve students' learning output of java programming language. However, we are still not sure about the theories that account for the difference. In the future, we will investigate related theories, such as learning styles [31] and cognitive theory [32], to explain the cause of the differences further.

7   Conclusion

This paper shows that the industry-standard Java development tool-Eclipse IDE- can be used by non-computer undergraduates as a development tool. Students have no difficulty in transferring from c language development IDE to Eclipse because they are familiar with the process of developing with IDE. The interactive and visual tool, BlueJ, can be adopted as a teaching tool to facilitate students' understanding of object-oriented concepts. However, BlueJ should not be used as a Java developing tool and students should be familiar with professional tools for software development [23].

Although BlueJ is helpful for students in understanding object-oriented concepts, more learning strategies should be adopted to improve students' ability of object-oriented software design. Otherwise, students who are familiar with the C language would like to write object-oriented programs with procedure paradigms instead of using object-oriented principles [30]. Therefore instructors should be aware of the phenomenon mentioned by researchers.

Acknowledgements This research is supported by the Talent Development Project from the Ministry of Education of Chinese of the People's Republic of China (grant: No.201802048023) and the Education Reform Project from Beijing Institute of Graphic Communication (grant NO.22150120030).

References

1. Holm EA, Williams JC, Herderick ED, Huang H (2020) Additive manufacturing trends: artificial intelligence & machine learning. Adv Mater Processes 178(5):32–33
2. Habibollahi Najaf Abadi H, Pecht M (2020) Artificial intelligence trends based on the patents granted by the united states patent and trademark office. IEEE Access 8:81633–81643
3. Horváth R, Javorský S (2014) New teaching model for Java programming subjects. Procedia Soc Behav Sci 116:5188–5193
4. McCracken M et al (2001) A multi-national, multi-institutional study of assessment of programming skills of first-year CS students. In: Working Group Reports from 6th Annual Conference on Innovation and Technology in Computer Science Education, ITiCSE-WGR 2001, pp 125–140. Canterbury, United kingdom
5. Emurian HH, Holden HK, Abarbanel RA (2008) Managing programmed instruction and collaborative peer tutoring in the classroom: applications in teaching Java™. Comput Hum Behav 24(2):576–614
6. Kolling M (2016) Educational programming on the Raspberry Pi. Electronics 5(3)
7. Velazquez-Iturbide JA, Hernan-Losada I, Paredes-Velasco M (2017) Evaluating the effect of program visualization on student motivation. IEEE Trans Educ 60(3):238–245
8. Hauswirth M, Adamoli A (2013) Teaching Java programming with the Informa clicker system. Sci Comput Program 78(5):499–520
9. Cunningham HC, Liu Y, Zhang C (2006) Using classic problems to teach Java framework design. Sci Comput Program 59(1):147–169
10. Abenza PPG, Olivo AG, Latorre BL (2008) VisualJVM: a visual tool for teaching Java technology. IEEE Trans Educ 51(1):86–92
11. Fenwick JB Jr, Norris C, Barry FE, Rountree J, Spicer CJ, Cheek SD (2009) Another look at the behaviors of novice programmers. SIGCSE Bulletin Inroads 41(1):296–300
12. Machanick P (2007) Teaching Java backwards. Comput Educ 48(3):396–408
13. Szelenyi F, Zecca V (1991) Visualizing parallel execution of FORTRAN programs. IBM J Res Dev 35(1–2):270–282
14. Hummel SF, Kimelman D, Schonberg E, Tennenhouse M, Zernik D (1997) Using program visualization for tuning parallel-loop scheduling. IEEE Concurr 5(1):26–40
15. Vilela PRS, Maldonado JC, Jino M (1997) Program graph visualization. Softw Pract Experience 27(11):1245–1262
16. Coronado E, Mastrogiovanni F, Indurkhya B, Venture G (2020) Visual programming environments for end-user development of intelligent and social robots, a systematic review. J Comput Lang 58
17. Haginiwa T, Nagata M (1995) Visual environment organizing the class hierarchy for object-oriented programming. IEICE Trans Inf Syst E78-D(9):1150–1155
18. Edwards S (1999) Visual programming: tips and techniques—using VisualAge for Java. In: Proceedings of the Conference on Technology of Object-Oriented Languages and Systems, TOOLS, pp 413
19. Bruno EJ (2005) NetBeans 4.1 & Eclipse 3.1. Dr. Dobb's J 30(8), 14–23
20. Geer D (2005) Eclipse becomes the dominant Java IDE. Computer 38(7):16–18
21. Catal C, Sevim U, Diri B (2011) Practical development of an Eclipse-based software fault prediction tool using Naive Bayes algorithm. Expert Syst Appl 38(3):2347–2353
22. Surla BD (2013) Developing an Eclipse editor for MARC records using Xtext. Softw Pract Experience 43(11):1377–1392
23. Kölling M, Quig B, Patterson A, Rosenberg J (2003) The BlueJ system and its pedagogy. Comput Sci Educ 13(4):249–268
24. Ragonis N, Ben-Ari M (2005) A long-term investigation of the comprehension of OOP concepts by novices. Comput Sci Educ 15(3):203–221
25. Van Haaster K, Hagan D (2004) Teaching and learning with BlueJ: an evaluation of a pedagogical tool. Issues Inf Sci Inf Technol 1:455–470

26. Kelleher C, Pausch R, Kiesler S (2007) Storytelling alice motivates middle school girls to learn computer programming. In: 25th SIGCHI Conference on Human Factors in Computing Systems 2007, Association for Computing Machinery, pp 1455–1464. United states

27. Cohen M (2013) Uncoupling alice: using alice to teach advanced object-oriented design. ACM Inroads 4(3):82–88

28. Zschaler S, Demuth B, Schmitz L (2014) Salespoint: a Java framework for teaching object-oriented software development. Sci Comput Program 79:189–203

29. Perez-Schofield BG, Ortin F (2019) A didactic object-oriented, prototype-based visual programming environment. Sci Comput Program 176:1–13

30. Westfall R (2001) Hello, world considered harmful. Commun ACM 44(10):129–130

31. Sener S, Çokçaliskan A (2018) An Investigation between multiple intelligences and learning styles. J Educ Train Stud 6(2):125–132

32. Will P, Rothwell A, Chisholm JD, Risko EF, Kingstone A (2020) Cognitive load but not immersion plays a significant role in embodied cognition as seen through the spontaneous act of leaning. Q J Exp Psychol 73(11):2000–2007

Interdisciplinarity of Foreign Languages Education Design and Management in COVID-19

Rusudan Makhachashvili⏺, Ivan Semenist⏺, and Dmytro Moskalov⏺

Abstract The transformative quality of the knowledge economy of the XXI century, the development of the networked society, emergency digitization due to the COVID-19 pandemic measures have imposed new interdisciplinary demands on the Arts and Humanities university graduates' skills and competencies, according to dynamic needs of the job market. The inquiry focus is the comprehensive diagnostics of the development of interdisciplinarity, multipurpose orientation across social domains, and universality of education design and skillsets of students of European (English, Spanish, French, Italian, German) and Oriental (Mandarin Chinese, Japanese) Languages major programs through the span of educational activities in the time-frame of COVID-19 quarantine measures of March 2020 to March 2021. A computational framework of foreign languages education interdisciplinarity is introduced in the study. The survey analysis is used to evaluate the dimensions of interdisciplinarity, universality, and transdisciplinarity, informed by the interoperability of soft skills and digital skills for Foreign Languages Education across contrasting timeframes and stages of foreign languages acquisition and early career training in the COVID-19 paradigm. The inquiry findings disclose: evaluation of interdisciplinarity of foreign languages training across a framework of social and cognitive dimensions; assessment of interdisciplinary and universal skills, crucial for successful professional application of foreign languages graduates; estimated needs of interdisciplinary retraining of Foreign Languages majors to meet dynamic job market requirements of the twenty-first century.

Keywords Interdisciplinarity · Foreign Languages Education (FLE) · Digital literacy · Education design

R. Makhachashvili (✉) · I. Semenist · D. Moskalov
Borys Grinchenko Kyiv University, Bulvarno-Kudryavska-st., 18/2, Kyiv, Ukraine
e-mail: r.makhachashvili@kubg.edu.ua

I. Semenist
e-mail: i.semenist@kubg.edu.ua

D. Moskalov
e-mail: d.moskalov@kubg.edu.ua

1 Introduction

1.1 State-of-the-Art Overview

Transformative shifts in the knowledge economy of the XXI century, Industry 4.0/5.0 development and elaboration of networked society, emergency digitization due to quarantine measures has imposed pressing revisions onto interdisciplinary and cross-sectorial job market demands of Arts and Humanities university graduates' skillsets, upon entering the workforce. This, in turn, stipulates reevaluation of the interdisciplinary approaches to comprehensive professional competences in foreign languages acquisition, education, and application.

Theoretical problems of holistic, multidimensional modeling of reality and its separate spheres are directed by the deterministic interaction of objects, signs of their reception and interpretation (in the field of individual and collective consciousness), embodiment, consolidation and retransmission of the results of interaction of these systems of features.

Conditions for the development of modern globalization civilization determine the expansion and refinement of the paradigm of views on the theoretical principles of determining the groundwork and characteristics of the consolidation of the world order, its perception in culture, collective social consciousness and natural language.

The universality of language in this respect is accessed through is the concept of the logosphere, synthetically perceived as (1) the plurality of language units, which are conditionally exhaustive phenomenological realizations of abstract and empirical elements of different spheres of life [4, 23]; (2) the zone of integration of thought, speech, and experience continuums of cultures [5, 16, 26]; (3) the plurality of culturally relevant universal meanings and signs—semiosphere [27]; (4) a plurality of transcendent spiritual meanings—pneumatosphere [14].

Foreign Languages Acquisition on university-level major programs is a rigorous process that involves different stages and a regimen of activities, communication types and competences across interconnected domains [24, 25]. Interdisciplinarity and ubiquity (universality) of Foreign Languages Education (FLE) in the twenty-first century, therefore, is informed, in crucial ways, by intellectualization and amplified information capacity of human activities in general. Thus, the intellectualization of modern global culture determines a qualitatively new approach to understanding the processes of parallel development of human activities and cognitive (intellectual) experiences. That is the origin and methodological premise of the concept of "noosphere". Noosphere is the unity of "nature" and culture, especially from the moment when the intellectual culture reaches (by force of influence on the biosphere and geosphere) the power of a peculiar "geological force" [41].

The noosphere is defined as the current stage of development of the biosphere, associated with the emergence of humanity in it [16, 41], and is interpreted as part of the planet and planet ambient with traces of human activity.

The integral real component of the Noosphere is identified as the Technosphere—a set of artificial objects (technologies) created by the humankind, and natural objects changed as a result of technological activity of humankind [28]. In turn, Computer Being (computer reality, cyberspace) is a complex, multidimensional sphere of synthesis of reality, human experience and activity mediated by the latest digital and information technologies; technogenic reality, a component of the technosphere of existence [17, 28].

2 Study Design and Objective

The outlined pre-existing studies paradigm and the cognitive (Noosphere) premise of Foreign Languages Education informed the following methodological dimensions, this paper sets out to disclose: (1) the *interdisciplinary* dimension of FLE, disclosed through the mutual transformative potential of information and modern technology, as "knowledge in a scientific sense can lag only slightly behind this world transformation because knowledge becomes transformed in the process" [17]; (2) the *universal* dimension of FLE, disclosed through the pervasive, ubiquitous nature of humanitarian and linguistic (especially multi-cultural) knowledge applicability, as "science and technology revolutionize our lives, but memory, tradition and myth frame our response" [32]; (3) the *interoperable* dimension of FLE, informed by the underlying anthropocentrism of linguistic knowledge and skills, providing the interface for development and application of skills and activities across different domains, as "a human is a nexus of existential horizons" [22].

The result of a fundamental Technosphere shift in the sphere of Foreign Languages Education, induced by the COVID-19 pandemic development and enhanced by continuous iterative digitalization measures, was the need to take quick comprehensive action [29, 36] in order to achieve such desirable results: To activate comprehensive interdisciplinary skillsets, otherwise latent or underutilized in the Foreign languages educational process; To enhance the scope foreign language communication skills beyond the domains traditionally reserved for Liberal Arts education; To boost ICT competence and digital literacy of FLE stakeholders, to meet the requirements of COVID-19 job market and workplace.

Informed by the scope of methodological dimensions, the study objective is to critically review the applied cases and best practices in development of interdisciplinarity, universality and multipurpose orientation of content design, management and projected skillsets of European (English, Spanish, French, Italian, German) and Oriental (Mandarin Chinese, Japanese) Languages major programs through the span of educational activities in the time-frame of COVID-19 quarantine measures of March 2020 to March 2021. The survey analysis method is used to diagnose the dimensions of interdisciplinarity, universality and transdiciplinarity, informed by the interoperability of soft skills [2, 8, 9, 11, 19, 31, 37, 42, 43] and digital communication skills [3, 10, 12, 13, 27, 29, 39, 40] for Foreign Languages Education across contrasting timeframes and stages of foreign languages acquisition and early career training.

The study of groundwork principles of universality and interdisciplinary of Foreign languages professional training and linguistic education in general is a parcel of the framework project TRANSITION: Transformation, Network, Society and Education [29, 30]. The inquiry main findings disclose: assessment of interdisciplinary and universal skills, crucial for successful professional development overall; systematization and assessment of interdisciplinary and custom professional skills, enhanced by higher linguistic education; evaluation of a linguistic training universality/versatility by stakeholders; estimation of a linguistic training interdisciplinarity by stakeholders; cross-sectorial and interdisciplinary social spheres estimated most accommodating or lucrative for a foreign languages education skillset; estimated needs and avenues of interdisciplinary upskilling or retraining by Foreign Languages majors to meet dynamic job market requirements in the XXI century.

3 Findings and Discussion

4 Interdisciplinary Groundwork of Foreign Languages Education Design

The following grid of groundwork concepts is applied to profile the Foreign Languages Education (FLE) in different disciplinary dimensions (Fig. 1): INTERDISCIPLINARITY; UNIVERSALITY; INTEROPERABILITY.

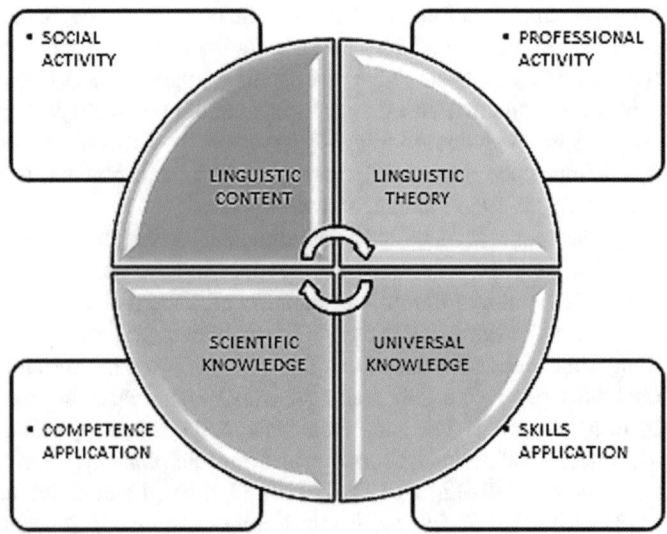

Fig. 1 Interoperablity model for FLE

The meaning of INTERDISCIPLINARITY is synthesized for the purpose of this study as an agglomeration of two or more fields of knowledge into one scope/goal of study, inquiry or activity [6, 15, 18, 21].

UNIVERSALITY is generally understood as a property of object or state **to "exist** everywhere **(ubiquity), or involve everyone"** [7]. In the context of this study we suggest to attribute the property of universality/ubiquity to social activity, vocational activity and professional performance.

The concept of INTEROPERABILITY is disclosed across different approaches [20, 33–35] as a characteristic of an object, product or system, that allows its interface to be comprehensible, to work with other objects, products or systems.

As applied to Foreign Languages Education, the concept of interoperability represents the property of functional, dynamic interconnectivity between the source and target domains of linguistic content, linguistic theory content, related areas of scientific and universal knowledge, and domains of professional and social application. Degrees of interoperability help define the measure of interdisciplinarity and universality of activities, skills and competence applications of FLE stakeholders (Fig. 1).

The generic concept of multiple disciplinarity [1, 38] comprises, in its turn, of a framework of interconnected notions (Fig. 2): Multi-disciplinarity; Interdisciplinarity; Transdisciplinarity.

Multi-disciplinarity, thus, is understood as a multitude of fields of knowledge, that comprise the scope of understanding a certain object, problem or area of inquiry.

Interdisciplinarity in this respect is interpreted as the interconnectivity of multiple spheres of knowledge that comprised the content of a problem or area of inquiry.

Trans-disciplinarity, subsequently, is perceived as a transcendent product of merging multiple interconnected knowledge domains.

Interdisciplinarity in FLE in general is, therefore, postulated in this study as a computational framework of interconnected types of disciplinarities (Fig. 3).

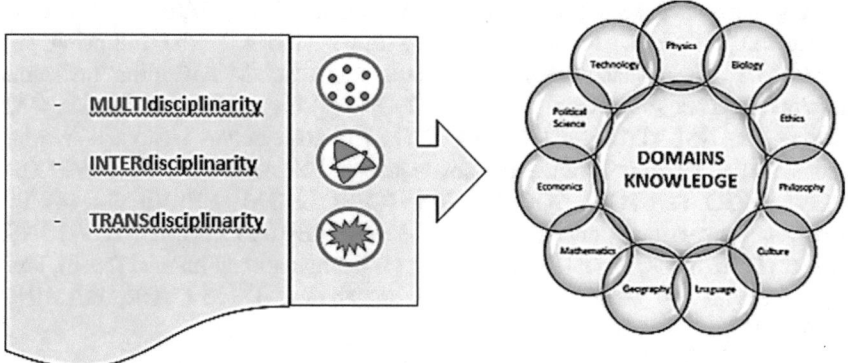

Fig. 2 Multiple disciplinarity model for FLE

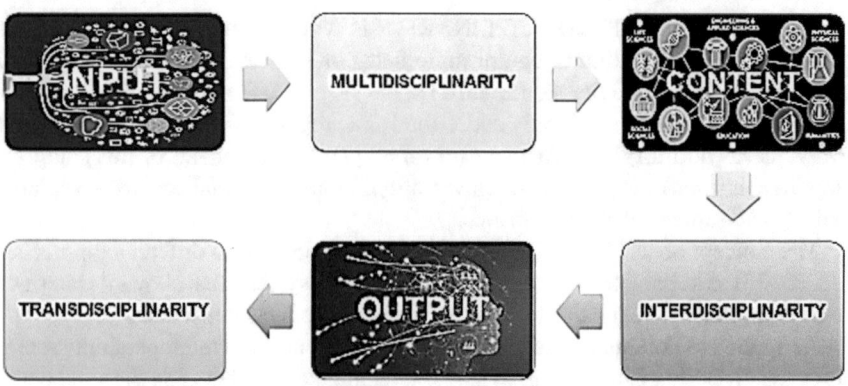

Fig. 3 Computational framework of multiple disciplinarities in FLE

Multidisciplinary **input** into the education design and content in the form of data, information and facts across different source domains of human knowledge in order (1) to constitute the thematic content of language acquisition; (2) to constitute the semantic referents of linguistic units; (3) to constitute the vast framework of reference and contexts for communicative application.

Interdisciplinary connections of the educational **content** for FLE—internal interconnectivity of theoretical and applied disciplines, external interconnectivity of FLE content with non-related areas of human knowledge (computer science, physiology, anthropology, philosophy etc.).

Transdisciplinary **output** in the transcendent nature target knowledge domains and universal applicability of skills, training and outlook of the FLE professionals upon graduation.

Interdisciplinary and transdisciplinary skills ensure *universal* applicability of FLE majors on the job market across various spheres of social activity.

Actual job market demands for FLE graduates in the year 2020 (benchmarking conducted across national and international hiring platforms—LinkedIn, Indeed.com, Work.ua, Jooble.org, include the positions in the following key areas: (1) Teacher of language/literature, corporate coach/MOOC tutor/curriculum developer/teacher (negotiation)—EDUCATION; (2) Translator, proofreader, CAT editor—TRANSLATION, COPYEDITING; (3) Researcher (scholar)—writing grants and grant applications, linguist-expert—RESEARCH AND DEVELOPMENT, NGO SECTOR; SOCIAL SERVICES; LEGAL SERVICES; (4) PR manager, Copywriter, Content manager, SMM—MEDIA COMMUNICATIONS; ADVERTISING, CONTENT-CREATION; (5) Computational linguist (NLP), lexicographer, applied terminologist, digital humanities—IT SECTOR, GAMING INDUSTRY.

5 Interdisciplinary, Universal and Interoperable Dimensions of FLE: Survey Results

The survey analysis is applied for in-depth diagnostics of professional competence and projected employability of in-training linguistic specialists. The inquiry seeks to identify various groups of applied skills, digital skills, and interdisciplinary soft skills, customized for up to date Foreign language university programs.

The **survey sample** consists of 447 respondents across 4 years (Freshman to Senior) of the Bachelor's programs in European (Spanish, Italian, French, English) and Oriental (Mandarin Chinese, Japanese) languages in universities of Kyiv, Ukraine.

The **design of the online survey** included the diagnostics of the following parameters: (1) assessment of multidisciplinary factors that inform Foreign Languages Education; (2) transdisciplinary social domains, estimated most accommodating for a foreign languages education skillset application; (3) transdisciplinary employment options and projected universal career paths for a foreign languages major in Ukraine; (4) systematization and assessment of interoperable skills, enhanced by higher philological education; (5) estimation of linguistic training interdisciplinarity by stakeholders; (6) evaluation of linguistic training universality/versatility by stakeholders; (7) estimated needs and avenues of upskilling or retraining by Foreign Languages majors to meet dynamic job market requirements.

The diagnostics is structured according the computational framework of the interdisciplinarity of INPUT = > CONTENT = > OUTPUT of FLE curriculum.

The following types of questions were used for the diagnostics: multiple-choice; Likert scale score; short answer.

Diagnostics of multidisciplinary INPUT, that allowed to identify informed choice of career paths in FLE, yielded the following results across the board (Fig. 4):

1st year students **highest** scoring target domain that informed the choice of FLE: (1) **mastering a foreign language of preference (27%)**; (2) a career in business sector (26%); (3) a career in education (25%).

1st year students **lowest** scoring target domain that informed the choice of FLE: (1) **to be a translator (2%)**; (2) a career in IT sector (23%); (3) a career in public service (24%).

4th year students **highest** scoring target domain that informed the choice of FLE: (1) **a career in public service (31%)**; (2) **a career in IT sector (24%)**; (3) a career in business sector (24%).

4th year students **lowest** scoring target domain that informed the choice of FLE: (1) to be a translator (1,2%); (2) a career in education (20%); (3) **mastering a foreign language of preference (24%)**.

Consequently, it is evident that mastery of a foreign language, regardless of application domain and sphere of education have gained the multi-disciplinary significance with the FLE majors through the span of recent 5 years (2017–2021). Computer science and IT domains have dropped in multi-disciplinary potential level of FLE INPUT by the year 2020.

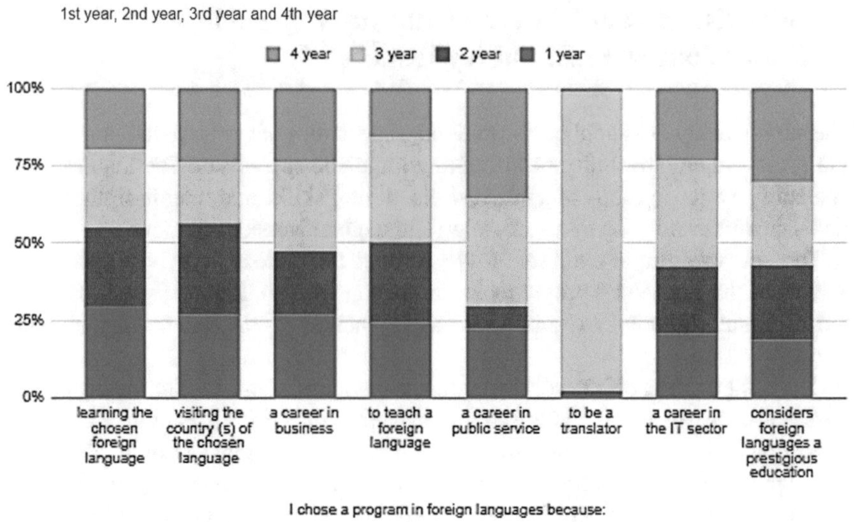

1st year, 2nd year, 3rd year and 4th year

■ 4 year   ■ 3 year   ■ 2 year   ■ 1 year

[bar chart with y-axis from 0% to 100% in increments of 25%, and x-axis categories: learning the chosen foreign language; visiting the country (s) of the chosen language; a career in business; to teach a foreign language; a career in public service; to be a translator; a career in the IT sector; considers foreign languages a prestigious education]

I chose a program in foreign languages because:

Fig. 4 Mutidisciplinary choice of career paths in FLE

Projected transdisciplinarity of FLE career avenues according to content analysis of responses is shaped around target domains of FLE applicability (estimated according to verbal predicates used) for INPUT timeframe (1st and 2nd year students—Fig. 5) yielded the following results: translation services; working abroad (focus on work in China); education:

Comparison and contrast with multidisciplinary INPUT is mostly consistent with the original choices for the career path (due to a limited time span of the FLE education).

Projected transdisciplinarity of FLE career avenues according to content analysis of responses is shaped around target domains of FLE applicability (estimated according to verbal predicates used) for OUTPUT timeframe (3rd and 4th year students—Fig. 6) yielded the following results: translation services; working abroad (no specific focus); business/corporate sector; language as a work tool:

Comparison and contrast with multidisciplinary INPUT indicates that respondents didn't originally plan for career options in translation, but changed their mind due to interdisciplinary potential of the FLE content.

Fig. 5 Content analysis of transdisciplinarity of FLE career avenues (1st and 2nd year students)

Fig. 6 Content analysis of transdisciplinarity of FLE career avenues (3rd and 4th year students)

Interdisciplinary social domains most accommodating or lucrative for foreign languages education are estimated as follows: *Private sector (business); Public sector (civil service, public education, state social sector, etc.); Foreign economic activity; Industry; IT sector; Law/legislature; Agriculture; Volunteering; Finance.*

The distribution of transdiscilplinary potential of FLE across different social domains is estimated in the following quantitative limits (Table 1):

The distribution of transdiscilplinary potential of FLE across different social spheres across different years of study is estimated as follows (social domains ranked by the highest score 5)—Fig. 7:

1st year respondents' highest ranking social domains of FLE applicability are: Private sector (business)—80%; Industry/Law/Social services—67,60%; Finance—63,50%.

4th respondents' highest ranking social spheres of FLE applicability are: Agriculture—67,10%; IT sector—57,10%; Finance/Law—55,80%.

The highest average scoring transdisciplinary domains of FLE application are the Private business sector (77%), Public service sector (68,65%) and Foreign economy sector (58,62%). IT sector is estimated among top 5 domains (48,5%). However, the score range of Foreign economy sector demonstrates the highest standard deviation

Table 1 Social spheres most accomodating or lucrative for a foreign languages education

What are the social spheres most accommodating or lucrative for a foreign languages education?	Standard deviation	Mean
Private sector (business)	0,06,579,133,682	77%
Public sector (civil service, public education, state social sector, etc.)	0,01,368,393,218	68,65%
Foreign economic activity	8,65	58,62%
Industry	0,01,785,182,064	20,88%
IT sector	0,0,627,430,275	48,48%
Law/legislature	0,02,823,450,903	21,93%
Agriculture	0,03,168,990,375	7,15%
Social services	0,04,191,882,036	50,48%
Volunteering and NGO	0,03,523,137,806	33,95%
Finance	0,0,459,585,411	32,08%

1 year, 2 year, 3 year and 4 year

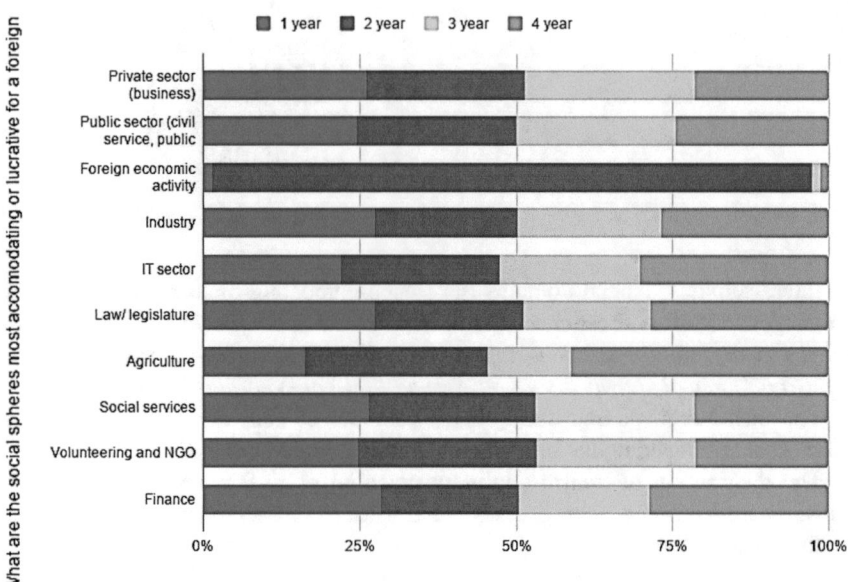

Fig. 7 Transdiscilplinary potential of FLE across 1–4 years of study

(8,65), indicative of an assessment gap as this domain as reliably transdisciplinary, informed by FLE content.

Assessment of skills in FLE across different levels of study yielded the evocative results, as to the potential of foreign languages education for enhancing interoperability of different types of soft and professional skills (Figs. 8 and 9). Key interoperable (soft) skills, across different skills frameworks, identified as enhanced by FLE are as follows: *New knowledge creation; Innovative and adaptive thinking; Interdisciplinary connections; Social intellect; Emotional intellect; Digital literacy; Cognitive management; Makering outlook; Cross-cultural communication; Collaboration; Communication; Enterpreneurship; Creativity; Critical thinking; Innovativity; Leadership; Problem solving; Team-work; Facilitation/mediation; Coordination.*

1st year respondents—score means of top ranking interoperable skills enhanced by FLE (INPUT)—Fig. 8: Emotional intellect—3,239,263,804; Cross-cultural communication—3,222,891,566; **Communication—3,272,727,273; Creativity—3,259,259,259;** Critical thinking—3,217,391,304; Innovativity—3,234,567,901; **Problem solving—3,306,748,466; Team-work—3,253,012,048; Digital literacy—3,251,533,742.**

4th year respondents—score means of top ranking interoperable skills enhanced by FLE (OUTUT)—Fig. 9: **Innovative and adaptive thinking—4,533,333,333;** Social intellect—4,378,378,378; Emotional intellect—4,386,666,667; **Digital**

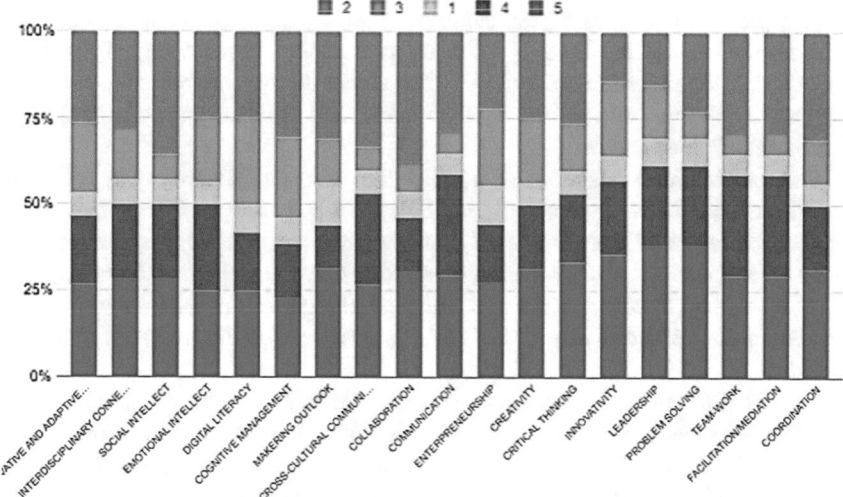

Fig. 8 1st year—skills enhanced by FLE

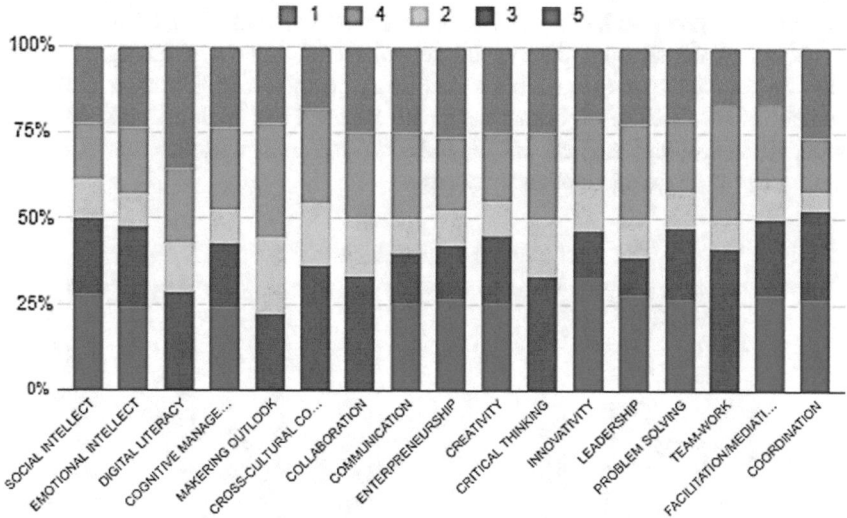

Fig. 9 4th year—skills enhanced by FLE

literacy—4,405,405,405; Collaboration—4,306,666,667; **Communication—4,608,108,108; Creativity—4,493,333,333; Critical thinking—4,608,108,108; Innovativity—4,506,666,667; Problem solving—4,493,333,333.**

The consistent interoperable skills [8, 19, 42], acquired through FLE, regardless of multi-disciplinary input or transdisciplinary output estimate are: communication, emotional intellect, creativity, problem solving and innovation. Digital literacy

[3, 10] proper features as a prominent interoperable skill, acquired through FLE, with respondents of the 1st and 4th years, presumably, because digital literacy is perceived in the timeframe of 2020–2021 as a core literacy, instrumental to foreign languages education and instrumental to application of other types of soft skills of the communicative nature.

Groundwork understanding of FLE design interdisciplinarity was estimated across such qualitative parameters (Fig. 10): *Mastery of several foreign languages; History, culture, and literature; Philosophy, psychology, sociology of law; Applied skills (programming, statistical analysis, mathematical modeling); Digital language data processing; Teaching skills and educational materials development; National and international historical-political and economic context; Trends of the globalized world.*

For 1st year students such parameters of FLE interdisciplinarity rank highest: **Mastery of several foreign languages—80,80%; History, culture, and literature—81,20%; Teaching skills and educational materials development—60%;** National and international historical-political and economic context—50%; *Digital language and data processing—43%.* For 4th year students such parameters of FLE interdisciplinarity rank highest: **Mastery of several foreign languages— 62,20%; History, culture, and literature—64,90%;** *Applied skills (programming, statistical analysis)—35,71%;* Trends in the globalized world—32,60%.

Polyglocy, domains of history, culture and literature, as well as digital sphere and IT are a consistent average interdisciplinary priority of FLE content. However, students of the 2020 of enrollment estimate National and international historical-political and economic context of FLE and education domain studies over globalization studies as an interdisciplinary priority.

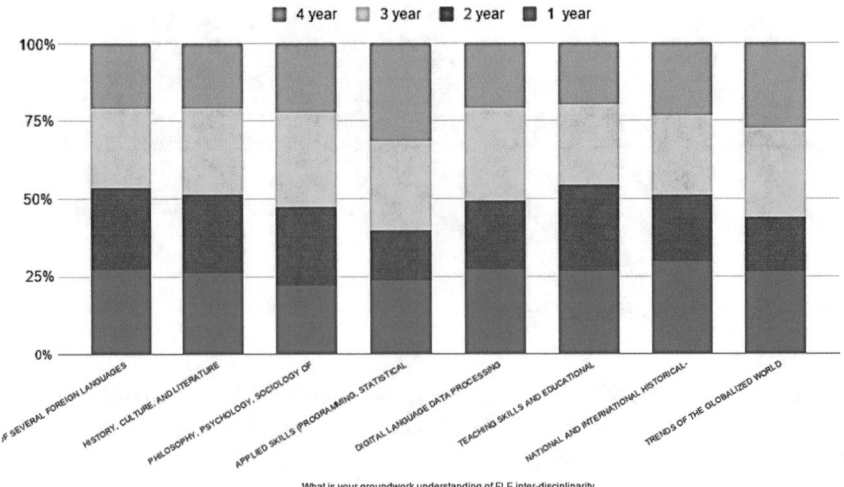

Fig. 10 Interdisciplinarity of FLE assessment

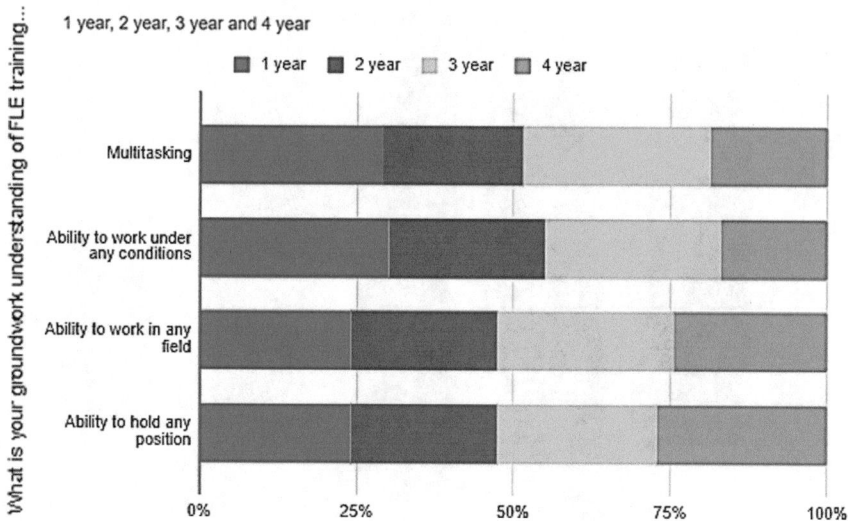

Fig. 11 Universality of FLE assessment

Groundwork understanding of FLE design universality was estimated across such quantitative parameters (Fig. 11): *Multitasking; Ability to work under any conditions; Ability to work in any field; Ability to hold any position.*

For 1st year students such parameters of FLE interdisciplinarity rank highest: Multitasking—45%; Ability to work under any conditions—41,4%; **Ability to work in any field—72,8%**; Ability to hold any position—39,1%.

For 4th year students such parameters of FLE interdisciplinarity rank highest: Multitasking—28,8%; Ability to work under any conditions—23,3%; **Ability to work in any field—74%**; Ability to hold any position—43,80%.

Respondents of both INPUT and OUTPUT timeframes overwhelmingly estimate universal applicability of foreign languages education as ability to work in any professional field or domain (72,8% and 74% of respondents respectively).

The estimation of the need for interdisciplinary upskilling or reskilling (Fig. 12), having completed a higher educational program in FLE is determined across such parameters: FLE is universal and quite sufficient; Desire to master related humanities sphere (psychology, law, culture-studies, international relations, sociology, history, philosophy; Desire to master an applied/technical specialty (computer science, economics, engineering, human health); Believe that FLE alone is not enough for a successful life.

1st year respondents identify the need to reskill or upskill after completion of the Bachelor's program in FLE as follows: Want to **master related humanities sphere** (psychology, law, culture studies, international relations, sociology, history, philosophy)—**57,60%**; FLE is universal and quite sufficient—30%; Want to master an **applied / technical specialty** (computer science, economics, engineering, human health)—18,20%. For FLE INPUT timeframe (1st and 2nd year students) dominant

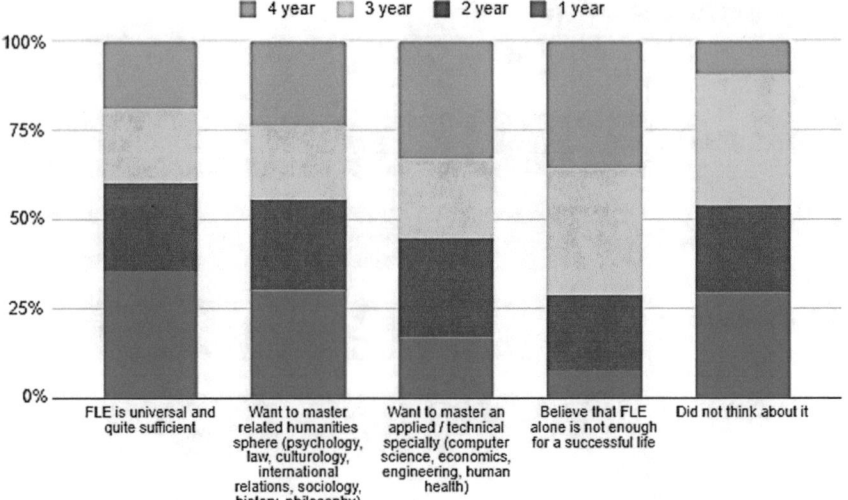

Fig. 12 Interdisciplinary upskilling or reskilling in FLE assessment

is the identified need to interdisciplinary **upskill** in related humanities/Liberal Arts disciplines, which is consistent with multi-disciplinary understanding of FLE.

4th year respondents identify the need to reskill or upskill after completion of the Bachelor's program in FLE as follows: Want to **master related humanities sphere** (psychology, law, culturology, international relations, sociology, history, philosophy)—**45,30%**; Want to master an **applied/technical specialty** (computer science, economics, engineering, human health)—36%; Believe that FLE alone is not enough for a successful life—36%. For FLE OUTPUT timeframe (4th and 3rd year students) dominant is the identified need to both interdisciplinary **upskill** in related humanities/Liberal Arts disciplines and to cross-sectorially **reskill** in a technical/computer science field, which is consistent with transdisciplinary understanding of FLE.

6 Conclusions

The comprehensive diagnostics of the dimensions of interdisciplinarity, universality and transdisciplinarity of foreign languages education disclosed the interoperability of soft skills and digital communication skills across contrasting timeframes and stages of foreign languages acquisition by students of different years of the Bachelor's program and early career training.

Digital domain, digital communication and digital literacy are assessed as interoperable parameters that inform interdiciplinarity of foreign languages education in the timespan of the last 5 years (2017–2018).

The computational framework approach allows to reliably estimate the multiple disciplinarity ratio of the foreign languages education workflow (input—content—output).

Multidisciplinary **input** of FLE is dominated by the mastery of a foreign language, regardless of application domain and sphere of education. Interdisciplinarity of FLE is estimated as interconnectivity of such core domains of knowledge: history, culture, and literature; national and international historical-political and economic context; programming, digital language data processing.

The dominant interoperable skills, acquired through FLE, are: communication, emotional intellect, creativity, problem solving and innovation. Digital literacy features as a prominent interoperable skill, facilitating the application of other types of soft skills of the communicative nature.

The priority avenues of interoperable skills development and expansion of FLE universality in professional application include inter-disciplinary up-skilling across adjacent Liberal Arts and Social sciences domains, and transdisciplinary re-skilling across cross-sectorial domains, not immediately connected to language acquisition and communication (hard sciences, computer science, engineering, economics). These findings are indirectly corroborated by the diagnostic of transdisciplinary potential of FLE applicability across different social domains (Private business sector, Public service sector, Foreign economy sector, Finance, IT sector).

The survey results inform the derivation of the following recommendations for FLE: Critical review of the curriculum content to accommodate the dynamics of multi-disciplinary input expectations of the FLE stakeholders; Review and update of the FLE curriculum content interconnectivity and learning outcomes to accommodate the interoperable interface of skills, customized to facilitate professional activity and language application in the intensely digitized world; Derivation of a flexible model of FLE content upgrade to meet the dynamic transdisciplinary requirements of the job market and enhance universality of professional application for foreign languages majors.

The survey results will be furthered and elaborated in assessment of interdisciplinary and interoperable digital skills adaptability for separate groups FLE stakeholders, according to roles and tasks performed in the language acquisition workflow, as well as according to age and entry digital literacy level. The perspective of the study is in scaling the inquiry to estimate the parameters FLE interdisciplinarity and universality for separate groups of source and target languages acquired, as well as to diagnose interdisciplinary trends of FLE across countries of Asia and countries of Europe.

References

1. Alvargonzález D (2011) Multidisciplinarity, interdisciplinarity, transdisciplinarity, and the science. Int Stud Philos Sci 25(4):387–403
2. Abbott S (2013) The glossary of education reform. http://edglossary.org/hidden-curriculum. Accessed Jul 2020
3. American Library Association (2020) Digital literacy. https://literacy.ala.org/digital-literacy
4. Bakhtin MM (1979) Aesthetics of verbal creativity, Art
5. Barthes R (1968) Elements of semiology, Hill and Wang
6. Callaos N, Marlowe T (2020) Inter-disciplinary communication rigor. Rigor and inter-disciplinary communication: intellectual perspectives from different disciplinary and inter-disciplinary fields. TIDC, LLC, pp 4–29
7. Cambridge Dictionary (2020) CUP. https://dictionary.cambridge.org
8. Davies A, Fidler D et al (2011) Future work skills 2020, institute for the future for university of phoenix research institute. https://www.iftf.org/uploads/media/SR-1382A_UPRI_future_work_skills_sm.pdf
9. Dos Reis A (2015) To be a (Blended) teacher in the 21st Century—some reflections. Int J Res E-learning 1(1):11–24
10. DQ Global Standards Report (2019) World's first global standard for digital literacy, skills and readiness launched by the coalition for digital intelligence. https://www.dqinstitute.org/
11. Eduventures (2020) TechLandscape. https://encoura.org/2020-eduventures-tech-landscape-heres-what-to-expect/
12. European Commission (2020) Digital competence 2020. https://ec.europa.eu/jrc/en/digcomp/digital-competence-framework
13. European Commission (2020) European e-competence framework guideline. https://www.ecompetences.eu/
14. Florensky P (1988) Namehail as a philosophical proposition. On the name of God. Studia Slavica Hung, Budapest 34(1–4):40–75
15. Frodeman R (ed) The oxford handbook of interdisciplinarity (2nd edn) OUP
16. Gachev G (1993) Humanistic commentary to natural science. Issues Lit 11:71–78
17. Heim M (1993) The Metaphysics of virtual reality. Westport Publishers, LA, p 278
18. Holbrook JB (2013) What is interdisciplinary communication? Reflections on the very idea of disciplinary integration. Synthese 190(11):1865–1879
19. Hymes DH (1972) "Communicative competence", sociolinguistics: selected readings, Harmondsworth: Penguin, pp 269–293
20. Interoperability Working Group (2020) Definition of Interoperability. http://interoperability-definition.info/en/
21. Jacobs JA, Frickel S (2009) Interdisciplinarity: a critical assessment. Ann Rev Sociol 35:43–65
22. Khoryzhy S (1997) Notes on ontology of virtuality. Issues Philos 6:53–58
23. Kranz W (ed) (1996) Die Fragmente der Vorsokratiker, Zürich: Weidmann
24. Legal Act of Ukraine (2019) On higher education. https://zakon.rada.gov.ua/laws/show/1556-18#Text
25. Legal Act of Ukraine (2019) On standard of higher education in specialization field 035 "Philology". https://mon.gov.ua/storage/app/media/vishchaosvita/zatverdzeni%20standarty/2019/06/25/035-filologiyabakalavr.pdf
26. Losev A (1993) "Philosophy of the Name", Being. Name. Cosmos. M: Thought, pp 613–801
27. Lotman Y (2000) Semiophere. SPb: Art
28. Makhachashvili R (2020) Models and digital diagnostics tools for the innovative polylingual logosphere of computer Bbeing dynamics. Italian-Ukrainian contrastive studies: linguistics, literature, translation. monograph. Peter Lang GmbH Internationaler Verlag der Wissenschaften, Berlin, pp 99–124
29. Makhachashvili R, Semenist I, Bakhtina A (2020) Digital skills development and ICT tools for final qualification assessment: survey study for students and staff of European and oriental philology programs. Electronic scientific professional journal "Open Educational E-Environment of Modern University", vol 9, pp 54–68

30. Makhachashvili R, Semenist I (2021) Interdisciplinary skills development through final qualification assessment: survey study for European and oriental languages programs. In: Proceedings of the 12th International Multi-Conference on complexity, informatics and cybernetics, IIIS, 2021, pp 144–152

31. Morze N, Makhachashvili R, Smyrnova-Trybulska E (2016) Communication in education: ICT tools assessment. In: Proceedings from DIVAI, Sturovo: University of Nitra, pp 351–354

32. Schlesinger AM (2020) Papers manuscripts and archives division. The New York Public Library. http://archives.nypl.org/mss/17775#overview

33. Shannon CE (1948) A mathematical theory of communication. Bell Syst Tech J 27(3):379–423

34. Slater T (2013) Cross-domain interoperability. Netw Centric Oper Ind Consortium—NCOIC. www.ncoic.org

35. Slater T (2012) "What is Interoperability?", Netw Centric Oper Ind Consortium—NCOIC. www.ncoic.org

36. Taleb N (2010) The black swan: the impact of the highly improbable, 2nd (edn) Penguin, London

37. Taniguchi T (2021) Classification of educational skills for university students in computer programming classes. Int J Inf Educ Technol 11(7):313–318

38. The Digital Divide (2020) Project overview. https://cs.stanford.edu/people/eroberts/cs181/projects/digital-divide/start.html. Accessed Oct 2020

39. Torre I, Łucznik K, Francis KB, Maranan DS et al (2020) Openness across disciplines: reflecting on a multiple disciplinary summer school. In: Open(ing) Education: Theory and Practice, Brill, pp 300–328

40. UNESCO (2018) ICT competency framework for teachers. https://unesdoc.unesco.org/ark:/48223/pf0000265721

41. Vernadsky V (1991) Scientific thought as a planetary phenomenon, Moscow, Academia

42. World Economic Forum (2020) The future of jobs report. http://www3.weforum.org/docs/WEF_Future_of_Jobs_2020.pdf

43. Wulf G, Shea G (2002) Principles derived from the study of simple skills do not generalize to complex skill learning. Psychon Bull Rev 9:185–211

A Survey of Postgraduates' MOOC Learning Satisfaction Based on the Perspective of User Experience

Xiaoxue Li, Xinhua Xu, Shengyang Tao, and Chen Sheng

Abstract Based on the educational practice of graduate high-level talents training, this paper investigates the satisfaction of graduate MOOC learning from the perspective of user experience. Taking 312 graduate students as the research object, through the significance test, analysis of variance and correlation test of the questionnaire data, it is found that students' overall satisfaction with MOOC learning is relatively low, and their satisfaction with learning guidance, learning facilities and learning effect is roughly the same; learners of different genders have significant differences in satisfaction with management services, girls' satisfaction is higher than that of boys; there is a significant difference between the overall satisfaction and learning effect, the first year graduate students have the highest satisfaction, the third year graduate students is the lowest; the satisfaction of literature and history students is the highest, and the satisfaction of art and sports students is the lowest; the satisfaction of each part is positively correlated with the overall satisfaction. On the basis of the above conclusions, this research puts forward suggestions to strengthen the construction of instructional services, enhance the construction of MOOC evaluation system, improve MOOC management service and the level of MOOC learning support.

Keywords User experience · Graduate Students · MOOC · Satisfaction

1 Statement of Problem

MOOC has swept the world since it was proposed in 2012. It aims to break the resource barriers between universities and allow more people to have access to the wonderful courses of famous schools and teachers. Through online communication and discussion with teachers or peers, they have cognitive existence. The existence of teaching and the existence of society provide new possibilities for modern personalized education. However, with the continuous development and derivation of the

X. Li · X. Xu (✉) · S. Tao · C. Sheng
School of Computer and Information Engineering, Hubei Normal University, Huangshi, China
e-mail: xuxinhua@hbnu.edu.cn

© The Author(s), under exclusive license to Springer Nature Singapore Pte Ltd. 2022 257
E. C. K. Cheng et al. (eds.), *Artificial Intelligence in Education: Emerging Technologies, Models and Applications*, Lecture Notes on Data Engineering and Communications Technologies 104, https://doi.org/10.1007/978-981-16-7527-0_19

MOOC platform and MOOC courses, problems such as the high elimination rate of the courses and the low social recognition of the teaching quality have emerged one after another [1]. Studies have shown that the dropout rate of MOOC learners is generally 85–95%, and the dropout rate is relatively high [2]. What kind of MOOC platform, courses and teaching methods can win the satisfaction of learners, so as to realize the real improvement of the quality of large-scale online open courses is an important subject of the current MOOC construction. The learner is the direct user and perceiver of the MOOC platform, and their satisfaction with the MOOC learning experience will directly affect their continuous use of the platform and learning effectiveness. Therefore, a learner satisfaction survey can also help the MOOC platform to further optimize its design provides reference and reference for teachers to develop MOOC courses and design teaching.

On February 4, The Ministry of Education of the People's Republic of China issued the Guidance on the organization and management of online teaching in general colleges and universities during the prevention and control of epidemic situation. The requirements of "stop teaching and stopping without stopping learning" made online education suddenly become the storm center of this time. MOOC, as a mature online education platform, can bridge the resource gap between universities [3]. Through the course content learning, online discussion and other links, MOOC can provide high-quality education resources and opportunities for the course learning for the majority of learners. Under the catalysis of novel coronavirus pneumonia, the feeling and satisfaction of MOOC platform has become the key factor that influences the effectiveness of online teaching.

On the basis of building a moderately prosperous society in an all-round way and fighting a decisive battle against poverty, the party and the state urgently need to cultivate a large number of high-level talents with both political integrity and ability. Postgraduate education plays an important role in cultivating innovative talents, improving innovation ability and cultivating reserve talents for academic research [4].

Postgraduate education is the main way to cultivate high-level innovative talents, and it is quite different from undergraduate survival at the level of training methods and professional education. The deep learning of subject professional knowledge is exactly the full embodiment of modern personalized education. Catalyzed by the new crown pneumonia epidemic, the MOOC platform has become the primary choice for teachers' online education. Learners' experience and satisfaction with the MOOC platform have become key factors affecting the effectiveness of online teaching. Therefore, whether the course content and learning support services in the MOOC platform for graduate students can satisfy people, and how to design and select MOOC content suitable for graduate students are key issues that need to be resolved.

2 Related Research Review

For the research of MOOC, different scholars have conducted rich and in-depth discussions from their own perspectives, and many scholars have also begun to examine the quality of MOOC from the perspective of user satisfaction. Learner satisfaction refers to the psychological feeling of happiness, pleasure or disappointment that students produce by comparing their gains with their previous expectations after learning [5]. Yingchuan [6] found that the interaction, course resources and support services in MOOC learning have a significant positive impact on learners' satisfaction, while the online learning burden of learners has a significant negative impact. Xu [7] et al. established a model for evaluating the learning support service of MOOC based on the satisfaction of MOOC learning support service, and pointed out that the construction of MOOC should be strengthened from the aspects of MOOC learning guidance, assessment and evaluation system, and learners' learning adaptability.

Some scholars also took satisfaction as the starting point, and deeply analyzed how to make MOOC sustainable and excellent operation and development, and how to design a successful MOOC course. He [8] analyzed the operating mechanism of MOOCs in the United States and borrowed from its successful experience. He pointed out that my country's MOOC services and learning should start from opening up free resources, issuing certificate of completion, and accrediting credits, and then moving towards a real degree. Disruptive innovation. Ling [9] took the ChinaX course in the edX learning platform as an example, analyzed the elements of it, and found the key elements for successful MOOC design and eleven specific methods.

The term "user experience" is defined by the International Organization for Standardization (ISO 9241-210) in 2010, that is, people's cognitive impressions and responses to the products, systems and services that they use or expect to use [10]. The definition points out that user experience is all the user's feelings before, during and after using a product, system or service, including emotions, beliefs, preferences, cognitive impressions, physical and psychological reactions, behaviors, and achievements. The current application of user experience in the education field is very weak, and the research on the satisfaction of MOOCs based on user experience is even rarer. Therefore, the research on the satisfaction of postgraduates' MOOC learning from the perspective of user experience is not only conducive to improving the satisfaction of MOOCs. The quality of the course platform is conducive to the effective choice of the platform and the course for learners.

As can be seen from the above, at present, in the field of MOOC, it mainly studies the satisfaction of all MOOC learners in terms of platform use experience and teaching mode, and there is no satisfaction survey specifically for graduate students. Therefore, based on the requirements of various massive open online course platforms in terms of curriculum content and learning support services, this study constructs a questionnaire from the perspective of user experience, aiming to provide guidance and help for graduate students to choose and use massive open online course platform in the future, and inspire the designers of massive open online course platform to better improve various functions and further enhance user satisfaction. At the same time, the satisfaction research on the case of graduate students using massive open online course platform also provides a good online learning resource and environment for the professional construction of graduate students in China's universities, and promotes the training of graduate students.

3 Research Design

3.1 Survey Design

Based on the combing and summarization of domestic and foreign related litera-ture on MOOC learning satisfaction, referring to the more mature satisfaction scales at home and abroad, combined with the characteristics of the scale and openness of MOOC, the indicator design and dimension division are carried out from the perspective of user experience. Draw on the empirical research on learning satis-faction of related MOOC [11–13], synthesize the operable questions in different questionnaires, draft the first draft of the questionnaire, and then delete it under the guidance of relevant experts and teachers. The choice of overlapping and overlapping expressions, unclear connotations, ambiguities in topics, etc., was finally formed in this study based on the user experience of the graduate student MOOC learning satisfaction survey questionnaire (see appendix). This questionnaire is divided into three parts: the first part is the basic situation of the survey object, the second part is the students' cognition and use of the MOOC, and the third part is the satisfaction measurement of the students who have attended the MOOC. Table, the scale has six dimensions: overall satisfaction, learning guidance, learning facilities, learning methods, management services and learning effects. See Table 1 for the connotation of specific dimensions. The five point Likert scale was used to measure the students' subjective evaluation and judgment in the questionnaire, that is, very agree, relatively agree, general, relatively disagree and very disagree, with the values of 5, 4, 3, 2 and 1 respectively.

3.2 Data Sources

With the help of the questionnaire star platform and online filling, the distribu-tion objects are selected through the combination of stratified sampling and random sampling, and the graduate students of all grades and majors in Hubei Normal Univer-sity are selected. In order to ensure the data validity of the questionnaire, add "have you ever participated in MOOC" in the basic information module of the respon-dents one question, to ensure that in the effective questionnaire, all learners have had the experience of MOOC learning, to prevent data interference. After ques-tionnaire collection and data cleaning, 312 valid questionnaires were obtained. The

Table 1 Survey questionnaire of graduate students' MOOC learning satisfaction based on user experience

Dimension	Connotation	Specific Indicators
Overall satisfaction	Overall satisfaction with the MOOC learning process	Are you satisfied with the MOOC courses you have taken; Are you willing to continue to use the MOOC for learning; Are you willing to recommend to your relatives and friends to use the resources on the MOOC for learning
Guidance	Through face-to-face tutoring, audio-visual tutoring, online tutoring Q&A, individual tutoring and other teaching activities, guide and guide students to learn and master knowledge independently; equipped with qualified teachers	The MOOC's introduction to the course is very clear and detailed; The MOOC does a good job of answering questions; The interaction with the teachers solved my problem; The school or research institution where the MOOC teaching team is located has a good reputation and a strong faculty
Learning facilities	Provide online learning platform and multimedia learning auxiliary resources	The MOOC platform has various types of courses; The provided videos, courseware, tests and other independent learning resources are abundant; The functions of the network teaching platform, virtual forum and other facilities can meet my learning needs; The design of the MOOC platform Very satisfied (such as interface design, navigation design, etc.)
Learning method	Cooperative learning with other learners is conducive to mutual motivation, mutual learning, knowledge mastery and ability improvement, and is of great significance to the completion of studies	Are you willing to use the Internet to communicate with classmates and teachers and share ideas; Whether interaction with students in forums or other virtual communities solves the problem; Frequent exchanges between teachers and students, students and students in learning forums, and whether the teaching atmosphere is good; Evaluate whether the activity is a very good way of learning

(continued)

Table 1 (continued)

Dimension	Connotation	Specific Indicators
Management service	Provide information, related consulting services, documents, and systems that learners need in a timely manner	The consulting services provided by MOOCs that are related or unrelated to learning can help me relieve my work, life and psychological pressure; The learning tasks in the course learning process are clear and reasonable; The task learning volume and difficulty set are moderate; It can be timely and comprehensive Accurately provide learners with information about class opening, teaching, counseling, examinations, etc.; The assessment method of course scores is scientific and reasonable
Learning result	Monitor and evaluate students' self-learning process through normal homework and contact, final exams and other assessment methods, so as to encourage students to complete self-learning tasks	The usual homework and exercises in the MOOC platform are helpful to my understanding, mastery and application of knowledge; The final exam on the MOOC platform helps me to systematically master knowledge and improve application ability; My MOOC homework is of high quality; My MOOC learning efficiency is better; MOOC learning has expanded my knowledge; MOOC learning has boosted my interest in learning

Table 2 Research sample composition information

Basic information	Classification	Number of samples	Basic information	Classification	Number of samples
Gender	Male	97	Major	Literature and history	82
	Female	215		Science and Engineering	148
Grade	Graduate freshmen	106		Economy and management	47
	Graduate Sophomore	168		Art	35
	Graduate students in the third grade	38	Have you ever studied MOOC	Yes	298
Number of MOOC courses completed	One or less	114		No	14
	Two to three	141	Type of master's degree	Academic master	192
	More than three	57		Professional master	120

research objects are oriented to different disciplines, including 82 samples of literature and history, 148 samples of science and engineering, 47 samples of economics and management, and 35 samples of art. In the survey sample, the ratio of male to female is 97: 215, and there are 106, 168 and 38 questionnaires for graduate students of grade one, grade two and grade three respectively. The survey group basically covers graduate students of various grades and training types. Combined with the actual ratio of male to female and the distribution of graduate students of different qualities in Hubei Normal University, it can be determined that the sample selected in this study is representative. The specific research sample composition information is shown in Table 2.

3.3 Research Hypothesis

In order to understand the overall satisfaction of graduate students to study in massive open online course and the partial satisfaction of various functional services provided by massive open online course platform, the following assumptions are made according to the basic structure of the questionnaire constructed above:

H1: there is a significant positive correlation between learning guidance and MOOC learning satisfaction.

MOOC guidance can help students quickly grasp the core and key points of course learning, and it is the "steering wheel" and "roadmap" in the process of students' learning and learning. Therefore, the quality of MOOC guidance affects students' learning persistence and interest. It is very necessary to give timely, comprehensive and in-depth guidance to the students. The guidance can guide the students to preview the knowledge learning content in advance, enhance their self-confidence, and lay the foundation for the follow-up formal course learning and the realization of teaching objectives.

H2: there is a significant positive correlation between learning facilities and MOOC learning satisfaction.

The types and functions of learning facilities, the types and contents of teaching resources in MOOC learning platform are important factors that affect the satisfaction of learners in the learning process. Excellent learning facilities can help learners maintain or improve their good experience and mood in learning. Bassi et al. [14] early survey of Online MBA students also shows that course content plays a key role in learners' MOOC learning satisfaction, and provides a method to test in the new MOOC environment.

H3: there is a significant positive correlation between learning style and MOOC learning satisfaction.

The discussion in the learning forum and the communication and interaction with other learners represent a key feature of the implementation of MOOC on all major platforms. The communication and cooperative learning with teachers and other learners can ensure the smooth and continuous learning of students to a certain extent, which is conducive to the supervision and incentive of learning among learners. In fact, early research on online learning does support the positive role of learner learner interaction. Therefore, the influence of learning partners in MOOC will have a positive impact on the learning satisfaction of MOOC.

H4: there is a significant positive correlation between management service and MOOC learning satisfaction.

Whether learners can get timely troubleshooting and related documents and system requirements in MOOC learning is an important factor affecting the learning experience, and also an important method to enhance learning enthusiasm. Therefore, only by clearly mastering the curriculum plan, evaluation criteria and activity task requirements of MOOC learning, can learners make corresponding preparation and safety in order to complete the course smoothly.

H5: there is a significant positive correlation between learning effect and MOOC learning satisfaction.

After the completion of MOOC learning, learners' understanding, mastery and application of knowledge determine the learning effect. Through the arrangement of homework and exercises, problem discussion, final examination and other methods to monitor and evaluate the learning effect of learners, and encourage students to complete the learning task immediately, which helps learners to systematically master knowledge and improve application ability, enhance the learning effect, and improve the learning satisfaction of MOOC.

Table 3 Reliability and validity of the questionnaire

Variable	Cronbach α value	Adaptation index	Detection value
Overall satisfaction	0.714	P value	***
Guidance	0.752	Chi square / degree of freedom	2.481
Learning facilities	0.801	REMEA	0.074
Learning style	0.835	CFI	0.911
Management service	0.794	IFI	0.907
Learning effect	0.766	SRMR	0.062

Note *P < 0.05, **P < 0.01, ***P < 0.001

4 Research Results and Analysis

4.1 Questionnaire Reliability and Validity Test

Since the first two parts of this questionnaire are the collection of basic information parameters of graduate students, the third part is related to the evaluation index of MOOC learning satisfaction, so this study uses Cronbach α coefficient to test the reliability of the third part of the questionnaire (see Table 2). After calculation, the Cronbach α coefficients of each variable are greater than 0.7, indicating that the reliability of the questionnaire is good. Similarly, in terms of validity, Confirmatory Factor Analysis (CFA) was used to fit, and robust maximum likelihood method was used to estimate parameters. The overall fitting results are shown in Table 3: $\times 2/DF = 2.548$, CFI $= 0.911$, IFI $= 0.907$, all greater than 0.9, It shows that the improvement degree of the questionnaire model is better; the RMSEA is 0.074, less than 0.8, which indicates that the gap between the theoretical model and the saturated model is ideal without calculating the number of samples and the complexity of the model. The SRMR is 0.062, less than 0.8, which indicates that the total residual error of the standardized model is small. From this point of view, the questionnaire model design fit well, and the questionnaire structure validity is high.

4.2 Data Processing and Analysis

Analysis of Learners' Overall Satisfaction

In this paper, there are six dimensions in the satisfaction scale of students who have been to massive open online course, and the learner satisfaction scale mainly contains 26 items, which examines the learners' true feelings about overall satisfaction, guidance, learning facilities, learning methods, management services and learning effects.

The survey results in the dimension of overall satisfaction show that 33% (N = 103) of the students are satisfied with the MOOC courses they have studied, 25% (N = 79) of the students are very willing to continue to use MMC to study, and 14% (N = 44) of the students are very willing to recommend the resources in the MOOC classes to their relatives and friends (see Table 4). Most graduate students are satisfied with their MOOC learning experience, so they are willing to let themselves and their friends and relatives continue and join the MOOC course in the future study process. Because of the different majors and interests of each learner, the types of MOOC courses they choose are different. It can be seen that most courses in MOOC platform are of high quality, and learners have better user experience, which is recognized by the vast majority of students.

Analysis of Significant Differences in Different Demographic Characteristics

The average values of overall satisfaction, guidance, learning facilities, learning style, management service and learning effect were 3.2451, 3.4875, 3.3698, 3.3112,

Table 4 Distribution of learners' overall satisfaction

Dimension	Connotation	Specific indicators	Very agree	Relatively agree	General	Relatively disagree	Very disagree
Overall satisfaction of learners	Overall satisfaction with massive open online course's learning process	1. I am very satisfied with the courses I have taken in massive open online course	28	103	149	24	8
		2. I am willing to continue to study in massive open online course	79	113	94	19	7
		3. I would like to recommend to my relatives and friends to use the resources in the class to study	44	98	125	36	9

Table 5 Satisfaction of different gender learners to MOOC learning

Variable	Gender	N	Mean value	Standard deviation
Overall satisfaction	1	97	3.297	0.514
	2	215	3.258	0.732
Guidance	1	97	3.495	0.695
	2	215	3.364	0.582
Learning facilities	1	97	3.743	0.563
	2	215	3.668	0.441
Learning style	1	97	3.625	0.748
	2	215	3.549	0.723
Management service	1	97	3.411	0.631
	2	215	3.387	0.625
Learning effect	1	97	3.523	0.485
	2	215	3.626	0.634

Note 1 for male and 2 for female

3.5216 and 3.4329 respectively, and the standard deviations were 0.8513, 0.8742, 0.8921, 0.9125, 0.8326 and 0.8323 respectively. It can be seen that learners' overall satisfaction with MOOC learning is relatively low, among them, the satisfaction of MOOC learning style is the lowest, and the satisfaction of management service is the highest. SPSS was used to calculate the mean value, independent sample t-test and variance analysis of the overall satisfaction, learning guidance, learning facilities, learning style and learning effect of MOOC from the gender and grade dimensions. The results showed that there was no significant difference in the overall satisfaction, learning guidance, learning facilities, learning style and learning effect of different gender learners, but there was significant difference in the management service of MOOC ($t = -2.168$, SIG $= 0\ 047$) (see Table 5). There are significant differences in overall satisfaction and learning effect among different grades.

Then from the analysis of variance of different professional learners (including literature and history, science and engineering, economy and management, art and sports) on the overall satisfaction, guidance and other indicators, it is found that in addition to guidance and management services, there are significant differences in the overall satisfaction, learning facilities, learning methods and learning effects of MOOC among different professional learners (see Table 6).

Correlation Test Between Each Dimension and the Overall Satisfaction of MOOC Learning

The data of satisfaction and the statistical results of guidance, learning facilities, learning mode, management service and learning effect were input into SPSS software for Spearman correlation analysis. The data in Table 7 are obtained (here, take the guidance and overall satisfaction as an example):

Table 6 Variance test of learning satisfaction of different majors in MOOC

		Sum of squares	df	mean square	F	Significance
Overall satisfaction	Between groups	4.158	2	2.369	6.416	0.004
	Within group	123.121	309	0.329		
	Total	127.279	311			
Guidance	Between groups	1.268	2	0.852	1.268	0.157
	Within group	149.365	309	0.369		
	Total	150.633	311			
Learning facilities	Between groups	4.135	2	2.845	6.398	0.015
	Within group	145.324	309	0.159		
	Total	149.459	311			
Learning style	Between groups	4.587	2	2.125	4.262	0.013
	Within group	179.465	309	0.459		
	Total	184.052	311			
Management service	Between groups	1.264	2	0.744	1.059	0.067
	Within group	157.236	309	0.215		
	Total	158.500	311			
Learning effect	Between groups	4.787	2	2.394	2.398	0.016
	Within group	165.325	309	0.529		
	Total	170.112	311			

Table 7 Correlation between guide and overall satisfaction figure descriptions

Project		Guidance	Overall satisfaction
Guidance	Spearman correlation	1	0.742**
	SIG. (bilateral)		0.005
	Number of samples	312	312
Overall satisfaction	Spearman correlation	0.742**	1
	SIG. (bilateral)	0.005	
	Number of samples	312	312

Note **Significantly correlated at the 0.01 level (two-sided)

It can be seen from Table 5 that the correlation coefficient between the guided learning and the overall satisfaction is 0.742; the probability of the correlation coefficient test is $P = 0.005 < 0.01$; the two asterisks (**) beside the correlation coefficient in the table indicate the significance at 0.01 level, the null hypothesis can be rejected, and the alternative hypothesis can be accepted: there is a significant correlation between guided learning and overall satisfaction. In the same way, we analyzed the correlation between data from other dimensions and the president's satisfaction, and found that the satisfaction of learning facilities, learning methods, management services, and learning effect dimensions are positively correlated with overall satisfaction, among which learning methods and school effects are related to overall satisfaction. The degree is a significant positive correlation.

5 Conclusion and Suggestion

Based on the survey and analysis of 321 postgraduates in various grades and majors of Hubei Normal University, this study can draw the following conclusions: First, the overall satisfaction of the graduate students in MOOC learning is relatively low. Among them, the satisfaction with management services is the highest, and the satisfaction with learning is the highest. The satisfaction of the method is the lowest, and the satisfaction with the tutoring, learning facilities, and learning effect is roughly the same. Second, from the specific perspective of each dimension, learners of different genders have significant differences in the satisfaction of management services. Girls' satisfaction is higher than that of boys. From the grade dimension, the variance analysis of overall satisfaction and learning guidance reveals that learning There are significant differences in overall satisfaction and learning effects. The first degree students have the highest satisfaction, and the third degree students have the lowest. Using variance analysis to test the differences in the satisfaction of different professional learners, it is found that learners of different majors are generally satisfied with the MOOC There are significant differences in degree, learning facilities, learning methods, and learning effects. On the whole, the degree of satisfaction of the learners of literature and history is the highest, and the satisfaction of the learners of art and sports is the lowest. And because the overall satisfaction is composed of five parts of satisfaction, such as tutoring and learning facilities, the correlation test shows that the satisfaction of each part is positively correlated with the overall satisfaction.

Further analysis of the questionnaire reveals that most learners are less satisfied with the communication and discussion with teachers and other learners in the MOOC learning process, especially the "interaction with teachers solves my problem" in the learning part. The problem of "problems" and the "frequent exchanges between teachers and students, students and students in the learning forum, and a good teaching atmosphere" in the learning style are the least satisfactory. This shows that the teacher-student, student-student interaction in the MOOC learning process has an important impact on learners' learning satisfaction. On the other hand, it also shows that the interactive design of the MOOC needs to be improved. In response to the above conclusions, this study puts forward the following recommendations.

5.1 Strengthen the Service of Guided Learning Construction

The introduction of MOOC course guidance and teachers' timely guidance on learners' difficult problems, inappropriate learning methods or personal learning needs play a very important role in ensuring students' smooth learning, continuous learning and learning efficiency. Because the average satisfaction of students on MOOC guidance is generally low in this study, and through the correlation analysis, the relationship between guidance and total efficiency is obtained Therefore, strengthening MOOC guidance service can improve students' overall satisfaction to a certain extent. First of all, teachers should try their best to teach students effective methods of MOOC learning, so as to reduce students' ineffective learning caused by using wrong learning methods, thus affecting learners' MOOC learning satisfaction. Therefore, teachers should properly communicate with students to stimulate students' continuous learning motivation. Secondly, the discussion area or forum of MOOC is the main interaction between learners, teachers and learners Space, students can ask questions according to their own learning situation or discuss and answer the questions of other learners, and teachers' timely answering can also optimize students' learning experience. In the current teaching practice, QQ group is the most active platform for learner interaction [15]. However, the interaction of MOOC is usually asynchronous, and the delay of information acquisition is also an important factor affecting learners' satisfaction. Therefore, MOOC can create a real-time discussion area, carry out real-time Q & A in the form of video or voice, enrich and improve learners' cognition of learning guidance service, and improve students' learning efficiency. Finally, it is necessary to cultivate and form a good interaction atmosphere, so that learners can communicate with each other To share resources, learn together, and jointly promote the formation of excellent learning groups.

5.2 Strengthening the Construction of MOOC Evaluation System

According to the nature of the course and the type of the content, there are some differences in the current evaluation and scoring standards of MOOC for learners, which are generally composed of the following parts: watching video progress, unit test results, unit homework results, discussion, final test results. From this we can see that every time learners complete the corresponding part of the task, they will have the corresponding results. In the future, first, we should strengthen the construction of the evaluation system of MOOC from the content and form of evaluation or test, and improve the scientificity and quality of MOOC evaluation; second, teachers should carefully choose the appropriate assessment methods according to the types and characteristics of courses, add rich types of homework and test questions, optimize the components of assessment standards, and add appropriate evaluation methods In order to keep the freshness of students' study, we should pay more attention to the assessment.

5.3   Improve MOOC Management and Service

This study found that the management service satisfaction is slightly higher than other parts and is positively correlated with overall satisfaction. This requires us to continuously improve the MOOC management level: One is that the teaching side should give learners timely and accurate learning reminders. At present, there are a few courses that give certain reminders during the course, for example, send emails to inform learners before uploading each unit video. Most courses only send e-mails to remind learners before the course starts. It is necessary to send e-mails to remind learners in time during the course, which helps learners to reduce missed course learning due to other reasons. The second is that relevant institutions should establish a credit transfer mechanism as soon as possible. A prominent problem in the development of MOOCs in China is the relatively high dropout rate [16]. To solve this problem, the key lies in the certification system. The third is that the teaching party should provide good consulting services. Many MOOC platforms and courses need to be improved in terms of providing consulting services to learners. The teaching party shall provide learners with comprehensive consulting services, including course registration, course resources, course learning and assessment, etc., and open a variety of consulting service methods, such as consultation through telephone, email, and virtual communities. Promote the enthusiasm of learners.

5.4   Improve MOOC Learning Support Level

In some MOOC courses, the phenomenon of network failure and malfunction of multimedia equipment such as audio and projection are common. Good interface design and clear navigation system are important guarantees for learners to study smoothly. Therefore, the design and production of learning platforms and more collaboration among learners should be strengthened. A good MOOC learning platform can not only enhance students' interest in learning, but also enhance the quality and effectiveness of students' learning. In addition, good collaboration between learners is also very important for improving MOOC learning effects, and learners should be fully guided to conduct in-depth communication and knowledge construction. At present, MOOC is still in the initial stage of development, and the interaction between learners and learners is still very imperfect. An important reason is that the

MOOC's learning group is unstable. There may be students joining the MOOC at any time, and there may also be students who drop out at any time. This affects the stability of the interaction between MOOC learners. Communication and discussion between learners can also play an important role in the construction of knowledge. Teachers and platforms should encourage active interaction between learners, give full play to the role of learners' collaborative learning, and continuously improve the effectiveness of MOOC learning.

References

1. Yonglin Y, Shirong Z, Tao D et al (2014) From "MO Class" to "Small and Micro Class", look at the application of big data in teaching [J]. Mod Educ Technol (12):45–51
2. Dan H (2013) MOOC: subversion and innovation-a summary of the 4th "China distance education young scholars forum" [J]. China Dist Educ 11:5–14
3. Ministry of education "guidance on the organization and management of online teaching in Colleges and universities during the period of epidemic prevention and control", 2020.02.05
4. Jinping X (2020) Made important instructions for postgraduate education [EB/OL]. http://www.cdgdc.edu.cn/xwyyjsjyxx/sy/syzhxw/284888.shtml
5. Huafei W, Wenmin F (2005) Research on the content system of university customer satisfaction [J]. J Liaoning Educ Admin Coll 22(9):28–29
6. Yingchuan Z (2018) A survey of the satisfaction degree of MOOC and its influencing factors of college students [J]. Higher Educ Res 39(02):73–78
7. Xu F, Xiangping C, Reformxue Y (2016) Satisfaction research of MOOC learning support services: based on the structural equation model [J]. Open Educ Res 22(05):76–85
8. He B, Xinhui H (2020) Why the American MOOC degree program succeed: a probe into the operating system and mechanism [J]. Mod Dist Educ Res 32(03):60–68
9. Ling Q, Xia W, Mingmei Z (2015) Analysis of the key elements and strategies for the successful design of MOOC——taking Harvard University's ChinaX course as an example [J]. Educ Res 36(08):23–29
10. ISO 9241-210.Ergonomics of Human-System In-teraction-Part 210: Human-Centred Design for Interactive Systems [S] (2010)
11. Wenjiao Q, Guang Z, Wei L, Shiling JH (2017) Empirical study on the cognition and satisfaction of MOOC among college students [J]. Chin Univ Teach (08):87–91
12. Guofeng L (2015) The construction of flipped classroom of multi-evaluation literature retrieval course combined with MOOC [J]. Libr Sci Res 23:7–14
13. Li, Shengbo, Li C, zhengqinhua, (2016) A survey of MOOC curriculum design in China [J]. Open Educ Res 2:46–52
14. Anzaroot S, McCallum A (2013) UMass citation field extraction dataset. http://www.iesl.cs.umass.edu/data/data-umasscitationfield. Accessed 27 May 2019
15. Qinhua ZL (2016) An Empirical Study on the relationship between interaction center and interaction quality in MOOC. Audio Vis Educ China 2:58–63
16. Defang L (2016) The key to breaking the high dropout rate of MOOC lies in the certification system [EB / OL]. http://www.finance.ifeng.com/a/20160620/14507778_0.shtml. Kleinberg JM (1999) Authoritative sources in a hyperlinked environment. J ACM 46(5):604–632. https://doi.org/10.1145/324133.324140

A Goal Analysis and Implementation Method for Flipped Classroom Instructional Design

Bo Su, Wei Zhang, Yang Ren, and Le Qi

Abstract When developing teaching plans, the goals are often set in line with one's expectations but are not very actionable when implemented, and it is difficult to form a system for post-implementation evaluation. In this paper, by drawing on the teaching goal analysis technique to decompose the desired teaching goals into operable steps and quantifiable indicators in a step-by-step and planned manner, a simple measurement and data collection system is built on this basis, and the data analysis of the collected data is used to guide the adjustment and implementation of the teaching plan to ensure the steady achievement of the established teaching goals, and a complete roadmap for the dynamic implementation of teaching is constructed.

Keywords Objective analysis technique · Operability · Data analysis · Flipped classroom · Learning assessment

1 Presentation of the Problem

In the design of teaching objectives, we often use words such as emotion, attitude and value, which appropriately express our teaching objectives, but at the operational level, there is often a lack of achievable paths and tools to assess the stepwise progress of the objectives, making the same course teaching implementation plan too discrete for different instructors to implement, and it is difficult or even impossible to measure the effectiveness of teaching for the set objectives. It is difficult, if not impossible, to measure the effectiveness of instruction in achieving stated goals.

For example, in the reform of classroom teaching to achieve "three things", i.e., soulfulness, wariness, and fun, many of them design and realize "increasing" the fun of the teaching process as an important goal of instructional control1, and observe their teaching design methods. It can be found that the means to achieve this goal vary greatly, but all claim that their design is to achieve the fun of teaching. By

---

B. Su · W. Zhang · Y. Ren · L. Qi (✉)
Aviation Maintenance NCO Academy, Air Force Engineering University, Xinyang, China
e-mail: qixiaole@buaa.edu.cn

© The Author(s), under exclusive license to Springer Nature Singapore Pte Ltd. 2022 273
E. C. K. Cheng et al. (eds.), *Artificial Intelligence in Education: Emerging Technologies, Models and Applications*, Lecture Notes on Data Engineering and Communications Technologies 104, https://doi.org/10.1007/978-981-16-7527-0_20

	Number of questionnaires	Third-party evaluation	The feeling of the participants
Table 1 Questionnaire results concerning understanding of the requirement of fun	20	12	8

implementing anonymous sampling questions to a unit of front-line faculty, it was shown in Table 1 that opinions varied greatly.

Although not explicitly stated in the curriculum documents, different perspectives must lead to differences in implementation, and ultimately the degree of satisfaction of the original intent of the design of the classroom revolution of fun is bound to be different. Combining the original design intention of the flipped classroom and the curriculum team's understanding of fun in the teaching revolution, we analyze and elaborate on the prescribed actions in the design of teaching objectives, and give our views and suggestions for the reference of our colleagues. It is important to emphasize here that the ideas in this paper are only one way to try out under the banner of the classroom revolution and a design idea that our curriculum team is prepared to implement, and does not deny the validity and effectiveness of other design ideas.

2 Goal Analysis of Flipped Classroom Learning-Fun

2.1 Goal Analysis

In the process of preparing the flipped classroom lesson on the Stores Management System, we started with a goal analysis. The broad steps were divided into extraction, ranking, and integration. The steps of the goal analysis algorithm were referred to the goal analysis method in the book Goal Analysis by instructional design guru Robert Mager [1], which has the advantage of making qualitative goals actionable, with the following principles and steps.

1. Write down the goal.

   The teaching should be fun.

2. Exhaustive enumeration allows you to agree on all elements in the target:

   All students attend to the teaching easily and keep so for a longer and longer time.
   Students should be delighted during the process.
   Participation of the classroom activities should be active and effective.
   Response frequency to teachers' questions should increase.
   Students spend more time after class on the subject.
   Questions regarding the teaching content from students increase steadily.

Table 2 Sorting target elements

No.	Target elements for *Fun*
1	Increasing attending time
2	Active participation in discussion and rasing questions
3	More time spent on study both in class and after class
4	Respond to teachers and students more and to their points
5	Keeping a positive attitude
6	Get prepared for class
7	…

3. Sort the items in step 2 as is shown in Table 2.

1. A one-sentence description of each final clause after sorting.

In this step, we need to come up with a complete statement for each performance to describe the properties of the target descriptions. The properties may include the nature, quality, quantity and other features related. You may review them until you are satisfied with them all. Our example of this step is clearly shown in Table 3.

2. Test the integrity of statements.

To test the integrity, we may ask that suppose one has obtained or shown each of the proposed performances listed in earlier steps, whether or not you will consider the goal has been achieved. If you answer yes, then the analysis is done.

Table 3 Summarizing sorted target elements

No.	Target elements for *Fun*	Acceptable nature in one sentence
1	Increasing attending time	Average students spend more time attending to the teaching in class
2	Active participation in discussion and raising questions	Discussion participations and questions from students keep on increasing in number
3	More time spent on study both in class and after class	Everage time spent on the taught materials before, during and after teaching is increased
4	Respond to teachers and students more and to their points	Analysis during or after the teaching shows students display mastery of the content during interactions
5	Keeping a positive attitude	Students are happy during the process or at least delighted
6	Get prepared for class	Students spend time preparing for the teaching and interactions during the class
7	…	…

2.2 Goal Implementation

As it is generally accepted in academic and practical circles in disciplinary learning, cooperative learning can be designed to yield better learning outcomes than individual learning through the use of good group contingencies, and through cooperative learning. It can enhance positive relationships among learners, not only allowing them to demonstrate more social skills, but also sustaining learning that can develop higher-order thinking skills in terms of behavioral performance and can also increase the instructor's control over classroom management as well as increase motivation to learn [2]. Therefore, for the chapter on the basics of stores management system, we flipped the classroom session to cooperative learning, combined with the fun requirements of the classroom revolution.

The purpose of education is to create thinkers. Quality questioning will promote teacher-student synergy, which in turn will increase student engagement and learning. Quality questioning begins with recognizing the complexity of cognitive abilities and focuses attention, stimulates thinking, and leads to learning at different cognitive levels [3]. Therefore, we use quality questioning as an important tool for interesting teaching and learning. During the instructional design process, 6P framework are used to maximize its effects4, including prepare the question, present the question, prompt student thinking, process student responses, polish questioning practices and partner with students. The framework incorporates key questioning practices and the corresponding behaviors of high levels of student engagement in learning. These practices are derived from extensive knowledge based on questioning and student-teacher interactions. Each of the core practices influences and is influenced by the other five core practices, and they constitute the process of quality questioning.

The interesting classroom process allows participants to gain a deep understanding of the role, nature and necessary morphological composition of the suspension management system. After the online free study, the groups were able to learn through offline cooperative groups where each member was able to clearly answer the different sides (validity dimensions) of the three questions above.

Cooperative learning may consist of thinking about and answering the following questions.

1. What two aspects of equipment are required for an operational aircraft to attack a target (guide participants to analyze weapons and fire control systems)

The Fun objective can be reflected from the following indicators if there are:

Increase in the number of participants in the discussion.
Increase in speech length.
Increased relevance of the presentation to the topic.
Increased participation of students who are usually more passive.

Exemplar Fun activity can be designed as follows.

For example, students are guided to simulate an aircraft attacking a target by throwing a piece of chalk into a trash can, which leads to weapons and fire control systems.

2. If you are asked to design the fire control system which two aspects should be considered at least (guide the participants to analyze the hard and soft parts, i.e. targeting and weapon management, to introduce the concept of stores management system)

The Fun objective may use the same indicators. And the fun activities design may include guiding participants to carefully observe and writing down the whole process of a student throwing a piece chalk into the garbage can designed to the analysis of the aiming and throwing process, which is close to the teaching objectives.

Clearly summarize the essence of the weapon management system through group discussion of the topics assigned by the teacher.
Clearly summarize the morphological composition of the weapon management system by discussing the topics assigned by the teacher in groups.
Learning assessment through survey questions.
Improve the design of the offline part of the flipped classroom based on the evaluation results until the standards are met.

In a learning-oriented classroom, tasks are the means to accomplish learning objectives and indicate the form in which they are reached. In a learning-oriented classroom, mistakes and making errors are considered opportunities for learning. Teachers encourage students to ask questions about their learning, and quality questioning drives learning for both teachers and students. Goals of Fun can only be achieved through a learning-oriented classroom, designed as above as an example.

3 The Role of Learning Assessments and Their Proper Implementation

In a flipped classroom model, students spend valuable time in the classroom focusing on active learning, working together on difficult issues in the learning topic, and gaining deeper understanding [4]. This is a macro goal of instructional design, but is there a "ruler" that we can use to measure the effectiveness of learning in the classroom to "make learning more explicit, provide autonomy, and focus on learning itself"? This "ruler" is the learning assessment.

To conduct a scientific assessment of learning ability, it is first necessary to establish an observational measurement system that addresses the goals of instruction. In the case of non-commissioned officer professional education, where the instructor-to-cadet ratio is relatively high, the use of technological means combined with traditional means allows for the construction of a measurement system with a high cost-to-efficiency ratio appropriate to the curriculum. The significance of the measurement system is to establish a data-driven dynamic feedback system for teaching and learning through data collection, which allows observable key indicators on both sides of the instructional spectrum to be adjusted toward the goal through continuous adjustments.

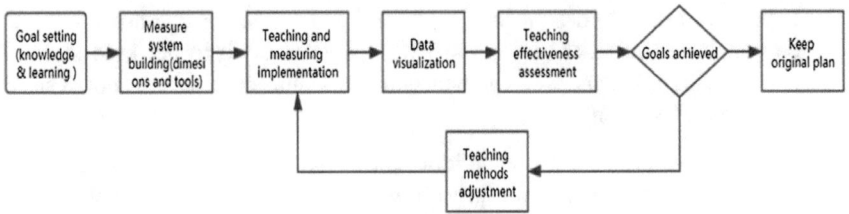

Fig. 1 Dynamic adjustment process of learning assessment implementation

For the current courses in the curriculum group, we have designed a generic assessment process implementation plan as follows.

As it can be seen from Fig. 1 that the implementation plan is actually a closed loop control engineering format.

In this loop, the very first step is to set a clear goal or a goal setting, both objective goals and process goals.

After the goal setting is determined, a measurement system is to be established, including the dimensions to measure and tools used to realize the measurement process.

Once the measurement system is established, teaching and measuring can be carried out so as to implement the planned procedure of teaching and to collect necessary data.

Educational data only do not provide values directly. That is why data visualization is needed to make the data easier to interpret and use.

Teaching effectiveness is expressed as casual relations in the form of logic [5]. There is plenty of research related that can be used for that purpose.

After the assessment is done, the temporal relations of the current effectiveness and that of the past can be compared to see if adjustment is needed or not, i.e., the trajectory of our teaching effectiveness can be observed and monitored to lay a solid foundation for future adjustment of teaching details.

Combining the above analysis, a knowledge objective and a formative objective were selected specifically to introduce the following case study.

3.1 Define the Objectives

Understand that the "stores management system" is the bridge and hub between the pilot and other airborne equipment and the airborne weapons. The goal of the itinerary is to be able to speak positively based on the discussion and the instructor's prompts, and to speak in a way that leads to the understanding of "bridges and hubs". Since we are taking fun as the goal of the study and the indicator of improvement, the learning assessment is designed more for this goal as well. However, the overall goal of the course, which is the basis and vehicle for achieving interest, must also be fully and effectively reflected in the design.

3.2 Define Measurement Dimensions and Instruments

Mathematical methods are used to describe and value student learning behaviors. Whereas learning measurement focuses on objectively quantifying the extent to which students achieve instructional goals, learning assessment focuses on how valuable the results of the measurement are. The fundamental difference between measurement and assessment of learning is whether or not learning behaviors are valued. In the teaching and learning process, the two are not opposites, but are closely related, with assessment generally being based on measurement, but measurement itself not being a substitute for assessment.

Integrating the specific descriptions of the teaching objectives obtained through the goal analysis in the previous section, the measurement indicators were set as follows. Since we are to carry out the teaching in a flipped class setting, more focus is placed on cooperative learning and interactions related activities.

Measurement of knowledge objectives: accuracy of responses to questions following cooperative learning.

Process Objective Measure: The number of people involved in the discussion of responses, the extent of coverage and accuracy during cooperative learning.

3.3 Teaching Implementation, Simultaneous Implementation of Measurement Records, and Knowledge Objective Measurement

Due to the presence of learning assessments, the teaching activities are actually divided into two major parts: teaching implementation and measurement implementation. This solution aims to find a generic and feasible flipped classroom implementation method, and therefore uses a combination of real-time and non-real-time methods for measurement implementation to save manpower. Real-time measurements consist of the teacher's recording of the quality parameters of student participation in the interaction, including the relevance of responses, initiative, frequency, and number of active questions. Non-real-time measurement recording consists mainly of on-site video equipment recording the interaction process, which is combined with real-time recording by measurement analysts in the classroom to compile and derive measurement data sets. The two activities, instructional activities and learning assessment data measurement, are sometimes serial and sometimes parallel, and change depending on the specific design of the instructional activity, requiring a more tedious process of conceptualization and rehearsal at the beginning of the instructional design, and no major increase in workload at a later stage once the demonstration is successful.

3.4 Data Visualization

Data visualization, is the scientific and technical study of the visual representation of data. Among other things, this visual representation of data is defined as a kind of information extracted in some summary form, including various attributes and variables of the corresponding unit of information [6]. It is a concept that is in constant evolution and its boundaries are constantly expanding. It refers mainly to technically advanced technological methods that allow visual interpretation of data through representation, modeling, and display of stereoscopic, surface, attribute, and animated displays using graphics, image processing, computer vision, and user interfaces. Data visualization covers a much wider range of technical approaches than special technical approaches such as stereo modeling.

Data visualization aims primarily at communicating and communicating information clearly and effectively with the help of graphical means. In order to effectively communicate ideas, aesthetic form and function need to go hand in hand to achieve insight into a rather sparse and complex data set by visually communicating key aspects and features. In our case, The data were recorded in an EXCEL sheet and displayed visually using Excel or Tableau software [7].

As is shown in Fig. 2, number of participants in classroom discussion increased steadily as time went by, indicating that the teaching design is fun to students and they are willing to be involved in the designed activities.

Assuming that the above results do not prove that the designed instructional activities have a direct causal effect on participants' engagement, the present Fig. 3 further reveals from the data the increasing relevance of student responses and discussion content in the instructional activities.

Similarly, the monotonic increase in the time trajectory plot, Fig. 4 of the number of active discussions once again proves that there is indeed a causal relationship with a high correlation between our teaching activities and students' positive responses.

As is shown in Fig. 5, for students who were less motivated in classroom activities, we also captured their progress in non-real-time classroom monitoring, and we monitored their cumulative time in classroom activities increasing in the middle

Fig. 2 Demonstration of number of participants temporal trajectory

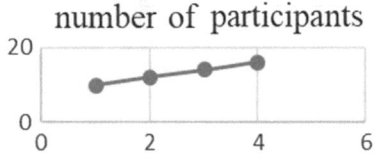

Fig. 3 Demonstration of response content relevance temporal trajectory

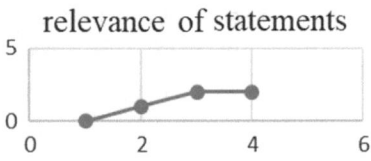

Fig. 4 Demonstration of active discussion temporal trajectory

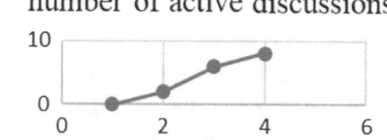

Fig. 5 Demonstration of average length of speech temporal trajectory

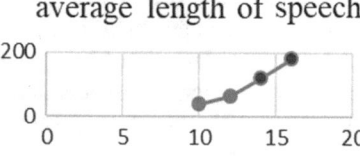

and later part of the course, indicating that this group of students had significantly increased their classroom participation and fully demonstrated that our teaching was fun indeed.

3.5 Data Analysis and Prediction

The method used for data analysis is mainly visual inspection, which shows the results of the vector and quantitative assessment of the effect by comparing it with previous participant performance. The baseline and baseline logic in health assessment is to be used here.

Baseline logic can be defined as a response pattern whose values measured over a period of time show very little variation as the basis of experimental reasoning for behavioral analysis, containing prediction, validation, and replication.

Stable state strategy consists in exposing the subject to the same known state repeatedly before exposure to the next state, while minimizing or controlling for any influence factor that is not relevant to the behavior, in order to obtain a stable response pattern.

The raw information obtained by adding variables to the continuous measurement behavior and measuring the behavior when the variables are not present is called the baseline [8]. From a purely scientific or analytical point of view, the primary purpose of establishing a baseline is to use the subject's performance without the presence of the variable as an objective basis for detecting future effects caused by the presence of the variable.

3.6 Adaptation of Teaching Implementation

As you can see from the chart above, if the data show that, in the case of our classroom implementation, if the established indicators are steadily improving in the direction we expect, then our teaching is effective relative to our goals, and if the data show that the desired goals are not changing as expected, then it is important to analyze the various aspects of course preparation, and then make and adjust to implement monitoring and data analysis again.

A functional relationship is presented when observed behavioral changes can be attributed to specific manipulations imposed on the environment. Systematic manipulation of the subject's environment (self-variant) so that predictable behavioral changes (dependent variables) are reliably and repeatedly produced demonstrates experimental control. In order for this attribution to be appropriate, the researcher must control for two sets of environmental variables. First, the researcher must present, remove, and/or vary the value of the dependent variable to control for the independent variable. Second, the researcher must control all other aspects of the experimental environment, i.e., the intervention variables, to maintain their consistency and to prevent unplanned environmental changes. These two operations provide a second definition of experimental control—the precise manipulation of the independent variables and the maintenance of consistency with respect to aspects of the experimental environment [9]. This lays the foundation for our adjustment of our teaching strategies.

4 Conclusion

A flipped classroom instructional design with learning-oriented quality questioning and cooperative learning as the primary means was adopted through a goal analysis approach for several elements of the classroom revolution. The analyzed instructional goals were used as a criterion for the design of a learning-based evaluation system that uses a combination of real-time and non-real-time measures of learning behavior without increasing the workload of instructional preparation. Data were visualized to demonstrate the temporal changes of key indicators using baseline logic and visual analysis to flexibly adjust the instructional design accordingly. The experimental data show that the method used is an operational and feasible generalized paradigm for flipped classroom instructional design, which can be used by peers for reference when teaching similar courses.

Although there is no direct implementation of traditional online teaching, under the advocacy of the teaching revolution in schools, we actively review the online and offline contents of the flipped classroom of the curriculum team, and implement and design it according to the direction and goals of the reform. The design and determination of objectives, the establishment and implementation of the measurement system, the collection and visualization of teaching process data, the dynamic adjustment scheme of teaching process based on data visualization analysis, and the implementation of learning assessment are an attempt to implement the teaching revolution. The following is a summary of our main findings.

1. The need for operationalized definitions of instructional goals can be obtained through goal analysis techniques.
2. Affective attitudes and values can have exogenous forms of measurement dimensions that can be quantified by designing observational systems for measurement.
3. Targeted collection and analysis of teaching process data are useful in guiding the achievement of established teaching objectives.

References

1. Mager RF (1972) Goal analysis [M], pp 71–80
2. Henley A et al (2017) Function-altering effects of rule phrasing in the modulation of instructional control. Anal Verb Behav 33:24–40
3. Walsh JA, Sattes BD (2005) Quality questioning: research-based practice to engage every learner [J]
4. Strelan P et al (2020) The flipped classroom: a meta-analysis of effects on student performance across disciplines and education levels. Educ Res Rev 30:100314
5. Alberto P, Troutman AC (2013) Applied behavior analysis for teachers [M]. Pearson, Upper Saddle River, NJ
6. Junfang G (2010) Research on the visualization of educational information data [D]. Shanghai Normal University
7. Hoelscher J, Mortimer A (2018) Using Tableau to visualize data and drive decision-making [J]. J Acc Educ 44:49–59
8. Christ TJ, Christ JA (2006) Application of an interdependent group contingency mediated by an automated feedback device: an intervention across three high school classroom [J]. Sch Psychol Rev 35(1):78–90
9. Leitch DB (2018) Applied behavior analysis for homeschoolers

Instruction Models of Located Cognition and Their Effectiveness Using Flipped Learning in Initial Training Students

Fabiola Talavera-Mendoza, Fabián Hugo Rucano Paucar, Rolando Linares Delgado, and Ysabel Milagros Rodríguez Choque

Abstract The current moments lead us to search for different strategies and modalities for the teaching-learning process, where face-to-face work has prevailed for a long time until the arrival of the pandemic (Covid-19) to migrate to a ubiquitous job. The objective of the study is to analyze the relationship that exists between the instructional cognitive model and its implications of using flipped learning as a methodology to acquire meaningful learning where teachers assign instruction through a recorded video. This study was carried out with a quantitative approach, with a correlational descriptive design, 60 students from a public university voluntarily participated, the same ones who are finishing their higher studies and were selected with an intentional non-probabilistic sampling, the instruments used were two surveys, one for flipped learning and the other for cognitive instructional design. The results reveal a level of high cultural relevance in the learning environment developed by the teacher who uses flipped learning with university students. It is concluded that from the perception of the students there is a positive correlation and that it is conditioned that flipped learning will be successful or adequate to the extent that the cognitive instructional approach is efficient.

Keywords Instructional model · Flipped learning · Ubiquitous environments

F. Talavera-Mendoza (✉) · F. H. R. Paucar · Y. M. R. Choque
Universidad Nacional de San Agustin de Arequipa, Arequipa, Perú
e-mail: ftalaveram@unsa.edu.pe

F. H. R. Paucar
e-mail: frucano@unsa.edu.pe

Y. M. R. Choque
e-mail: yrodriguezc@unsa.edu.pe

R. L. Delgado
Universidad Tecnologica del Perú, Cercado de Lima, Perú
e-mail: c16200@utp.edu.pe

E. C. K. Cheng et al. (eds.), *Artificial Intelligence in Education: Emerging Technologies, Models and Applications*, Lecture Notes on Data Engineering and Communications Technologies 104, https://doi.org/10.1007/978-981-16-7527-0_21

1 Introduction

We are facing a need to seek the autonomy of our students through action and reflection, in ubiquitous environments, which lead students to create their own spaces to analyze, infer and create new knowledge. Therefore, universities are not alien to this process, they have found the need to look for ways to interact with students, generating relevant changes or looking for turning points about the use of technology and the sustainability of work with an appropriate methodology [1]. In this way, distance-learning brings with it, challenges of how to organize learning environments that are open, flexible, autonomous, and realistic within the instructional model, guided by self-directed learning that allows self-regulation [2].

Then cognitive instructional design in virtual training takes on value and relevance by linking two dimensions: technological and pedagogical. In the case of the first one, it implies the possibilities and limitations for the management of a virtual platform, software applications, multimedia resources, etc.; and the second related to the knowledge of user characteristics, analysis of the objectives and/or competencies of virtual training, with start, development and closure activities including the evaluation of processes and learning outcomes [3].

In this way, it is important to know-how, these academic activities are being carried out virtually in universities accustomed to face-to-face work, the way students perceive and engage in the effort of teachers to prepare their instructional teaching material, which must be assimilated by the student, taking into account that the instructional way of offering knowledge has changed, is thus emerging as a need to know the effectiveness of the cognitive instructional model in flipped learning. It leads to transforming classroom space into a dynamic and interactive environment for the application of concepts and their creative application of content [4, 5]. Therefore, teachers do not need to give full lectures in class but should be reversed by freeing up that time to participate in practical activities such as discussions, group discussions, use of material; that lead to collaborative, argumentation, and reflection activities [6], where they learn by doing [7], so that students anytime, anywhere can access information [5, 8]. Besides, a flip-out environment encourages students to participate in more collaborative and project-based activities [5, 6], accompanied by formative assessments that include questionnaires, tests, and discussion forums. Where the role of the teacher opens up to observe, diagnose and provide feedback to students in real-time [9, 10].

The studies carried out and analyzed slipped around two aspects: (a) understanding the perspectives and attitudes of students and teachers about classrooms and (b) the behaviors of using it. About the first point, the works highlighted their importance around feedback, self-learning, and the thoughtful thinking of flipped, in this perspective we sought to know the effects of online academic help and student participation in achieving self-efficacy and self-directed learning [8]. On the other hand, the role of the teacher in narrated lectures is added, promoting teaching practices of greater influence for individualized learning [11]. In this sense, the analysis of previous knowledge and content is incorporated [10] and for greater internalization of flipped learning a model of appreciative research was used perceived from four phases: discovery, dream, design, and destination [12]. It was finally addressed from motivational characteristics and academic performance by significantly linking the value of homework, learning control, and self-efficacy [13].

The second aspect was directed to the work based on Ajzen's theory to explain and predict human behaviors based on intention, subjective norms, and behavioral control [14]. Also, learning styles instruments were added [15] and in another study, personality traits were incorporated to relate it to the level of satisfaction with flipped learning [4].

How can it be seen? Most of the articles published on flipped learning focus on effectiveness, efficiency, motivation, academic performance, and behavior. Not many studies of the cognitive regulations of the students and the relevant problems in the interactions have been carried out yet [16]. We will delve into the level of effectiveness in the knowledge and understanding of the instructional cognitive model mediated by flipped learning in the teaching process in initial training university students of a Faculty of Education.

2 Theoretical Framework

Flipped learning allows problem-solving and the development of critical thinking [6] and consists of four pillars: (i) Flexible environment, There is no time and place limit for teachers to prepare their instructional material and students have the opportunity to view and analyze the videos as many times as they require, (ii) Learning culture, teacher-student relationship take an active role in building and acquiring knowledge, (iii) intentional content, according to learning needs, the teacher organizes the class and tasks with an intentionality and (iv) professional educators, seek to generate spaces of critical and argumentative reflection in interactions with students to expand, clarify or deepen a knowledge [17]. Within this context, we have that students can become autonomous people in their decisions since there is flexibility at the time of learning and is subject to the rhythms and styles of learning, which can be assumed as an advantage and as a disadvantage related to economic resources for access to connectivity [17].

When teachers make their videos, they can ensure the perfect fit of content, rigor, and personal connections [7]. In reference, the level of efficiency and perception has been worked on taking into account nine dimensions (Socio-educational, motivation, interactions, autonomy, collaboration; deepening of content; problem-solving, class time, and qualifications) demonstrating the advantage and benefit of flipped learning [18]. In parallel with this study, the direct relationship between motivation and academic performance was found based on student interactions on the platform where you can measure the time students spend reviewing material and videos, visitation rate, time spent on each topic, as well as evaluating learning control about self-efficacy, intrinsic motivation, extrinsic motivation and value of the task [13]. In this line, we worked three moments in the university environment: Preparation, inter-action, and consolidation where each of them will be described with their components of interaction in this process of distance education.

3 Located Cognition

Therefore, cognition refers to the activity in the mind of the individual based on the activity, context, culture, and situations planned through social interaction. In this sense, the instructional design is based on a constructivist design that must consider interpersonal and gestural aspects in the presentation of materials for perception, as well as the enculturation process in which students gradually integrate their social practices into a community to build meaning and sense of what they have learned [2]. Furthermore, the expert is the one who models and promotes certain knowledge in the novice [19].

The cognition located does more than just challenge our ideas about what is considered "cognitive": it tells us that we could benefit from looking much more closely at the behavior. Behavior is considered an opportunity to learn, for example, in exploring or searching for information, or a cognition output [20]. The design of learning environments currently leads us to propose the increase of student partic-ipation in tasks oriented to decision-making, problem-solving, and intervention in practice; that is, instructional systems are required in constant dialogue with the student so that they can update their progress, performance, and expectations. The designers of the instruction will also be the students themselves [21]. In this way, there is a direct connection between the theory of located learning and cognitive learning [22].

Faced with the need to carry out a remote work by COVID-19, the university's curriculum programs face a common problem, how to articulate the contents to highly significant, contextualized actions, to involve students in an agile and collaborative way for an emotional and welfare regulation that lead to autonomy, with the applica-tion of critical, reflective and conscious tasks. Therefore, it is considered important to base the practice on the development of cognition located that have a common goal based on processes of belonging, participation, and practice [23]. Also, it seeks to promote apprenticeships with scaffolding, mutual negotiation of meanings, and the construction of knowledge [2].

The instructional approach, also known as instructional design, is a discipline that links learning theory with educational practice, is interested in understanding and improving the teaching-learning process, constitutes a conducive space for students to obtain computer resources and teaching aids to interact and carry out activities aimed at established educational goals and purposes [5].

Natural environments influence natural cognitive states through real sensory inter-actions, sociocultural perception learned through incorporated practices is related to sociocultural expectations [24]. To work in authentic scenarios, two dimensions will be taken into account: (i) Cultural relevance, using aspects relevant to their culture as examples (discussions, demonstrations, illustrations), and (ii) Social activity, with a participatory and collaborative, tutored work for the solution problems (debates, class discussion, guided discovery) [2, 25].

Six instructional approaches are presented: (1) Decontextualized instruction, the teacher is the protagonist and transmitter of the information associated with the traditional approach, in which abstract and decontextualized readings are usually provided, (2) Collaborative analysis of invented data, based on exercise task, where content and data are alien to students' interests, (3) Understanding-based instruction to relate concepts and procedures, (4) Collaborative analysis of relevant, student-centered, and real-life data that seeks to induce reasoning and critical discussion together, (5) Situated simulations that promote the motivation of simulated problems or cases taken from real life, to develop the type of reasoning and mental models of ideas and concepts and (6) Learning in situ, seeks to develop skills and knowledge of the profession, solving social problems or of the community of belonging. It emphasizes the usefulness or functionality of what has been learned and learning in real settings [2] (Fig. 1).

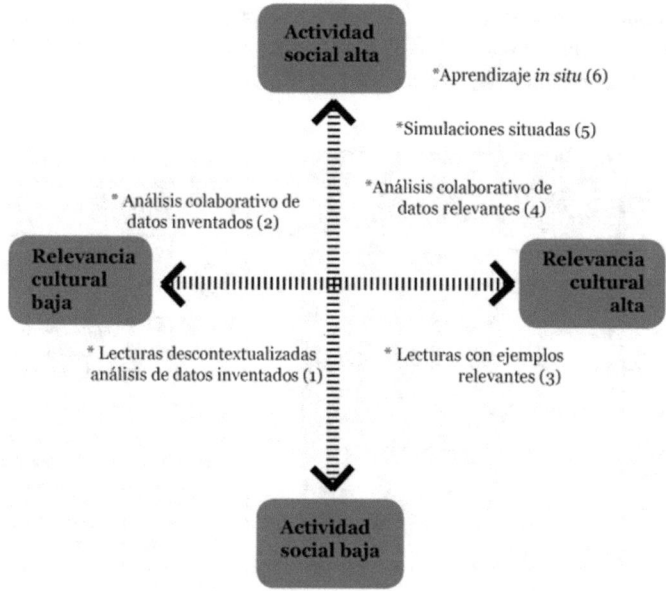

Fig. 1 Instructional approaches to located cognition (*Source* Taken from Derry et al. [26])

4 Method and Materials

The research carried out led us to measure the effectiveness of the cognitive instructional design based on inverted learning in the Thesis Seminar Subject, during the semester I—2020, 60 university students participated, the sampling was intentional, using the inclusion criteria with students who They actively participated in the flipped learning methodology and exclusion students enrolled for the second time in the subject were not considered. It was developed in a quantitative approach, which is a sequential, probative, orderly, and rigorous process [26], to arrive at general explanations [27]; the design was descriptive correlational. Participation was voluntary and anonymous. All the forms were sent online in Google forms and they have the respective permissions for being minors. The study was implemented in 16 weeks.

Procedures:

In this line, three moments were worked in the university environment: Preparation, interaction, and consolidation where each of them is described with their interaction components in this process of ubiquitous education (Fig. 2).

The five components that were carried out in the development of the learning sessions contain elements of: (Thematic planning, discussion of concepts, accompaniment, modeling, and formative evaluation) with activities that allow understanding the essence of each moment and the conscious evaluation and reflective of the opportunities generated in it, as detailed below (Fig. 3).

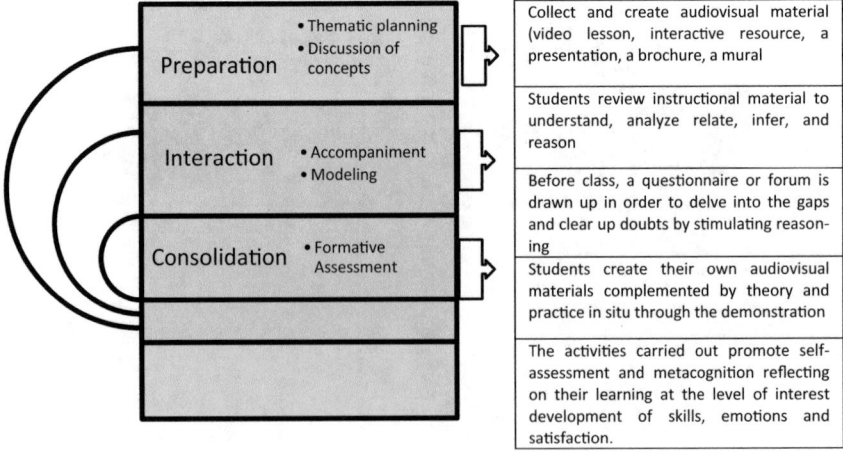

Fig. 2 Flipped learning moments (*Source* self-made. *Elaboration* Own)

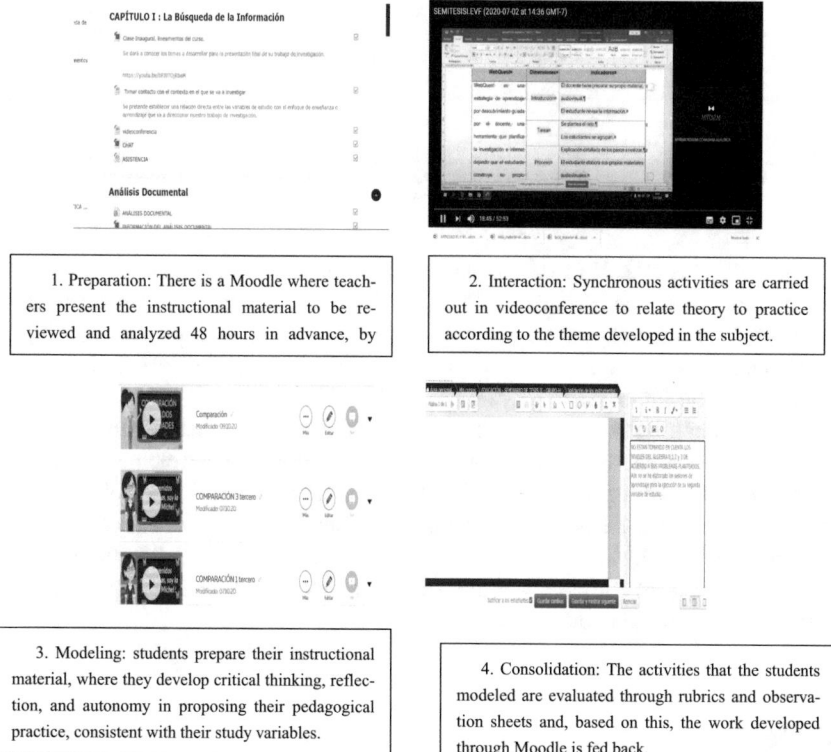

Fig. 3 Flipped learning model (*Source* Moodle DUTIC)

5 Instruments

The research used two questionnaires, one on cognitive instructional design and the second to learn about teaching in flipped learning, the completed questionnaires were encoded in a spreadsheet. A coding system was used with 05 Likert alternatives from 01 to 05: Totally agree, very agree, agree, disagree, and totally disagree. For the instructional approach, 06 dimensions and 24 indicators were considered: Collaborative analysis of invented data (03 indicators), Decontextualized instruction (06 indicators), Simulated situations (04 indicators), On-site learning (03 indicators), Understanding-based instruction (03 indicators) and Collaborative analysis of relevant data (05 indicators). For flipped learning 05 dimensions and 25 indicators were considered: Modeling (05 indicators), Discussion of concepts (04 indicators), Feedback (05 indicators), Thematic planning (05 indicators), and Evaluation (06 indicators) [28].

The psychometric validity of the content of the survey has been carried out through expert judges. The Aiken coefficient V has been applied to the experts' responses. The criterion of 10 expert judges was used, for the variable cognitive instructional

Table 1 Reliability statistics (*Elaboration* Own)

	Cronbach's alpha	Cronbach's alpha based on standardized elements	No. of elements
Located instructional approach	909	918	24
Flipped learning	975	975	25

design that has a total V of 0.986, and for the flipped learning variable you have a total V of 0.986, which is valid over 90%. For the internal consistency of the research instrument: an instructional approach to cognition located and flipped learning the Cronbach Alpha was used and answered by the 60 students, the data analyses were processed in the SPSS V.25 program (Table 1).

The Cronbach alpha applied to all 24 items of the cognition instructional approach variable was 90.9% and the reliability index based on standardized elements is 91.8%. In the Flipped Learning variable applied to the 25 items, Cronbach's alpha is 97.5% and the standardized element-based reliability index also shows us a high-reliability level of 97.5%; both research instruments are validated on 90% reliability by being at an excellent level and demonstrating that the degree to which the questionnaire elements relate to each other.

6 Results

The use of the Cognitive Instructional Approach allowed this study to reveal the scores of students' statements according to the different dimensions considered by this methodology and establish their relationship with Flipped Learning the factors representing the opinion of the group of participants with similar perceptions were extracted, thus discovering the different thought patterns expressed in their assessments.

This table shows the different dimensions, with the minimum, maximum, half, and deviation values, which are obtained from the indicators that were worked on (considering very agree plus totally agree), where more than 50% of Students consider dynamic elements in the dimensions: instruction based on the understanding and collaborative analysis of relevant data, with averages of 31.33 and 31.40, and standard deviations of 2.51661 and 5.89915 respectively and the other dimensions are They are close to and close to 50% of the students, except for the dimension of a collaborative analysis of inverted data, which only 6.1% indicates that the students agreed and totally agreed, this is due to the indicators that were worked on as The activities or tasks provided by the teacher are poorly contextualized and unrealistic. Most of the assigned tasks are mechanical and do not contribute to their professional training and the development of the subject tends to exercise and memory work with very routine or basic activities. (Table 2).

Figure 4 shows us the percentages of the results applied to the theory of Derry,

Table 2 Descriptive statistics—cognitive instructional approach located (*Elaboration* Own)

Dimensions	Indicators	Minimum	Maximum	Half	Dev. deviation
Collaborative analysis of invented data	3	2,00	5,00	3,67	1,52,753
Decontextualized instruction	6	7,00	43,00	22,33	15,33,188
Simulated situations	4	21,00	37,00	29,25	7,93,200
On-site learning	3	29,00	30,00	29,33	0,57,735
Understanding based instruction	3	29,00	34,00	31,33	2,51,661
Collaborative analysis of relevant data	5	21,00	35,00	31,40	5,89,915

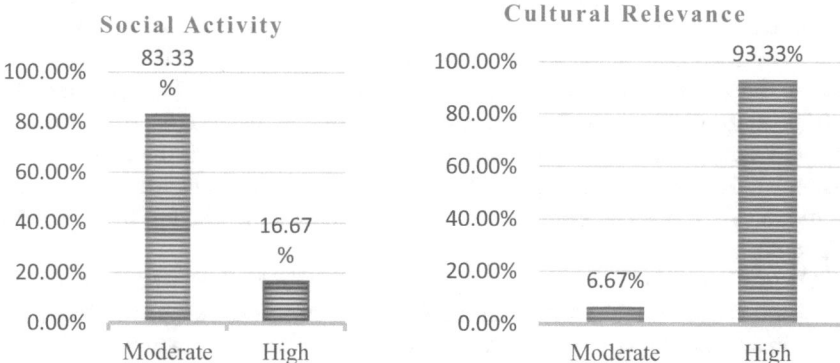

Fig. 4 Cognitive model of the instructional approach: social activity and cultural relevance (*Elaboration* Own)

Levin, and Schauble, whose starting point is the so-called motivational instruction considered through two dimensions. In social activity it shows us that 83.3% is moderate, due to the dimensions of a collaborative analysis of reversed data and decontextualized instruction that had minimal results and were affected the results of the other dimensions; and in cultural relevance, it shows us that 93.3% is high because the results of the minimum dimensions represented on regular average and the other dimensions yielded favorable results (Fig. 4).

In this table, you can see the different dimensions, with the minimum, maximum, half, and deviation values, the same ones that are obtained from the indicators that were worked. (considering the very agree plus the totally agree) with the participation of 60 students, we see that most of the results are favorable over 50% of the total; where the evaluation dimension has a higher result with an average of 34 and a standard deviation of 5.51362; in general we have a minimum standard deviation which indicates that the results are close to average (Table 3).

To choose the statistical test the normality test is first performed considering the database of the variable instructional approach and flipped learning product of the

Table 3 Descriptive statistics—flipped learning (*Elaboration* Own)

Dimensions	Indicators	Minimum	Maximum	Half	Dev. deviation
Modeling	5	24,00	32,00	28,80	3,11,448
Discussion of concepts	4	26,00	33,00	30,25	3,09,570
Accompaniment	5	28,00	32,00	30,40	1,81,659
Thematic planning	5	26,00	43,00	32,00	7,03,562
Evaluation	6	25,00	41,00	34,00	5,51,362

survey, considering 95% confidence, therefore the P-value or value of alpha is 0.05, with the application of the software SPSS V.25, as shown in the following table.

In the Table, the results of the normality test of both variables are presented, where data less than 50 take the Shapiro-Wilk statistical test. The Shapiro-Wilk results corroborate that the significance level of both variables is 0.019 and 0.163, that is, the variable cognitive instructional approach located, the P-value is 0.019 < 0.05, then the distribution of the study variable is not normal; and the study variable flipped learning, the P-value is 0.163 > 0.05, then the distribution of the study variable is normal, therefore, it is appropriate to apply one of the non-parametric tests that, due to its breadth of analysis, corresponds to the non-parametric Rho test by Spearman. (Table 4).

Correlation analysis is a statistical procedure that determines whether two variables are related or not, and Spearman's Rho correlation analysis indicates the strength and direction of the association between two statistical variables, the values of the independent variable, instructive approach, vary systematically concerning the values of the dependent variable Flipped Learning.

The following table presents the results and the correlation analysis between the variables calculated in the SPSS V.25 software.

According to the results on the calculation of the Correlation Coefficient, we have to: The cognitive instructional approach variable has a correlation level of 0.558 concerning the flipped learning variable; that is, the correlation is positive moderate, and significant at the level of 0.01 (bilateral).

Table 4 Normality tests (*Elaboration* Own)

	Kolmogorov-Smirnov[a]			Shapiro-Wilk		
	Statistical	gl	Sig	Statistical	Gl	Sig
Located instructive cognitive approach	211	24	007	897	24	019
Flipped learning	142	24	200[*]	940	24	163

*This is a lower limit of true significance
[a]Lilliefors significance correction

Table 5 Correlations (*Elaboration* Own)

			Instructional approach	Flipped learning
Spearman's Rho	Located instructive cognitive approach	Correlation coefficient	1,000	558[**]
		Sig. (bilateral)		005
		N	24	24
	Flipped learning	Correlation coefficient	558[**]	1,000
		Sig. (bilateral)	005	–
		N	24	25

Correlation is significant at the 0.01 level (bilateral)

The variable instructive approach to the flipped learning variable has a significance level with a P-value of $0.005 < 0.05$, i.e. we check the hypothesis raised by the research. In this sense, we can assert that flipped learning will succeed or be appropriate to the extent that the instructional approach is efficient and appropriate (Table 5).

7 Discussion

It has been found that the level of effectiveness in knowledge and understanding of the instructional cognitive model mediated by flipped learning in the teaching process in students has favorable results. It is denoted that mediation in the activities designed by teachers allows to create environments that facilitate knowledge-building processes and not using an appropriate instructional design in the virtual modality will not cause an impact on the training process [29].

According to the literature presented with it, as an innovative practice has allowed migrating from presence to ubiquitous spaces, showing satisfaction, motivation, efficiency, and better academic performance [5, 8]; that allows us to validate our study that does, presents a good instructional material that enhances cognitive abilities generates a positive relationship. Therefore, cognitive commitment is linked to overcoming course requirements and a preference for challenges [30]. As well as understanding that, if the inverted learning is well structured, it allows to reduce the cognitive burden [31]. This has been demonstrated and evidenced, since, the non-presence, carried with a good methodology is complemented favorably by the learning-teaching process, helping to provide sustainability for future experiences for the development and achievement of the competencies of the subjects.

Similarly, studies reveal high rates in collaborative work and situation-driven understanding-based instruction that allows complementarity between theory and practice in achieving common and group goals, as well as the achievement of self-regulated learning, by increasing individual learning time, modeling being another procedure for teaching, which leads to planning, execute and evaluate what has been learned, as long as it is taken to local contexts, thus being a strength within the obtained values. Corroborating that peer interaction and timely and evaluative teacher feedback influences the improvement of learning [30].

The predominance of the cognitive instructional approach was inclined towards high cultural relevance, students are in the process of enculturation and adaptation to this new mode of work where extrapolation and adaptive learning of content to reality are being gradually inserted.

An interesting finding is the prevalence of the collaborative analysis instruction model of relevant data followed by understanding, focusing on real contexts to solve problems and develop skills. In this way, the approach of instructional cognitive learning for virtual learning environments allows acquiring different levels of skills, strategies, and scaffolding necessary to generate learning [32], which can promote the decrease of academic performance of students [33].

8   Conclusions

We can conclude that this study has allowed us to identify the relationship between the cognitive instructional model and Flipped Learning, in the perception that university students have that they are preparing to teach, which is very positive. Corroborated in a study to measure cognitive load and the implementation of Flipped Learning turned out to be less than that of the training of a face-to-face class [31], as the review of the instructional materials in advance reduces the cognitive load. Being a highly significant factor, because the student has to read and understand previously to be able to apply the theory in practice and this process at the beginning demanded a lot of effort and later it became a good practice in the development of the subject.

The implications of this study are based on situated cognitive learning and collaborative instruction, which predominated from the perception of the students, with commitment and participation in the classroom being the basic aspects to generate positive learning, demonstrating a high cultural relevance, where the apprentices appropriate cultural practices and tools, giving importance to scaffolding, construction and mutual negotiation of the acquired learning.

Likewise, it is necessary to carry out more in-depth research on the cognitive instructional model and Flipped Learning to continue presenting scientific evidence that allows improving the work of university teachers in the teaching-learning process. As well as the attitude and emotional degree that these processes imply and the intrinsic cognitive load that students develop, taken to a field of personalization of learning with artificial intelligence.

References

1. Guadalupe Martinez-Borreguero JJ-V-N-C (2020) Desarrollo de una estrategia didáctica basada en objetos de aprendizaje para el mejoramiento del proceso educativo. Sustainability 2–23
2. Díaz Barriga Arceo F (2003) Cognición situada y estrategias para el aprendizaje significativo. Revista electrónica de investigación educative 5(2):1–13
3. Coll C, Mauri T, Onrubia J (2008) Los entornos virtuales de aprendizaje basados en el análisis de casos y la resolución de problemas. En Psicología de la educación virtual, editado por C. Coll y C. Monereo. España: Morata
4. Kim JY (2018) A study of students' perspectives on a flipped learning model and associations among personality, learning styles and satisfaction. Innov Educ Teach Int 55(3):314–324
5. Castro VFR, Castro MIR, Arias FJT, Jalca JEC, Pin ÁLP, Pilay YHC, Nazareno OEG (2019) El Flipped Learning, El Aprendizaje Colaborativo Y Las Herramientas Virtuales En La Educación, vol 43. 3Ciencias
6. Bergmann J, Sams A (2014) Flipped learning: Gateway to student engagement. International Society for Technology in Education, Arlington, VA
7. Fulton KP (2012) 10 reasons to flip. Phi Delta Kappan 94(2):20–24
8. Chyr WL, Shen PD, Chiang YC, Lin JB, Tsai CW (2017) Exploring the effects of online academic help-seeking and flipped learning on improving students'
9. Bauer-Ramazani C, Graney JM, Marshall HW, Sabieh C (2016) Aprendizaje invertido en TESOL: definiciones, enfoques e implementación. TESOL J 7(2):429–437
10. Kang K (2018) Exploring teaching and learning supporting strategies based on effect recognition and continuous intention in college flipped learning. J Korea Converg Soc 9(1):21–29
11. Brewer R, Movahedazarhouligh S (2019) Flipped learning in flipped classrooms: a new pathway to prepare future special educators. J Dig Learn
12. Sargent J, Casey A (2020) Flipped learning, pedagogy and digital technology: establishing consistent practice to optimise lesson time. Eur Phys Educ Rev 26(1):70–84
13. Xiu Y, Thompson P (2020) Flipped university class: a study of motivation and learning. J Inf Technol Educ Res 19:041–063
14. Umutlu D, Akpinar Y (2020) Effects of different video modalities on writing achievement in flipped English classes. Contemp Educ Technol 12(2):ep270
15. Yoshida H (2019) Flipped learning for pre-service teacher education: with focus on instructional design for elementary and secondary education. In: inted2019 Proceeding, pp 5718–5727. IATED
16. Zheng XL, Kim HS, Lai WH, Hwang GJ (2020) Cognitive regulations in ICT-supported flipped classroom interactions: an activity theory perspective. Br J Educ Technol 51(1):103–130
17. Üğüten SD, Balci Ö (2017) Flipped learning. J Suleyman Demirel Univ Inst Soc Sci 26(1)
18. Moreno-Guerrero AJ, Romero-Rodríguez JM, López-Belmonte J, Alonso-García S (2020) Flipped learning approach as educational innovation in water literacy. Water 12(2):574
19. Díaz Barriga F (2005) Principios de diseño instruccional de entornos de aprendizaje apoyados con TIC: un marco de referencia sociocultural y situado. Tecnología y comunicación educativas 20(41):4–16
20. Prtchard JD (2018) Situated cognition and the function of behavior. Comparat Cogn Behav Rev 35–39
21. Barriga FD (2005) Principios de diseño instruccional de entornos de aprendizaje apoyados con TIC: un marco de referencia sociocultural y situado. Revista Tecnología y Comunicación Educativa 5–16
22. Tsui P (2020) Sustainable development of hotel food and beverage service training: learning satisfaction with the situated cognitive apprenticeship approach. Sostenibilidad 12(5)
23. Niemeyer B (2006) El aprendizaje situado: una oportunidad para escapar del enfoque del déficit. Revista de educación 341:99–121
24. Schilhab TS (2019) Socio-cultural influences on situated cognition in nature. Front Psychol 10:980

25. Derry S, Levin JR, Schauble L (1995) Stimulating statistical thinking through situated simulations. Teach Psychol 22(1):51–56
26. Sampieri RH (2014) Definiciones de los enfoques cuantitativo y cualitativo, sus similitudes y diferencias. RH Sampieri. Metodología de la Investivación
27. Cascante LG (2013) Metodología de la investigación educativa: posibilidades de integración. Revista Comunicación 182–194
28. Ramírez H (2016) Los estilos de aprendizaje. Mercurio
29. Belloch C (2017) Diseño instruccional
30. Lo CK, Hew KF (2020) A comparison of flipped learning with gamification, traditional learning, and online independent study: the effects on students' mathematics achievement and cognitive engagement. Interact Learn Environ 28(4):464–481
31. Karaca C, Ocak M (2017) Effect of flipped learning on cognitive load: a higher education research. J Learn Teach Dig Age 2(1):20–27
32. García-Cabrero B, Hoover ML, Lajoie SP, Andrade-Santoyo NL, Quevedo-Rodríguez LM, Wong J (2018) Design of a learning-centered online environment: a cognitive apprenticeship approach. Educ Tech Res Dev 66(3):813–835
33. Eze TI, Obidile JI, Okotubu OJ (2020) Effect of cognitive apprenticeship instructional method on academic achievement and retention of auto mechanics technology students in technical colleges. Eur J Educ Stud

Ethical, Socio-cultural and Administrative Issues in Education of an AI Era

Controversial Issues Faced by Intelligent Tutoring System in Developing Constructivist-Learning-Based Curriculum

Yue Wang

Abstract This article aims at exploring the identified benefits and challenges when integrating Educational Artificial Intelligence (AIEd) into contemporary Constructive-Learning-Based Curriculum design. It is argued that the Intelligent Tutoring System, which lies at the centre of AIEd, could help enhancing learners' constructive learning experience by providing personalised and engaging learning environment, and it could promoting learners' twenty-first century skills during the learning process. However, biased data in the Intelligent Tutoring System may also give rise to learning recommendation bias and inequality. Also, the algorithms in Intelligent Tutoring System is uncapable of capturing unquantifiable learning information, and put its emphasize mainly on the quantifiable educational recommendations, that will deteriorate the construction of the Constructivist-Learning-Based Curriculum.

Keywords Curriculum · Constructivist Learning · Intelligent Tutoring System

1 Introduction

AI refers to "the theory and development of computer systems that able to perform tasks normally requiring human intelligence" [17]. According to United Nations Educational, Scientific, and Cultural Organization (UNESCO)'s 2019 reports, "[a]rtificial Intelligence (AI)-powered services have already become prevalent in human lives in many places, including the least developed countries." Even the least developed countries have attached great importance on AI in the Industrial Revolution 4.0 (IR 4.0), which is characterized by a fusion of technologies [33]. Policy makers designed national strategies to fully exploit this significant social and economic transformational opportunity. On the World Economic Forum, a whole set of skills that a high-quality human resource should equip to survive in the IR 4.0 was

Y. Wang (✉)
Nanfang College Guangzhou, Guangzhou, China
e-mail: yuewang606@163.com

proposed, and named as "twenty-first century skills". And AI-literacy is one of the skills suggested [22].

Policy documents and academic papers also suggest that AI has become one of the most influential factors to power the development and transformation of many industries, including agricultural, medical, financial, business, etc. [25]. One of the most affected area is the manufacturing industry, as AI can be used to optimize manufacturing supply chain by helping predicting the market demands [26]. Thus, to survive in a competitive market, many enterprises in the manufacturing sector has adopted AI-based systems and techniques, which led to an increasing demand for new types of human resources with AI-related skills, known as the ICT competencies (the ability of solving problems using Information and Communication Technology).

Moreover, there is a rocketing demand of human resource equipped with ICT skills in the service sector, which is the one that creates more jobs than any other sectors in the recent years [24]. For example, in 2018, the service sector created 46.3% jobs of the whole job market in China [31]. Therefore, both of the manufacturing and the service industries have attached great importance on the implementation of AI, which means a tremendous increasing demand of skills that are related to the application of AI in the industries.

Kohlberg [18] proposed that the main objective of education is to "grow children into productive citizens, and live a well rounded life in society." And the society for now has a goal about cultivating AI tech savvy labor [39]. It is crucial for educators to investigate how can the application of AIEd contribute to the cultivation of this new types of human resources, and to ultimately help them improve their employability and "live a well rounded life" in a society surrounded by AI.

AIEd is a powerful tool that can "make computationally precise and explicit forms of educational, psychological and social knowledge which are often left implicit." [17], while offering practitioners and learners with a "more fine-grained under-standing of how learning actually happens." It is therefore important to further explore the impact of AIEd, its advantages, the challenges it faced, and how it may influence the broader society.

This article aims to examine the AIEd tools' influences on curriculum design and it impact on the wider society. In the following part of the article, I will discuss the pros and cons of introducing AIEd tools in the contemporary curriculum. Then, I will explore the impact of AIEd tools on the society, especially on the education equality in developing countries. Finally, based on my analysis on the impact of AIEd, I will give some suggestions for teachers, learners, and educational policymakers and raise questions about AIEd that need to be further addressed.

2   AIEd: Tools and Technologies

AIEd is a system that brings together AI and education by using computer programmes to analyse, monitor, and assist the learning process [17]. This system helps develop a digital learning environment that adapts to the capabilities and needs of individual learners, known as the Adaptive Learning Environment. The technological tool that lies at the centre of AIEd is the **Intelligent Tutoring Systems (ITS)** [19]. ITS aims at providing a computational one-to-one tutor who offers tailored learning experience to individual learners. It provides the users with three types of services: (1) personalised learning activities, learning contents, assessments that fit the learners' learning process and cognitive needs; (2) timely feedback that allows the learners to have a clearer understanding of their learning stage, and to further better control their learning process; (3) suitable pedagogical approaches suggestion in accordance with learners' responses that can lead, challenge, and support learners [17].

2.1 ITS

ITS is established based on a knowledge-based model, which consists of models that represent various types of "knowledge" and algorithms that process the "knowledge". The "knowledge" can include many components, such as facts, rules and problem-solving methodologies. The knowledge-based model, ITS, consists three main types of "knowledge": (1) the knowledge of pedagogy, such as teaching methods and strategies; (2) the knowledge of a specific subject, such as history textbook knowledge; (3) the knowledge of the learner, such as their previous learning achievements, emotional states and etc.

ITS researchers have also introduced other models of "knowledge", such as metacognitive, physiological, psychological and etc. [17]. These models can cooperatively and critically process, evaluate, and compare the "knowledge" in the models with the information generated from the learners during the learning process, and provide more accurate recommendations to individuals.

2.2 Two Other ITS Tools

ITS is often supported by two other technological tools: **Intelligent Support for Collaborative Learning (ISCL)**, which contributes in alleviating interaction problems in the online learning context, and **Intelligent Virtual Reality (LVR)** that can augment immersive learning experiences. These AIEd tools work together to co-efficiently improve learner's constructive learning experience.

In the constructive learning environment, learners "may work together and support each other as they use a variety of tools and information resources in their guided pursuit of learning goals and problem-solving activities" [38]. Also, Constructive learning is to "construct knowledge individually and/or socially based on learners' interpretations of experiences in the world. Instruction is to engaging learners in meaning making (knowledge construction)" [16] (see Fig. 1).

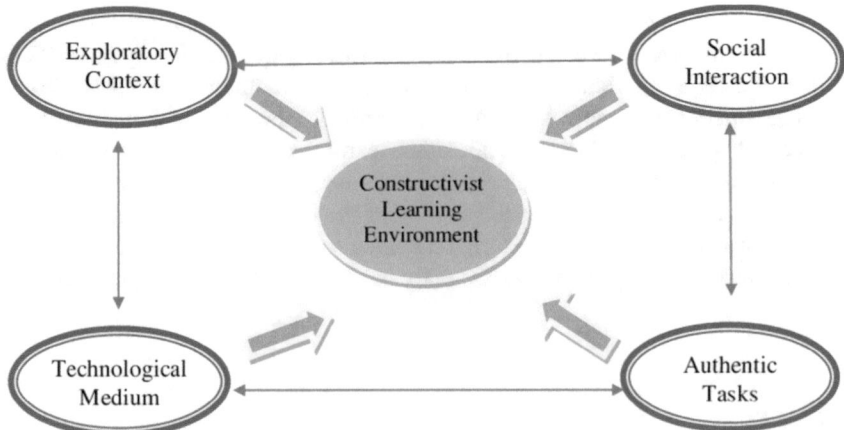

Fig. 1 Constructive learning environment

ISCL offers learners with necessary interactions by playing different roles in online collaborative activities. It provides a virtual peer that can interact with learners in the form of dialogue, introducing new ideas through the best cognitive way same with the learner's; It also provides an assistant for a human tutor, who can alert and inform human tutors when there is significant event happens in the collaborative activity, such as students going off the topic or repetitively misunderstanding concepts [28].

LVR allows learners to immerse themselves into the virtual world to experience while learning. It provides authentic immersive experiences that simulate some inaccessible aspect of the real world, for example, learners might be able to explore a nuclear power plant or biology structure in the virtual world (see Fig. 2).

These three AIEd tools will enable learners and teachers to identify the most efficient learning and teaching methods in different context. They will also shed light on how can other factors, such as environment, influence the efficiency of learning.

Fig. 2 The concept usage of intelligent virtual reality

3   AIEd: SAMR Model and ITS

Professor Mishra [23] proposed a technology integration teaching method, and Australian education expert Carrington [8] developed the Padagogy technology integration wheel. The most representative one is the SAMR model (Substitution-Augmentation-Modification-Redefinition) created by Dr. Ruben in 2008 to address the issue of how to select, use, and evaluate information technology in teaching. The aim of the SAMR model is to alter the learning experience [15].

The SAMR model is divided into four layers:

(a)   The bottom layer is the Substitution layer, the lowest level of technology integration application, which represents the technology used in teaching activities under the traditional education model;

(b)   the second layer is the Augmentation layer, the technology it uses has some enhancement effects;

(c)   the third layer is the Modification layer, the technology it uses allows to redefine the task itself;

(d)   the highest layer is the Redefinition layer, which uses technology that allows redefining learning tasks or creating new task technologies, so that students can display their learning results through various technical means.

The SAMR model is mainly used to evaluate the application value of technology in teaching. From Substitution, Augmentation, Modification to Redefinition, the value of technology may expand. Pfaffe [27] used the SAMR model as a framework for evaluating Mobile-learning activities and Dimensions of Learning, summarizing the specific applications of many mobile learning tools in the teaching process, and carrying out different levels of SAMR. There are five dimensions in a learning environment [40], which are Initiative, Collaboration, Constructivity, Authenticity and Goal Orientation. In Table 1, the author analyzes ITS's constructivity by applying SAMR model.

Table 1 Constructivity of ITS tools under SAMR evaluation model

	Substitution	Augmentation	Modification	Redefinition
Constructivity	ITS tools can only assess learners' knowledge level and provide learning goals	ITS tools provide learners with necessary learning content and searching tools to facilitate them build up the connection between background knowledge with new knowledge	ITS tools provide scaffolding tools and learning contexts for learners to construct knowledge	ITS tools support learns autonomously and unconsciously go through knowledge construction process that based on their former knowledge

4 Controversial Issues of Integrating ITS into Curriculum

In order to examine AIEd's influence on constructivist-learning-based curriculum, we should first discuss what are the key elements of a constructivist-learning-based curriculum. There are seven components in the curriculum: Learner, Learning Content, Learning Environment, Pedagogical Practices, Assessment, Teacher, Leadership and Management [20, 32], among which "Learner" is particularly addressed and put at the centre of the curriculum. Because the core of the constructive curriculum is to develop knowledge and specific skills for learners, that can help them adopt to the world outside of school [5, 11]. Therefore, I will particularly explore whether the adoption of ITS can enhance learner's learning experiences. Moreover, as discussed above, IR4.0 requires learners to have AI-literacy to better fit in the society surrounded by AI. Thus, I will also examine learning outcomes, to be specific, ITS's impact on learners' acquisitions to 21st-century skills.

4.1 Benefits of ITS on the Constructivist-Learning-Based Curriculum

Enhancing Learning Experience. ITS can greatly improve the personalisation of the learning experience by providing tailored one-to-one tutoring [17]. After entering ITS, the domain model will offer the learner proper learning content. Then the pedagogical model will provide initial learning activities while monitoring the whole activity process. Data generated from learner's interaction with the ITS, such as time-devoted on the activities and learner's emotional state, will be captured, processed, and then shared in the learner model with other two models which can thus adjust their recommendations accordingly. Gradually, with a richer and highly personalised database, ITS can adapt to the learner's interests, capabilities, needs, and learning styles.

ITS can also assist human teachers to make learning more personalised. AI can free teachers from routinely and time-consuming tasks such as record keeping and marking that usually use up teachers' energy [21]. Teachers could then have more time to lay emphasize on helping students, providing the ingenuity and empathy needed to understand student, and help students to solve their individual problems.

Personalised service can further improve the efficiency of learning. Bloom's [2] research studied the efficiency of one-to-one tutoring. It found that students who received personal tutoring scored two standard deviations higher than other students who didn't. This finding suggests that there is a positive correlation between person-alised learning with learning efficiency. Webster [36] also found that some measures can help optimize learner's learning efficiency, such as constant practice, real-life situation simulation during the learning process, timely feedback, and personalization. The increasing range of data capture devices—such as biological data, voice recognition, and eye tracking—will enrich the ITS with more indicators and evidence that can be used to accurately track and assess learners' learning activities. After collected massive data of the learner, ITS can analyse the data to conclude the learner's learning capability, cognitive level, and learning habits. Therefore, the learning activities and contents will be delivered to the learners through tailored learning content and practices with the most suitable teaching methods, which could make learning as an enjoyable and engaging experience for each learner.

Opponents may argue that collaborative learning can be especially difficult and inefficient in the context of online collaborations, where participants can not meet in person [13]. Yet with other assistant tools learning efficiency can be further improved through better collaboration and real-life situation simulation [36]. ISCL can help learners get better interaction in the best time in the online collaborative activities. It can also stimulate online collaborative activities' potential ability of promoting learning efficiency. And IVR has the ability to engaging learners and stimulating their learning interests, enabling learners to actively transfer what they have experienced and learned in the virtual world to the real world, and ultimately enhance educational outcomes.

Taken together, ITS can provide a personalized and engaging learning experience for learners. Moreover, with the ability to offer personalised service, it can help learners with special educational needs, motivating those who cannot attend schools, as well as supporting disadvantaged populations [30].

Cultivating 21st-Century Skills in the IR 4.0 Era. Cath et al. [9] suggested that the adoption of AIEd poses new opportunities to the contemporary curriculum that AIEd can help learners better fit the society surrounded by AI. Educators should work not only to familiarize them with AI technology but also to cultivate learners' 21st-century skills that are essential for one to thrive in the LR 4.0 era [7]. The World Economic Forum has proposed 21st-century skills which consist of 12 skills that can be divided into three categories, the following Table 2 gives a presentation of all skills.

Table 2 21st-century skills

Learning and innovation "The 4 C's"	Digital literacy	Career and life
Critical thinking& problem solving	Information literacy	Flexibility & adaptability
Creativity and innovation	Media literacy	Initiative & self-direction
Communication	ICT Literacy	Social & cross-cultural interaction
Collaboration		Productivity & Accountability
		Leadership & responsibility

The adoption of AIEd into the curriculum could directly contribute to the cultivation of those skills. Firstly, ITS puts the learners in control of their learning, which may help learners to develop self-direction skills. ITS also provides learners with tailored learning activities that post questions to challenge them to think deeper of the learning content. The automated tutoring allows learners to "trail and error" without embarrassment, which makes them more willing to try to solve the problems, and eventually improves their problem-solving skills [12]. Behaviouristic studies [1, 37] identified several factors associated with creativity cultivation, include being curious and questioning, being willing to explore. When learners immersed themselves into the IVR, they'll be able to explore the world and better satisfy their curiosity, which contributes to the cultivation of creativity. Most importantly, the more learners engage themselves into the AIEd system, the more familiar they get with the AIEd learning environment, which helps to enhance their Digital Literacy in all.

4.2 Challenges Faced by ITS in Constructivist-Learning-Based Curriculum

ITS technologies depend mainly on educational programmed computer software and algorithms to process data collected [3]. An algorithm is a set of instructions to perform on a particular group of data, therefore, algorithms generated from inaccurate or biased data will probably lead to incorrect output. Algorithms will greatly influence the accuracy of the analysis on learners' behaviour and the recommendations the ITS provides to the learners, it is vital to evaluate the quality of data and the algorithms. Incompetent algorithms will not be able to provide accurate personalised learning experience to individual learners, diminishing one of the biggest advantage of ITS in the contemporary curriculum, and causing lasting behavioral changes.

Biased Data. Before the ITS algorithms implemented into practice, it has to be trained on a certain data set [4]. If the training-dataset is biased towards a certain group, ITS may not be applicable to a different group, or the quality of its service might be compromised. For example, let's say a specific ITS algorithm is trained on a database of students from a Western European country, or from the most developed cities in China, then this ITS algorithm can not be directly applied to learners from other countries, or less developed areas in China. If been adopted uncritically, the ITS algorithm will have low reliability when reflecting the actual functioning of another learning context due to the biased database.

Moreover, due to the complexity of ITS technology, and the rather fixed data training process, it would be difficult for ITS users, namely, teachers and students, to modify the algorithm even if they have sensed its problems. This might harm learners' trust to the AIEd and diminish its value in the contemporary curriculum.

Potential Risks of ITS Algorithms. The educational concepts and curriculum knowledge of the AIEd algorithm developers can directly determine which data to be collected, how they will be analysed, what kind of educational issues to focus on, and how to evaluate the learning outcomes [34]. Currently, the AI developers in education aree mostly employed by the companies in the private sector [35]. Most of the AIEd algorithm designers have limited knowledge of the education and curriculum. They are not often familiar with the new educational needs, educational process, teaching methods, and the real needs of students and teachers. ITS developers might celebrate certain educational values when designing ITS algorithms. As a result, the ITS algorithm will inevitably be developed with various misunderstandings on education and biased educational ideas. As is discussed above, if the algorithms designed that favors some values over the others, the risk of wrong judgement will tremendously increase that can not reflect the real situation of education.

Furthermore, it is important to note that learning is a complex activity that could be influenced by both educational and non-educational factors, such as learners' socio-economic status, emotional status, etc. [14]. Although many countries have been making a great effort of collecting critical educational data, there is a huge gap between the full range of factors that ITS should concern and the factors that ITS have concerned. UNESCO [35] suggested that "The current AIEd system only includes education system as the only source of data relevant to learning provision". Other non-educational factors have not been integrated into the ITS system, even though they can also hugely affect the learning outcomes. For example, the socio-economic status has a positive correlation with learners' learning outcomes [14]. Including these non-educational factors into the ITS system can better understand how do these factors might influence learning efficiency. Learners' physical and mental health conditions can also impact on learning. Chou et al. [10] argued that bad emotional status might negatively impact on learners' performance at school. Data on autism and depression will shed light on the learners' inactiveness on learning activities. Without these data, when the depressed learners participated in the ITS system, their unusualal learning behavior might be misunderstood as a bad learning habit, and give wrong learning recommendations by ITS. Even some developers try to incorporate more variables, their accessibility is another issue. Personal data including history of mental illness, is difficult to be collected due to laws and regulations published to protect data security and personal privacy.

The accuracy and reliability of the algorithm may also be influenced by the low adaptability of the many models that ITS tends to contain. ITS developers try to estab-lish a more comprehensive system that by integrating more models which accommo-dates a variety of factors that influence learning. The models include the psycholog-ical model, meta-cognitive model, neurological model, etc. Although these models are used to analyse human behaviour, the actual problems that needed to be dealt with in the ITS system are normally broader than those handled by the original models. Because after adopted to the ITS system, the original models' application purposes, scenarios, and objects have to merge themselves with educational needs. When they are incorporated into the ITS system and applied to analyze the learning process, they may lead to analytical results that deviate greatly from the actual situation [34]. Based on the biased and inaccurate results, ITS may then provide inappropriate learning contents, pedagogical approaches, which may affect learners' learning outcomes.

Moreover, the contemporary curriculum normally consists of several education objectives that are not assessable [29]. For example, China's new curriculum reform aiming at building up national education values, such as fairness, justice, creativity. It demands secondary-school students to be cultivated as a person with morality, and achieve a good level of social and emotional growth. However, fairness, justice, morality, and emotional growth can not be easily quantified [6]. This is the unavoidable incompatibility between the algorithm and education. On the one hand, algorithm deals with problems with definite answers, which come from objective data analysis. On the other hand, education deals with human-related problems. The educational practices are full of uncertainties, and there is no clearly-defined evaluation for the morality growth of learners.

The inability of ITS system to build models to analyse this valuable but unquantifiable educational information will inevitably cause the ITS system biased put its emphasize on the quantifiable educational recommendations. For example, the AIEd will mainly recommend test-related practice, factual-knowledge and etc. The aim of developing learners' twenty-first century skills can be hard to achieve.

5   Conclusion

This article introduced the AIEd and its three main tools: ITS, ISCL, LVR, discussed the advantages that ITS can bring to the contructivist-learning-based curriculum: promoting a more personalised, engaging and effective learning experience to learners; and enhancing the cultivation of twenty-first century skills to prepare learners better fit the LR 4.0.

However, due to the ITS is based on the educational algorithms, the risks that might occur when evaluating learning process by applying algorithms had also been addressed. The biased database of the ITS algorithms can lead to inequality between learners. In the field of education, not all the influential factors of the learning outcome are quantifiable by algorithms. Even if the ITS tries to include as many as factor models' into the system to improve the inclusiveness of ITS and further improve its recommendation accuracy, there is a risk of low adaptability of the factor models. Because they were trained from different fields of study. Furthermore, besides data and algorithms in the ITS, the ITS developers normally have insufficient educational knowledge, which lead to the non-professional design of the ITS algorithms. This can bring troubles when teachers and learners use the ITS, as the ITS developers do not understand their real educational needs.

To address the disadvantages of ITS on the curriculum, two suggestions are given here: Teachers and students are suggested to participate in the design process of ITS products; When integrating more models into the ITS system, their adaptability should be tested by conducting sufficient empirical studies, both in the laboratory or in schools.

The promotion of ITS is widespread, it has the ability to promote education equality by providing more education opportunities to disadvantaged populations, and assist them to cultivate twenty-first century skills. But with the high-tech nature of ITS, the marginalized and disadvantaged population also suffered from insufficiency of educational infrastructures. Therefore, when policy makers designing ITS related policies, they should ask themselves more questions, such as: What are the most urgent infrastructures that need to be developed to better assist the spread of ITS?

References

1. Abdullateef E (2000) Developing knowledge and creativity: asset tracking as a strategy centerpiece. J Arts Manag Law Soc 30(3):174–192
2. Bloom BS (1984) The 2 sigma problem: the search for methods of group instruction as effective as one-to-one tutoring. Educ Res 13(6):4–16
3. Alcalá-Fdez J, Sanchez L, Garcia S, del Jesus MJ, Ventura S, Garrell JM, Otero J, Romero C, Bacardit J, Rivas VM, Fernández JC (2009) A software tool to assess evolutionary algorithms for data mining problems. Soft Comput 13(3):307–318
4. Allen WM (2007) Data structures and algorithm analysis in C++. Pearson Education India
5. Allen GC (2019) Understanding China's AI Strategy: clues to chinese strategic thinking on artificial intelligence and national security. Washington, DC: Center for a New American Security. Instruction as Effective as One-to-One Tutoring. Educational Researcher 13(6): 4–16
6. Beard D, Schweiger D, Surendran K (2019) Integrating soft skills assessment through university, college, and programmatic efforts at an AACSB accredited institution. J Inf Syst Educ 19(2):11
7. Bybee RW, Fuchs B (2006) Preparing the 21st century workforce: A new reform in science and technology education. J Res Sci Teach: Official J Natl Assoc Res Sci Teach 43(4):349–352
8. Carrington A (2016) Professional development: the padagogy wheel: it is not about the apps, it is about the pedagogy. Educ Technol Solut 72:54–57
9. Cath C, Wachter S, Mittelstadt B, Taddeo M, Floridi L (2018) Artificial intelligence and the 'Good Society': the US, EU, and UK approach. Sci Eng Ethics 24(2):505–528
10. Chou MH, Lin MF, Hsu MC, Wang YH, Hu HF (2004) Exploring the self-learning experiences of patients with depression participating in a multimedia education program. J Nurs Res 12(4):297–306
11. Cullen R, Harris M, Hill RR (2012) The learner-centered curriculum: design and implementation. John Wiley & Sons
12. Green A, Preston J, Janmaat J (2006) Education, equality and social cohesion: a comparative analysis. Springer
13. Gunawardena CN (1995) Social presence theory and implications for interaction and collaborative learning in computer conferences. Int J Educ Telecommun 1(2):147–166
14. Hartas D (2011) Families' social backgrounds matter: Socio-economic factors, home learning and young children's language, literacy and social outcomes. Br Edu Res J 37(6):893–914
15. Holmes W, Bialik M, Fadel C (2019) Artificial intelligence in education. Center for Curriculum Redesign, Boston
16. Jonassen D, Davidson M, Collins M, Campbell J, Haag BB (1995) Constructivism and computer-mediated communication in distance education. Am J Distance Educ 9(2):7–26
17. Luckin R, Holmes W, Griths M, Forcier LB (2016) Intelligence unleashed: an argument for AI in education. Pearson Education, London
18. Kohlberg L, Mayer R (1972) Development as the aim of education. Harv Educ Rev 42(4):449–496
19. Kay J (2012) AI and education: grand challenges. IEEE Intell Syst 27(5):66–69
20. Merritt BK, Blake AI, McIntyre AH, Packer TL (2012) Curriculum evaluation: linking curriculum objectives to essential competencies. Can J Occup Ther 79(3):175–180

21. McClintock R (1992) Power and pedagogy: Transforming education through information technology. Institute of Learning Technologies, New York
22. Markauskaite L (2006) Towards an integrated analytical framework of information and communications technology literacy: from intended to implemented and achieved dimensions. Inf Res: Int Electr J 11(3):3
23. Mishra P, Koehler MJ, Henriksen D (2011) The seven trans-disciplinary habits of mind: Extending the TPACK framework towards 21st century learning. Educ Technol 22–28.
24. Neumark D, Wall B, Zhang J (2011) Do small businesses create more jobs? New evidence for the United States from the national establishment time series. Rev Econ Stat 93(1):16–29
25. Porter ME, Heppelmann JE (2015) How smart, connected products are transforming companies. Harv Bus Rev 93(10):96–114
26. Perea-Lopez E, Ydstie BE, Grossmann IE (2003) A model predictive control strategy for supply chain optimization. Comput Chem Eng 27(8–9):1201–1218
27. Pfaffe LD (2017) Using the SAMR model as a framework for evaluating mLearning activities and supporting a transformation of learning. St. John's University (New York), School of Education and Human Services
28. Resta P, Laferrière T (2007) Technology in support of collaborative learning. Educ Psychol Rev 19(1):65–83
29. Stasko J, Badre A, Lewis C (1993) Do algorithm animations assist learning? An empirical study and analysis. In Proceedings of the INTERACT'93 and CHI'93 conference on Human factors in computing systems, pp 61–66
30. Sarkis H (2004) Cognitive Tutor Algebra 1 program evaluation, Miami-Dade County Public Schools. Lighthouse Point, FL: The Reliability Group
31. Shouhai D, Shan X (2015) Will the service sector promote China's long-term employment growth? China Econ 10(1):14
32. Scott D (2007) Critical essays on major curriculum theorists. Routledge, pp 9–10
33. Shahroom AA, Hussin N (2018) Industrial revolution 4.0 and education. Int J Acad Res Bus Soc Sci 8(9): 314–319
34. Tan WZ (2019) The algorithmic risk of artificial intelligence in education. Open Educ Res 25(6):20–30
35. UNESCO (2019) The challenges and opportunities of artificial intelligence in education. . https://en.unesco.org/news/challenges-and-opportunities-artificial-intelligence-education. Accessed 31 May 2020
36. Webster Sr, WG (1994) Learner-centered principalship: the principal as teacher of teachers. Praeger Publishers, 88 Post Road West, Box 5007, Westport, CT 06881
37. Waite SJ, Rea T (2007) Pedagogy or place? Attributed contributions of outdoor learning to creative teaching and learning
38. Wilson BG (1996) Constructivist learning environments: case studies in instructional design. Educational Technology
39. Xing B, Marwala T (2017) Implications of the fourth industrial age for higher education. The Thinker Issue
40. Xu P (2017) Research on the classification of educational APPs based on technology Integration Model Integration Models

Analysis of the Current Situation of Urban and Rural Teachers' Sense of Fairness in Online Education Equity

Qi Xu⑩, Xinghong Liu, Xue Chen, Han Zhang, and Min Pan

Abstract Online education brings new opportunities for the development of education equity. In order to reveal the inequities between urban and rural teachers in online education, we conduct an investigation on teachers in some regions of Hubei Province from the perspective of the sense of fairness in online education. Results show that groups with the best sense of overall fairness mainly come from urban and counties. Teachers in towns and rural areas have poor performance in terms of the sense of starting fairness, and there is a significant difference in the sense of starting fairness between teachers in urban and counties. Most teachers in urban and counties have strong abilities in technology application and platform management, there is a gap between urban and rural teachers in the digital teaching ability. Both the sense of starting fairness and the sense of process fairness have a significant impact on the sense of outcome fairness, and the sense of outcome fairness has the greatest impact on the sense of overall fairness in online education equity. urban and rural teachers show significant differences in the sense of outcome fairness. Region is the key to affect teachers' sense of overall fairness. The effect and quality of online teaching in various regions still need to be improved.

Keywords Urban and rural teachers · Online education · Education equity

1 Introduction

Online education promotes the development of education equity. Compared with traditional education, online education has the advantages of low cost, wide coverage, and freedom of time or space [1]. However, the practice of online education during the Covid-19 has exposed the unfairness of urban and rural education. There are still many inequities between urban and rural teachers. Many rural teachers have not received good technical training and it is difficult to carry out the teaching tasks, which leads to poor effect. Various problems may affect the healthy development

Q. Xu · X. Liu (✉) · X. Chen · H. Zhang · M. Pan
School of Computer and Information Engineering, Hubei Normal University, Huangshi, China
e-mail: 1582419651@qq.com

© The Author(s), under exclusive license to Springer Nature Singapore Pte Ltd. 2022   313
E. C. K. Cheng et al. (eds.), *Artificial Intelligence in Education: Emerging Technologies,*
Models and Applications, Lecture Notes on Data Engineering and Communications
Technologies 104, https://doi.org/10.1007/978-981-16-7527-0_23

of education and the realization of fairness and justice in the whole society. By investigating the sense of urban and rural teachers in online education equity, we analyze the unfairness of online education among urban and rural teachers, and lay the ideological foundation for the realization of education equity.

2   Concept of Education Equity

"Education equity" is a philosophical concept of different opinions. When individuals are in different environments, their understanding of this concept is inconsistent. Sociology advocates that education equity is equal educational opportunity. For example, Horace Man [2] first advocated equal opportunity education; Coleman [3] emphasized that the feature of equal education is free, that is, in a specific area or environment, children of different social backgrounds can receive the same educational opportunities and get free general curriculum education. His view coincides with that of Rawls's in substance. From the perspective of ethics, Rawls proposed that "equity" should be used to protect citizens' basic rights of freedom and the distribution of income and wealth. In Rawls's view [4], the fairest distribution should enable the most vulnerable groups in society to gain benefits. The OECD [5] has made a certain interpretation of education equity in order to assist its member countries in formulating education policies. They believe that education equity is a combination of fairness and inclusion. It is necessary to ensure that individuals are not affected by socioeconomic status, gender and other factors to receive education, and minimum standards must be established to ensure full coverage of education for all. Among the descriptions, Husen's three-stage theory of education "fair starting, fair process, and fair outcome" has the most extensive influence [6]: (1) starting fairness: individuals can overcome economics, family, and class influenced by other practical factors, and have equal rights and opportunities to receive education; (2) process fairness: emphasizing attention to the development needs of students; (3) outcome fairness: Individuals can finally obtain personalized education that meets their own characteristics, and their potential can be fully utilized. The three stages progress in sequence, and they jointly interpret the connotation of educational equity.

Many studies discuss education equity from the perspective of the student [7], but the role of the teacher is also important. For the teacher, few studies analyze the various inequities of them in online education. Under the normalization of the Covid-19, online education has gradually integrated into school education. As designers, instructors, managers and evaluators of online education, teachers have a profound influence on the future development of students [8]. In this study, In terms of starting fairness, teacher education equity means that teachers are not affected by economic status, living environment, gender, technology and other conditions, enjoy the opportunity to start online teaching, and obtain equal educational resources; In terms of process fairness, it means that every teacher can achieve from efficient teaching to high-efficiency teaching, including the ability to design and optimize the flow of teaching activities, effectively manage classroom discipline, seriously invest in online teaching process, provide guidance for students, make reasonable recommendations, and broaden students' horizons. In term of outcome fairness, it is that teachers can achieve personalized education within a certain area, improve online teaching ability, harvest good online teaching quality, and contribute to the construction of large-scale personalized education.

3 Research Design

By interpreting the connotation of online education equity, we design the questionnaire for urban and rural teachers' sense of fairness. We used random sampling to distribute questionnaires for three weeks in December 2020. A total of 400 questionnaires are planned to be distributed and 362 are recovered, of which 28 unqualified questionnaires are eliminated, and a total of 334 valid questionnaires are available, with an effective recovery rate of 83.5%. We use SPSS26.0 to conduct reliability and validity tests, cluster analysis, etc. to deeply explore the crux of urban and rural online education equity, and to clarify the online education inequity between urban and rural teachers. Specifically, this paper focuses on three sub-questions: **Question 1**: Do urban and rural teachers show differences in online education equity? **Question 2**: What are the characteristics of groups with higher, middle, and lower scores on online education equity? **Question 3**: What are the factors that have the greatest impact on online education equity of urban and rural teachers?

3.1 Variable Design

As shown in Table 1, the sense of the education equity is divided into 3 dimensions, which are divided into 7 sub-dimensions. Each sub-dimension has no less than 3 items for a total of 24 items, using the Likert five-level scale structure. The complete questionnaire is as follows: The first part is the basic information of the teacher, including his/her gender, age, etc.; the second part is the sense of equity survey.

3.2 Participant

We take teachers from Hubei Province who have some online education experience as the object, and use the elementary, junior, and senior schools in Huangshi as an example. Before the Covid-19, schools in this area had carried out training activities to improve the ability of teachers to apply information technology, and teachers had also participated in the training of "stopping class without stopping study" [9], so they had some network technology foundation and the ability to carry out online teaching. It provides guarantee for the investigation of online education equity in

Table 1 Variable design of the sense of equity in urban and rural online education

Dimension	Sub-dimension	Item dimension
Sense of starting fairness	Infrastructure	Network access, educational resources, hardware equipment, teaching platform, physical environment
	Technical Training	Self-training, school training, other forms of technical learning
Sense of process fairness	Online teaching attitude	Usefulness, ease of use, continuous use
	Online teaching management	Teaching time management, teaching plan management, teaching order management
	Online teaching behavior	Teaching content design, resource sharing, teaching interaction, home-school collaboration
Sense of outcome fairness	Online teaching quality	Teaching effect, teaching efficiency, teaching benefit
	Professional development	Knowledge, digital skills, teaching satisfaction

urban and rural teachers. Among them, 116 are males (34.73%) and 218 are females (65.27%); 6 teachers (1.8%) under 20 years old, 130 teachers (38.92%) 21–30 years old, 150 teachers (44.91%) 31–40 years old, 48 teachers (14.37%) over 41 years old; 120 teachers (35.93%) in towns, 80 teachers (23.95%) in rural areas, 46 teachers (13.77%) in county towns, and 88 teachers (26.35%) in urban areas; seniority is 30 (8.98%) for one year or less, 124 (37.13%) for 2 to 5 years, 146 (43.71%) for 6 to 10 years, and 34 (10.18%) for 11 years and above; 82 teachers (24.55%) teach Chinese, 124 teachers (37.13%) teach Mathematics, 54 teachers (16.17%) teach Foreign Languages, and 74 teachers (22.16%) teach other subjects; 106 teachers in elementary schools (4.26%), 138 teachers in junior schools (2.84%), 46 teachers in senior schools (59.66%), and 44 teachers in other schools (40.34%); there are 94 associates (28.14%), 174 undergraduates (52.1%), 26 graduates and above (7.78%), and 40 (11.98%) at other levels.

3.3 Reliability and Validity

In reliability test, we find that the overall reliability level of the questionnaire Cronbach Alpha is 0.952 (>0.8), and the reliability level of each dimension is >0.8. As shown in Table 2, the Cronbach Alpha of each sub-dimension is within the acceptable range (>0.7), indicating that the overall reliability of questionnaire is good.

Table 2 Research reliability and validity test

Dimension	Sub-dimension	Number of questions	Kronbach Alpha	KMO/p
Sense of starting fairness	Infrastructure	5	0.808	0.842***
	Technical training	3	0.700	
Sense of process fairness	Online teaching attitude	3	0.717	0.819***
	Online teaching management	3	0.723	
	Online teaching behavior	4	0.755	
Sense of outcome fairness	Online teaching quality	3	0.736	0.843***
	Professional development	3	0.767	
Sense of overall fairness		24	0.932	0.925***

*** means $p < 0.001$

The convergent validity of the questionnaire is determined by the standardized factor loading value of each variable. We find that the factor loading value of the item in each variable meets the requirements (>0.7). Through the method of "dimension reduction analysis", the result shows that KMO is 0.955 (>0.8), and Bartlett's sphericity test value reaches the level of significance of 0.000, indicating that each variable is suitable for factor analysis. Through principal component analysis and maximum variance method, a total of 7 factors are extracted. The cumulative variance explanation rate is 68.205%, indicating that the measurement of each variable in the questionnaire is effective and the overall validity of the questionnaire is good.

4 Result Analysis

4.1 Do Urban and Rural Teachers Show Differences in Online Education Equity?

As shown in Table 3. Teachers' sense of starting fairness shows a significant difference in each region with F of 25.26 ($p < 0.001$). Teachers in counties have the most positive evaluation of the starting fairness (M = 4.30), teachers in urban areas (M = 4.27) ranked the second, and teachers in towns (M = 3.88) and rural areas (M = 3.74) have lower evaluations. This shows that teachers in county and urban areas have greater advantages in network access and technical training. Teachers in counties are most satisfied with the quality and status of the access network, reflecting their ability to access the Internet smoothly and showing strong self-confidence in

Table 3 Variance analysis of teachers' online education fairness in each region

ANOVA		Mean	Standard deviation	95% confidence interval of the mean		Minimum	Maximum	F/p
				Lower limit	Upper limit			
Sense of starting fairness	Township	3.38	0.30	3.81	3.96	3.25	4.38	25.26***
	Rural	3.74	0.51	3.58	3.91	2.00	4.38	
	County	4.30	0.24	4.19	4.40	3.75	4.63	
	Urban	4.27	0.22	4.20	4.33	3.75	4.75	
Sense of process fairness	Township	3.58	0.35	3.79	3.97	3.11	4.44	12.93***
	Rural	3.84	0.51	3.67	3.40	2.33	4.44	
	County	4.17	0.35	4.02	4.32	3.00	4.78	
	Urban	4.24	0.22	4.18	4.31	3.67	4.78	
Sense of outcome fairness	Township	3.73	0.45	3.82	4.05	2.67	4.5	8.05***
	Rural	3.90	0.48	3.75	4.05	2.00	4.5	
	County	4.14	0.34	3.99	4.28	3.00	4.67	
	Urban	4.26	0.21	4.19	4.32	3.83	4.83	
Sense of overall fairness	Township	3.80	0.35	3.81	3.99	3.01	4.31	15.84***
	Rural	3.83	0.48	3.67	3.98	2.11	4.38	
	County	4.20	0.28	4.08	4.32	3.25	4.62	
	Urban	4.26	0.16	4.21	4.31	3.95	4.60	

*** means $p < 0.001$

technical training; rural teachers have the lowest scores on the starting fairness. It shows that the promotion of high-quality educational resources and advanced technology in remote areas is still insufficient, which undoubtedly has caused a serious obstacle to the sustainable development of the digital economy and society.

Teachers' sense of process fairness shows significant differences in F of 12.93 in each region ($p < 0.001$). The scores of urban teachers (M = 4.24), county teachers (M = 4.17), township teachers (M = 3.88), and rural teachers (M = 3.84) on the sense of process fairness shows a decreasing trend, indicating that teachers in urban and counties also have obvious advantages in the process of online education. Teachers with geographical advantages can actively respond to various problems in online teaching. It is difficult for teachers in towns and villages to successfully complete the tasks of online teaching. It is difficult for them to maintain the balance between online teaching management and evangelism. Coupled with being hindered by the fairness of the starting point, these teachers are facing numerous difficulties when developing online teaching.

Teachers' sense of result fairness shows a significant difference in each region with F of 8.05 ($p < 0.001$). Among them, the teacher's outcome fairness score: urban (M =

4.26) > county (M = 4.14) > township (M = 3.93) > rural (M = 3.90), which is very similar to the teacher's performance in process fairness. Teachers in towns and rural areas have low evaluations of the outcome fairness. Because there are fewer teachers in remote areas such as rural areas, it is difficult for them to control the quality of online teaching under the conditions of limited equipment, poor environment, and inexperience. It is difficult to obtain better professional development, which undoubtedly forms a huge contrast with teachers in urban and counties.

Teachers' sense of overall fairness shows a significant difference in F of 15.84 in each region (p < 0.001), and the comparison result: county (M = 4.20) > township (M = 3.90); Urban (M = 4.26) > township (M = 3.90); county (M = 4.20) > rural (M = 3.83); urban (M = 4.26) > rural (M = 3.83). That is, on the sense of overall fairness, urban teachers perform the best and rural teachers perform the worst. This difference runs through the equity of online education and is reflected in the sense of starting, process and outcome fairness.

4.2 What Are the Characteristics of Groups with Higher, Middle, and Lower Scores on Online Education Fairness?

The category that is much lower than the overall mean is defined as poor, and much higher than the overall mean is defined as excellent, and the interval between the overall mean ± 10% is defined as general [10]. The minimum distance between the initial centers is 2.503, and finally iterates 13 times to convergence, as shown in Fig. 1. The F test shows that all variables are significant (p < 0.001), so the clustering result is valid. The importance of each variable to the clustering result is ranked as follows: sense of overall fairness (F = 294.37) > sense of outcome fairness (F = 245.62) > sense of starting fairness (F = 162.73) > sense of process fairness (F = 171.28).

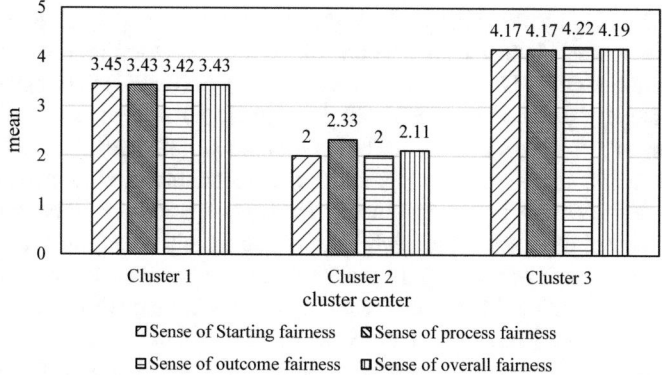

Fig. 1 Cluster analysis of urban and rural teachers' sense of fairness in online education

The teachers in cluster 1 are at the middle level, and the number of teachers in this category is the largest. Such groups are fairly distributed in rural areas, towns, counties, and urban areas. Most of them are novice teachers of Chinese (31.4%) and mathematics (42.9%). The teaching age is concentrated within 5 years (94.2%), and ages are 21–30 years old (85.7%). There are more female teachers (62.9%) than male teachers (37.1%), mainly in elementary (48.6%) and junior (37.1%) schools. Their highest academic qualifications are mainly colleges (62.9%) and undergraduates (20%). They scored the highest on the sense of starting fairness ($M = 3.45$), and the lowest score on the sense of outcome fairness ($M = 3.42$) and overall fairness ($M = 3.42$). Such teachers simply transplant offline teaching to online education. They have some digital technology foundation and can basically complete online teaching tasks, but the teaching effect still need to be improved. Due to the inertia of offline teaching, such teachers still use equal thinking and implement "undifferentiated teaching" for students in online teaching. The procedural teaching resources, single teaching activities, rigid teaching model, and lack of teaching evaluation have led to the fact that the teaching results have not reached their satisfaction.

The teachers in cluster 2 are at the lower level, and there are fewer teachers in this category. Most of them come from rural areas (98%). The proportion of male and female teachers is equal. They are 31–40 years old and have a teaching experience of 2 to 5 years. They mainly teach mathematics at the elementary level. Their highest education is mainly in colleges. Their performance in terms of the sense of the starting fairness ($M = 2$) and outcome fairness ($M = 2$) are both poor, and the sense of the overall fairness ($M = 2.11$) and process fairness ($M = 2.33$) is not optimistic. Since 2010, China has implemented the "National Training Program". The teaching level and quality of teachers in many remote schools have been improved. However, the old digital access equipment lacks updates and repairs, and it is difficult for such teachers to obtain advanced technology and high-quality resources. In addition, the low pressure for exams and admissions at the elementary school level may allow teachers to have more relaxed standards when judging the effects of online teaching, which is not conducive to their effective online teaching and their professional development.

The teachers in cluster 3 are at the higher level, such groups mainly come from urban and counties (60%). They account for 97% of the total number of teachers in urban and counties. There are more female teachers (65.6%) than male teachers (34.4%). The age group is mainly 31–40 years old (55.7%), and the teaching experience is mostly 6–10 years (55%). There are more people teaching other subjects (such as physics, chemistry, biology, information technology, etc.), accounting for 81.1% of the total number of other subjects. The school period is mainly junior school (42.7%). Their highest academic qualifications are mainly undergraduates (61.1%), and a large number of teachers with a graduate degree or above are also concentrated in this group, indicating that teachers with higher academic qualifications are better at using online teaching tools. Their evaluation of overall fairness ($M = 4.19$) is good, but their evaluation of starting fairness ($M = 4.17$) and process fairness ($M = 4.17$) is lower than their evaluation of result fairness ($M = 4.22$). Teachers of this category can generally take advantage of online teaching platforms and digital technology to interact with students in time and communicate effectively with parents. They have realized scientific and efficient classroom management, and have gained good teaching results.

4.3 What are the Factors that Have the Greatest Impact on Online Education Equity of Urban and Rural Teachers?

As shown in Table 4, Model 1, 2, 3, and 4 are based on individual characteristics variables, individual characteristics variables + sense of starting fairness, individual characteristics variables + sense of starting fairness + sense of process fairness, individual characteristics variables + sense of starting fairness + sense of process fairness + sense of outcome fairness as predictor variables, and linear regression analysis with sense of starting fairness, process fairness, outcome fairness and overall fairness as dependent variables.

The construction effect of Model 1 is good (p < 0.001), and the explainable variance reaches 30.9% (R > 0.5), so the regression equation can be obtained as: sense of starting fairness = 3.184 + (−0.045) * gender + 0.063 * age + 0.136 * region + 0.114 * seniority + (−0.006) * subject + (−0.002) * teaching stage + 0.069 * qualification. Urban and rural teachers' sense of starting fairness is mainly restricted by region. The second is seniority. The better the regional conditions and the older the school age, the more positive the assessment of starting fairness. Secondly, the starting fairness is inversely proportional to the variables of gender, subject, and teaching stage. That is, female teachers have a lower sense of starting fairness than male teachers. Teachers who teach science and engineering subjects have lower starting fairness scores than teachers who teach science and history subjects.

The construction of Model 2 is statistically significant (p < 0.001). The explainable variance of teacher characteristic variables and starting fairness to process fairness reached 74.3% (R > 0.5). Therefore, the regression equation is: sense of process fairness = 0.576 + 0.026 * gender + (−0.045) * age + 0.015 * region + 0.076 * seniority + (−0.017) * subject + (−0.019) * teaching stage + 0.026 * qualification + 0.825 * starting fairness. The variable that has the greatest impact on urban and rural teachers' sense of process fairness is the starting fairness ($\beta = 0.825$), indicating that the quality of digital access and technical support have a restrictive effect on online teaching. The second is the seniority ($\beta = 0.076$). When they are older, the more they have more teaching experience, the easier it is to carry out effective classroom management and knowledge transfer, which becomes one of their advantages in online education. The influence of age, subject, and teaching stage on the sense of starting fairness is negatively correlated, indicating that teachers' sense of process fairness gradually declines with age, which may be relevant to their job burnout and emotional indifference in teaching [11], When teachers teach subjects other than

Table 4 Regression analysis of online education fairness of urban and rural teachers

model	Dependent variable	Predictor variable	R	R2 /Adjusted R2	Change of R2	F/p
1	Sense of starting fairness	Qualification, Seniority, Gender, Region, Subject, Teaching stage, Age	0.582	0.339/0.309	0.339	11.625***
2	Sense of process fairness	Qualification, Seniority, Gender, Region, Subject, Teaching stage, Age, Sense of starting fairness	0.869	0.755/0.743	0.755	60.921***
3	Sense of outcome fairness	Qualification, Seniority, Gender, Region, Subject, Teaching stage, Age, Sense of starting fairness, Sense of process fairness	0.869	0.755/0.741	0.755	53.734***
4	Sense of overall fairness	Qualification, Seniority, Gender, Region, Subject, Teaching stage, Age, Sense of starting fairness, Sense of process fairness, Sense of outcome fairness	0.516	0.267/0.144	0.267	2.181/0.03

*** means $p < 0.001$

cultural courses, their sense of process fairness scores are lower. Because of their short class hours, the school easily ignores these groups, which has a certain impact on their active and effective online teaching. Teachers at higher levels have more work pressure, and teachers with higher levels of teaching score lower on the sense of process fairness.

The construction of Model 3 is statistically significant ($p < 0.001$). The explainable variance of teacher characteristic variables, sense of starting fairness, and process fairness to the result fairness reached 74.1% ($R > 0.5$), then the regression equation is: sense of outcome fairness = 0.413 + (−0.007) * gender + (−0.056) * age + (−0.006) * region + 0.108 * teaching age + 0.011 * subject + (−0.028) * teaching stage + (−0.026) * qualification + 0.329 * sense of starting fairness + 0.547 * sense of process fairness. The variable that has the greatest impact on urban and rural teachers' sense of result fairness is process fairness ($\beta = 0.547$), the second is starting point fairness ($\beta = 0.329$), and the effect of process fairness is more obvious. In addition, teaching age ($\beta = 0.108$), age ($\beta = 0.547$), teaching stage ($\beta = -0.028$), and qualification ($\beta = -0.026$) have successively weakened influences on the sense of outcome fairness, indicating that the teaching experience of a teacher is longer, when they are older, with lower education, and professors at lower levels, the higher their scores on the sense of outcome fairness. Region ($\beta = -0.006$) can not only directly affect the sense of outcome fairness, but also indirectly affect it through the sense of starting fairness, reflecting the key position of regional factor in the sense of fairness of urban and rural teacher education.

Similarly, the construction of Model 4 is statistically significant ($p < 0.05$). The interpretable variance of teacher characteristic variables, sense of starting, process, and outcome fairness to overall fairness is 14.1% ($R > 0.5$). The reason may be that the factors that affect the overall sense of fairness include more other variables, and its explanatory degree is expected to be further improved. The regression equation obtained is: sense of overall fairness = 66.951 + (−3.44) * gender + (−6.718) * age + (−2.309) * region + 4.022 * seniority + 1.795 * subject + 1.228 * teaching stage + 0.564 * qualification + (−0.626) * sense of starting fairness + (−1.572) * sense of process fairness + 14.772 * sense of outcome fairness. That is, the order of the influence of each variable on the overall sense of fairness of teachers is: result fairness ($\beta = 14.772$) > age ($\beta = -6.718$) > seniority ($\beta = 4.022$) > gender ($\beta = -3.44$) > region ($\beta = -2.309$) > subject ($\beta = 1.795$) > sense of process fairness ($\beta = -1.572$) > teaching stage ($\beta = 1.228$) > sense of starting fairness ($\beta = -0.626$) > qualification ($\beta = 0.564$). It shows that the sense of outcome fairness is the main factor that directly affects the sense of overall fairness. When teachers score higher in the sense of outcome fairness, they are more likely to perform better in the sense of overall fairness, which further shows that the sense of outcome fairness occupies an important position in the fairness of urban and rural teacher education.

5   Conclusion and Reflection

We analyze the urban and rural teachers' sense of fairness in online education equity, and find:

First, groups with the best sense of overall fairness mainly come from urban areas and counties. They are mainly in the junior schools, their age is between 31 and 40 years old, and the seniority is 6 to 10 years mostly. Urban teachers are more positive on the sense of overall fairness and the teaching quality is also better. This has a promoting effect on the development of online education on the surrounding towns or villages, but it may create some pressure on teachers in surroundings. The poor overall sense of fairness among teachers in rural areas is detrimental to the urban and rural online education across the country.

Second, teachers in towns and villages have poor performance in terms of the sense of starting fairness, and there is a significant difference in the sense of starting fairness between teachers in urban and counties. Teachers in towns and villages have long faced the difficulty of relatively short funds. It is difficult for them to access high-quality digital teaching equipment and their technical ability is relatively weak. The relative underdevelopment of towns and rural areas brings difficulties to the starting fairness in online education. The Covid-19 revealed that the country still has shortcomings in infrastructure construction. Education equity issues such as the low degree of co-construction and sharing and unequal distribution of educational resources still exist. Therefore, efforts should be made to improve the construction of informatization infrastructure and resources in remote areas, and provide technical support services according to the needs of teachers.

Third, most teachers in urban and counties have strong abilities in technology application and platform management, However, there is a gap between urban and rural teachers in the digital teaching ability. A large number of rural and township teachers only use basic information technology, and simply transplant offline teaching to online teaching. The seniority has a significant influence on the sense of process fairness. Some teachers with rich classroom teaching experience hold neutral or partial negative comments on the sense of process fairness, indicating that they have not fully mastered the methods of online teaching. Under the influence of the sense of starting fairness, it is difficult for township and rural teachers to quickly improve their online teaching level in short-term training and change their evaluation of process fairness. Affected by the seniority, the long-term traditional education of some non-novice teachers in remote areas has affected their teaching mode. In terms of online education adaptability, teaching interaction, resource compensation, and discrimination among students, it is difficult for them to reach a teaching level equivalent to that of teachers in urban areas and counties in the short term. In the normalization of the Covid-19, the construction of efficient teaching models and innovative teaching activities are important elements for teachers to achieve process fairness. Therefore, it is necessary to give full play to the cross-temporal and cross-regional advantages of the Internet, and urban and rural teachers should break through the communication barriers, conduct online teaching appraisal, technical exchanges, and explore new methods of online teaching together.

Fourth, both the sense of starting fairness and the sense of process fairness have a significant impact on the sense of outcome fairness, and the sense of outcome fairness has the greatest impact on the sense of overall fairness. The difference between urban and rural teachers' sense of outcome fairness is significant. Region is the key to affect the overall fairness of online education. The effect and quality of online education in various regions still need to be improved. Factors such as teacher's seniority, age and teaching stage have a significant impact on the sense of outcome fairness. Limited to the current undifferentiated information technology application ability training, many older primary teachers in rural and towns cannot keep up with the pace of training. Their lack of online teaching design organization experience has led to the emergence of a teaching style of "cramming method of teaching" or "the teacher saying counts". Therefore, there is an urgent need to organize and carry out hierarchical and classified training under the normalization of the Covid-19. It is necessary to formulate training plans with different goals and different contents in a targeted manner, and strengthen training assistance for poor areas and weak schools.

References

1. Dumford AD, Miller AL (2018) Online learning in higher education: exploring advantages and disadvantages for engagement[J]. J Comput High Educ 30(3):452–465
2. Kopan A (1974) Equality of educational opportunity. American University Press, America
3. Kahlenberg RD. Learning from James Coleman [EB/OL]. http://www.findarticles.com/p/art icles/mi_m0377/is_2001_Summer/ai_76812255
4. Frohlich N, Oppenheimer J, Eavey C (1987) Laboratory results on Rawls's distributive justice. Br J Polit Sci 17(1): 1–21. Accessed March 5, 2021. http://www.jstor.org/stable/193962
5. Bøyum S (2014) Fairness in education—A normative analysis of OECD policy documents. J Educ Policy 29(6):856–870. https://doi.org/10.1080/02680939.2014.899396
6. Husén T (1975) Social influences on educational attainment. Res Perspect Educ Equal 182–186
7. Xiaoyong Hu, Wei Xu, Yuxing Cao, Huanyun Xu (2020) The theory of equity in basic education in the new era of information promotion: intension, path and strategy. Electro-Educ Res 41 (09): 34–40
8. Hanwei D (2020) Ethical issues and coping strategies in online teaching. Mod Educ Technol 30(12):41–47
9. Min Z, Yongpeng C, Jun T (2020) Study on the practical effect of "stop-and-go" in digital campus experimental schools - Based on the survey and analysis of 107 schools in Hubei Province 30 (08): 106–112
10. Govender P, Sivakumar V (2020) Application of k-means and hierarchical clustering techniques for analysis of air pollution: a review (1980–2019). Atmos Pollut Res 11(1):40–56
11. Karthikeyan P (2016) role of teachers in developing emotional intelligence among the children. Shanlax Int J Educ 3(2):22–27

Construction of Multi-Tasks Academic Procrastination Model and Analysis of Procrastination Group Characteristics

Chao Zhou⬛, Jianhua Qu, and Yuting Ling⬛

Abstract Academic procrastination could affect the learning effect of students to a certain extent. In order to identify students with procrastination tendency in online courses, this paper constructs a Multi-tasks Academic Procrastination Model. Next, based on the learning log data of an online course, static procrastination indicators and fluctuating procrastination indicators are extracted to analyze students' procrastination behavior. Lastly, for finding out the behavior characteristics of different procrastination groups, this paper analyzes the procrastination behavior of learners through K-means clustering algorithm and visualization method. The results showed that the academic procrastination model with fluctuating procrastination indicators has better clustering effect than the academic procrastination model with static procrastination indicators, and online learners can be divided into three different procrastination groups, namely, severe procrastinators, general procrastinators and non-procrastinators. According to the characteristics of different procrastination groups, this paper gives some suggestions to reduce the occurrence of academic procrastination in online learning and improve the learning effect.

Keywords Academic Procrastination · Clustering Analysis · Group characteristic analysis · Visualization

1 Introduction

Academic procrastination refers to the behavioral tendency of learners to postpone one or more necessary learning tasks [1]. Due to the lack of effective supervision and management of online learning environment, and the lack of motivation for students' learning, academic procrastination has become the most common behavior in online learning [2, 3]. Relevant researches show that academic procrastination not only can measure students' self-management ability, but also has a negative impact on students' learning effect [4, 5]. Therefore, it is particularly important to build

C. Zhou (✉) · J. Qu · Y. Ling
Shandong Normal University, No. 1 Daxue Road, Jinan, China
e-mail: sdnu_zhouchao@163.com

© The Author(s), under exclusive license to Springer Nature Singapore Pte Ltd. 2022   327
E. C. K. Cheng et al. (eds.), *Artificial Intelligence in Education: Emerging Technologies, Models and Applications*, Lecture Notes on Data Engineering and Communications Technologies 104, https://doi.org/10.1007/978-981-16-7527-0_24

an effective academic procrastination model, explore the behavioral characteristics of procrastinators, and identify students who are prone to procrastination in course learning.

At present, the research on academic procrastination has attracted the attention of scholars at home and abroad. Researches on academic procrastination are mainly divided into two aspects: one is academic procrastination research based on questionnaire survey, and the other is academic procrastination research based on online course learning log. In the study of academic procrastination based on online course learning log, Xiong Luying et al. [6] used the total time of learners to complete all tests on the "365 University" platform as an indicator, constructed a procrastination model and analyzed the behavior of online learners in academic procrastination, which in turn helps learners make plans to reduce academic procrastination. Yang Xue et al. [7] constructed the academic procrastination model by using six indicators, such as homework completion time and mutual evaluation time, on Moodle platform, so as to identify academic procrastination and conduct online intervention, and ultimately reduce the number of students' procrastination. Danial Hooshyar et al. [8] established PPP (Prediction of Students' Performance Through Procrastination Behavior) model with homework publishing time, homework first viewing time, homework submission time and homework deadline as indicators, used K-means clustering algorithm to cluster students to obtain cluster labels, and then used a variety of classification algorithms to classify and predict. The final experimental results showed that PPP model can describe students' procrastination behavior well. Yeongwook Yang [9] extracted feature vectors representing students' homework submission patterns through homework publishing time, homework first viewing time, homework submission time and homework deadline to build a procrastination model, then used clustering method to classify students into three types of procrastination, and finally compared the classification effects of various classification methods to identify which method can best predict students' tendency to procrastinate.

Although the current researches have achieved some results, there are still the following two problems:

(1) The current researches are mainly based on the procrastination model constructed by a certain learning task, without considering the influence of multiple learning tasks on academic procrastination. For example, in the literature [6–9], they built a procrastination model only through tests or homework.

(2) At present, most of the indicators for measuring academic procrastination are static indicators, such as the number of late or unpaid homework, and less consider the impact of the fluctuation of procrastination data on academic procrastination. So, it is difficult to find that the degree of procrastination varies with time. For instance, Michinov et al. [10] found that the degree of procrastination of patients with high procrastination will gradually increase with the deepening of the course, while the degree of procrastination of patients with low procrastination will remain low throughout the course. Although AFTAB AKRAM et al. [11] considered the problem of procrastination tendency when

studying academic procrastination, the model was composed of 31 indicators, which made the model too complicated and made the experimental results difficult to understand.

In order to solve the above two problems and better analyze students' procrastination behavior, this paper constructs a Multi-tasks Academic Procrastination Model and adds an indicator to measure procrastination fluctuation to the model. On this basis, clustering analysis is carried out on the research data of procrastination behavior to better analyze the characteristics of procrastinating groups and find out the factors affecting the occurrence of procrastination behavior. And finally, three types of procrastination groups are obtained and the characteristics of procrastination groups are visually analyzed.

2 Research Question

In this study, we aim to build a Multi-tasks Academic Procrastination Model and add fluctuating procrastination indicators to the model. Through cluster analysis and visual analysis, the characteristics of procrastination groups are obtained, and the factors affecting procrastination behavior are found out and relevant opinions are given. The research question posted in this study is:

(1) Can the Multi-tasks Procrastination Model constructed in this paper better describe the online academic procrastination?
(2) What are the factors that affect different procrastination behaviors?

3 Model and Method

3.1 Multi-Tasks Academic Procrastination Model

At present, most studies on academic procrastination use static procrastination indicators. This paper constructs a Multi-tasks Academic Procrastination Model (called MTAP) on this basis, to describe academic procrastination more simply and comprehensively. The model mainly includes the following two types of procrastination indicators: static procrastination indicators (SPI) and fluctuating procrastination indicators (FPI). Among them, the SPI are used to describe the procrastination degree of completing a learning task, which mainly includes five parts: static procrastination of homework, static procrastination of test, static procrastination of sign-in, static procrastination of video and static procrastination of courseware. The FPI are used to express the change of procrastination degree in future learning, which mainly includes the procrastination fluctuation of five kinds of learning tasks: homework, test, sign-in, video and courseware (see Fig. 1).

A specific descriptions of the SPI and the FPI is given in Table 1.

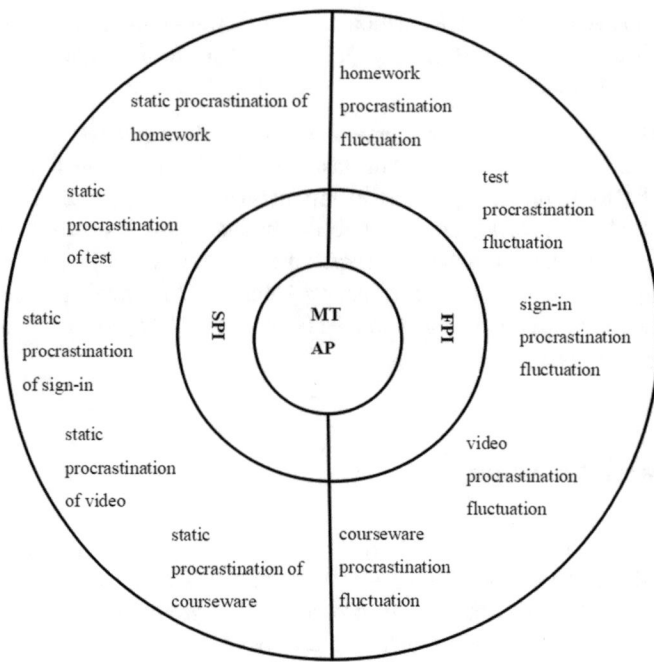

Fig. 1 MTAP

3.2 Extract Procrastination Indicators

Extraction of SPI

The current researches mainly extract the SPI through the deadline of learning tasks, and only measures the academic procrastination by the number of tasks completed late or overtime, which leads to the procrastination behavior of completing tasks at the last moment being judged as non-procrastination. In order to solve this problem, this paper ranks the total time spent on completing a task, and measures the procrastination degree of completing the task by ranking. The higher the ranking, the stronger the procrastination degree, thus we could obtain five static procrastination indicators: homework, test, sign-in, video and courseware.

Extraction of FPI

Although the SPI proposed in this paper can be used to analyze academic procrastination, it is difficult to reflect the fluctuation of academic procrastination. Therefore, in order to better describe the tendency of academic procrastination, this paper puts forward the FPI, which takes the standard deviation of the degree of procrastination to complete a task as the fluctuation of procrastination to complete the task. The larger the standard deviation, the greater the fluctuation of procrastination. The flow chart is shown in Fig. 2.

Table 1 Description of procrastination indicators

	Indicators	Meaning of indicators	Correlation variable
S P I	Static procrastination of homework	Rank of the total time taken to complete all homework	Homework publishing time Homework submission time
	Static procrastination of test	Rank of the total time taken to complete all the tests	Tests release time Tests submission time
	Static procrastination of sign-in	Rank of the total time taken to complete all sign-in	Sign-in release time Sign-in time
	Static procrastination of video	Rank of total time spent watching all videos	Videos release time Time to watch videos
	Static procrastination of courseware	Rank of total time spent browsing all courseware	Courseware publishing time Time to browse courseware
F P I	Homework procrastination fluctuation	Standard deviation of procrastination ranking when completing each homework	Rank of the time taken to complete each homework
	Test procrastination fluctuation	Standard deviation of procrastination ranking when completing each test	Rank of the time taken to complete each test
	Sign-in procrastination fluctuation	Standard deviation of procrastination ranking when completing each sign-in	Rank the time spent completing each sign-in
	Video procrastination fluctuation	Standard deviation of procrastination ranking when completing each video	Rank the time taken to complete each video
	Courseware procrastination fluctuation	Standard deviation of procrastination ranking when completing each courseware	Rank of the time taken to complete each courseware

3.3 Cluster Analysis

On the basis of the above, in order to get the characteristics of procrastination groups better, this paper makes cluster analysis on procrastination groups and divides them into different clusters. By analyzing the behaviors of different groups, this paper finds out the causes of academic procrastination, and puts forward relevant suggestions for teachers and students, so as to reduce the frequency of academic procrastination in online learning.

3.4 Visual Analysis

In order to show the clustering effect more intuitively and analyze the differences and connections of procrastination behaviors among different groups in more aspects,

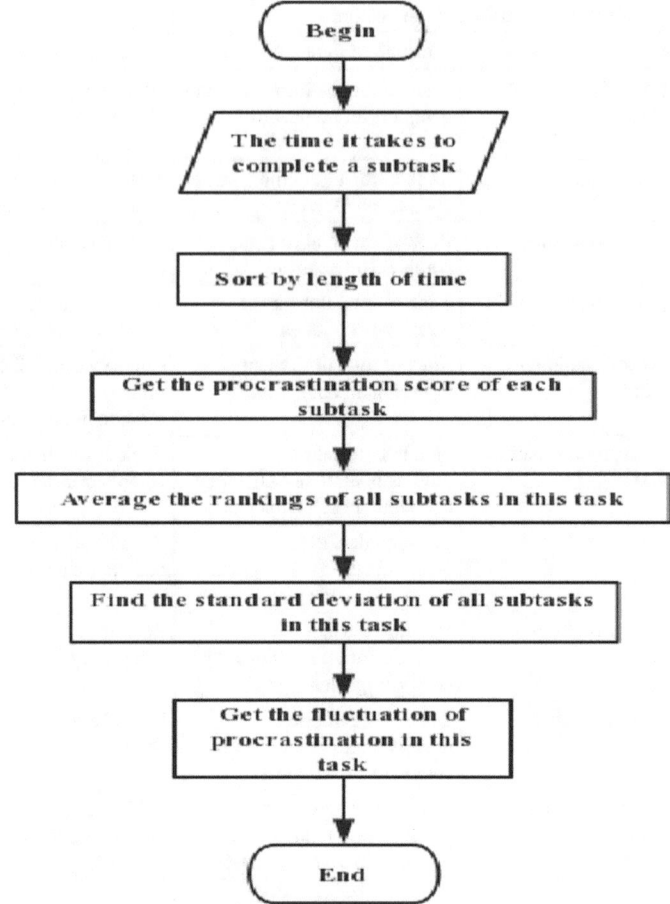

Fig. 2 Flow chart of calculation of the FPI

this paper firstly reduces the dimensions of the data of procrastination behaviors after clustering, then displays it by visual method, and finally visualizes the clustering centers between different groups. In this way, the differences of procrastination behaviors between groups can be shown more intuitively and the characteristics of procrastination behaviors of different groups can be analyzed intuitively.

4 Experimental Results

4.1 Data Set

In this paper, the study log data of 99 students in an online course is taken as the research object. After cleaning the data, the time of submitting homework for 19 times, submitting tests for 20 times, signing in for 30 times, browsing courseware for 32 times and watching videos for 154 times are finally obtained. On this basis, according to the indicator's calculation method proposed in 2.2 of this paper, two types of academic procrastination indicators are obtained, and the MTAP is constructed.

4.2 Comparison of Clustering Results

Two cluster experiments are conducted to verify that the proposed academic procrastination model is more suitable for cluster analysis. Experiment 1 is a clustering experiment based on the combination of the SPI and the FPI proposed in this paper. Experiment 2 only uses the SPI mentioned in this paper for clustering experiments. Experiment 1 and Experiment 2 both use K-means clustering algorithm to divide procrastination groups into K clusters. Because this paper wants to divide students into three types of procrastination groups, the final value of K is 3. In order to facilitate the observation of data distribution, this paper reduces the dimension of data, and the experimental results are shown in Fig. 3.

It can be clearly seen from Fig. 3A that the cluster boundaries are clear, there are few intersections around the center points of the clusters, and the classification of clusters are reasonable. The cluster boundaries in Fig. 3B are vague, the intersection around the center of the cluster is obvious, and the classification is not reasonable. Therefore, it can be concluded that the academic procrastination model proposed in this paper, which combines SPI with FPI, can better cluster the procrastination groups.

4.3 Evaluation of Clustering Effect

This paper calculates the Silhouette Coefficient and DB Index of the clustering experiments in Experiment 1 and Experiment 2 respectively (in Table 2) which could better illustrate that Experiment 1 has better clustering performance than Experiment 2.

By comparing the two kinds of clustering evaluation indexes, it can be found that the Silhouette Coefficient and DB Index of clustering using both the SPI and the FPI are better than clustering using only the SPI, which further proves that the MTAP proposed in this paper is more suitable for clustering analysis.

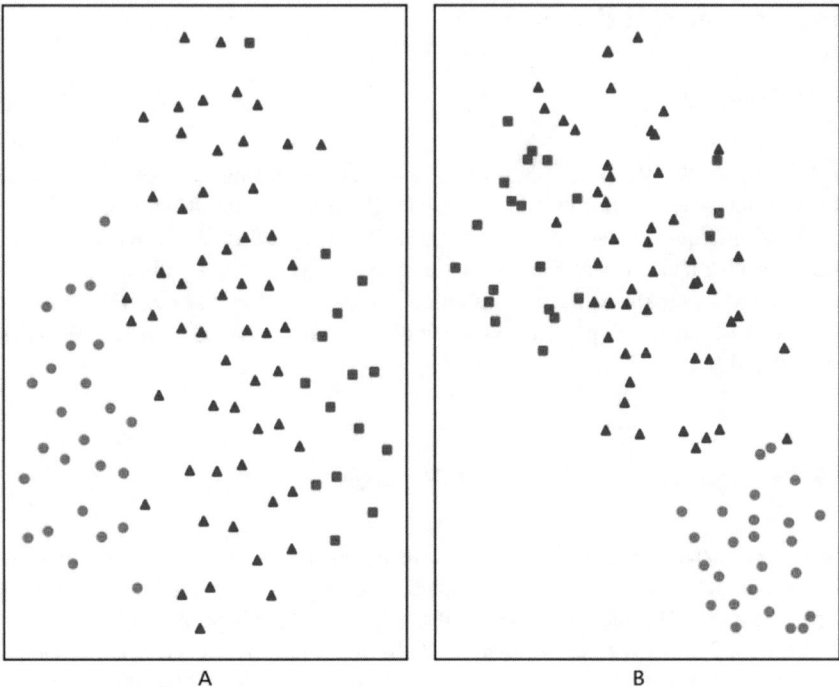

Fig. 3 Clustering results. Fig. A shows the clustering results of Experiment 1, Fig. B shows the clustering results of Experiment 2

Table 2 Silhouette coefficient and DB Index		Experiment 1	Experiment 2
	Silhouette coefficient	0.318	0.285
	DB Index	0.918	1.254

4.4 Visual Analysis and Suggestions

To better describe the characteristics of different procrastination groups and find out the causes of procrastination behavior, this paper visualizes the clustering centers between different groups based on the clustering experiment. The visualization results are shown in Fig. 4. The abscissa represents the attribute characteristics of procrastination behavior, and the ordinate represents the procrastination degree of each attribute characteristic among groups.

It can be seen from Fig. 4 that the three types of learners have obvious differences in task procrastination. According to the differences procrastination in task in homework, test, sign-in, courseware and video, they are divided into three types of

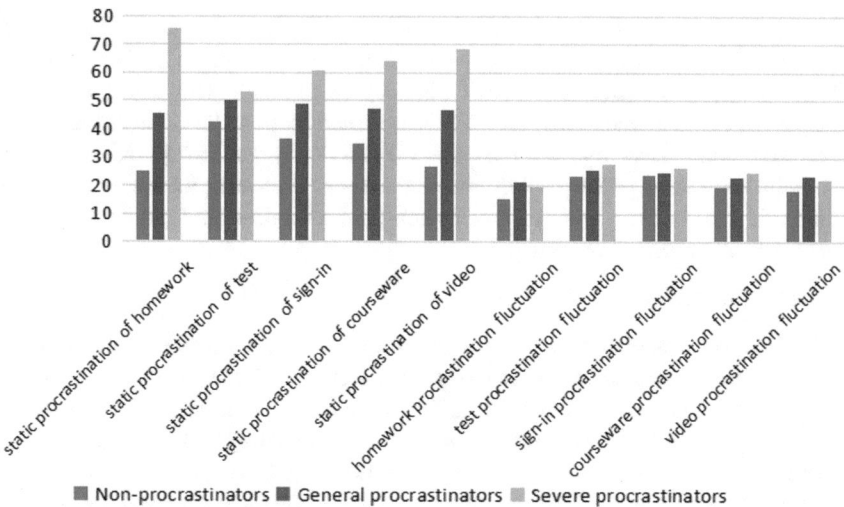

Fig. 4 Procrastination group characteristics

procrastination learners, namely severe procrastinators (cluster 1), general procrastinators (cluster 2) and non-procrastinators (cluster 3). The characteristics of each type of learners are described as follows:

1. Severe procrastinators-cluster 1

Severe procrastinators account for 26% of the total learners, that is, more than a quarter of the students have procrastination behavior. And their procrastination indicators values in completing homework, tests, sign-in, courseware and videos are the highest, which is delayed to a great extent. In addition, the fluctuation of procrastination degree is the biggest in completing the three tasks: test, sign-in and courseware. The procrastination degree is smaller than that of general procrastinators, but larger than that of non-procrastinators. In general, the procrastination degree of these learners is less volatile when it comes to completing homework and watching videos, which is high in most cases. Therefore, teachers should spend more energy on homework and videos, always urge severe procrastinators to complete homework submission and videos viewing as soon as possible, and actively guide them to complete the three tasks of tests, sign-in and courseware, so as to improve their interest in learning. It is hoped that this will convert them into general procrastinators or even non-procrastinators, thus improving their learning performance.

2. General procrastinators-cluster 2

General procrastinators account for 60% of the total learners, and their procrastination degree in completing the tasks assigned by teachers is lower than that of serious procrastinators, but higher than that of non-procrastinators. It shows that a large number of students still finish the learning tasks assigned by teachers seriously and have certain self-management ability and self-discipline ability. But the fluctuation of procrastination degree in completing both tasks of homework and videos is higher than the other two types of learners, which shows that students may change among these three types of learners because of difficulty and psychological factors when they complete these two tasks. Therefore, teachers should focus on this kind of learners in homework and videos, urge them to complete tasks early, and always guide and care about the knowledge problems they encounter when completing homework, actively guide this kind of learners to change to non-procrastinators, and reduce or control the transition to serious procrastinators.

3.   Non-procrastinators-cluster 3

14% of non-procrastinators are lower than the other two types of learners in terms of the degree of procrastination in completing tasks and the fluctuation of procrastination in completing tasks, indicating that a small number of students have strong interest in this course, keep great enthusiasm for completing tasks and have strong internal learning motivation. Therefore, teachers don't need to spend a lot of energy and time on such learners, and students can actively complete their learning tasks.

5   Conclusion and Future Work

In order to describe the academic procrastination in online learning more comprehensively and simply, this paper constructs an academic procrastination model which includes both the SPI and the FPI on the basis of related research. Through cluster analysis, students are divided into three types of academic procrastination groups, and the group characteristics of different learners are analyzed visually, so that the differences of the characteristics of different procrastination groups are obtained, and corresponding suggestions are put forward for teachers and students to reduce the occurrence of procrastination behavior. The results show that the clustering effect of the procrastination model proposed in this paper is better than that of the model containing only the SPI, and it can better describe the characteristics of procrastination. Among the three types of academic procrastination groups divided in this paper, the severe procrastinators have the highest degree of the SPI in completing various learning tasks, and have great procrastination, while the fluctuation of procrastination in homework and videos is small, which is a long-term procrastination student. However, the general procrastinators have great the fluctuation of procrastination in five learning tasks, and they have a great tendency to change into other two types of learners. Non-procrastinators have lower procrastination in both the SPI and the FPI. In the future research work, online academic procrastination model can be constructed based on multiple views of online learning behavior.

Acknowledgements This work was supported in part by the National Natural Science Foundation of China (No.61876101, 61802234) and the Innovation and Entrepreneurship Training Program for College Students of China (No. S202010445045).

References

1. Wan Wenjing, Xi Yong. Research on the cause of college students' academic procrastination in online learning: Qualitative analysis based on grounded theory (in Chinese) J. China Medical Education Technology, 2020,34(06):717-723.
2. Cerezo R, Sánchez-Santillán M, Paule-Ruiz MP, Núñez JC (2016) Students' LMS interaction patterns and their relationship with achievement: A case study in higher education. Comput Educ 96:42–54
3. Visser L, Korthagen F, Schoonenboom J (2015) Influences on and consequences of academic procrastination of first-year student teachers. Pedagog Stud 92:394–412
4. Tuckman BW (2005) Relations of academic procrastination, rationalizations, and performance in a web course with deadlines. Psychol Rep 96:1015–1021
5. Cerezo R, Esteban M, Sánchez-Santillán M, Núñez JC (2017) Procrastinating behavior in computer-based learning environments to predict performance: A case study in Moodle. Front Psychol 8:1403
6. Xiong luying, Guo xingjun, Jiang qi, Zheng qinhua. Study on Learner Academic Procrastination in Online Learning Environment (in Chinese) J. E-educ Res, 2019,40(12):29–35+50.
7. Yang Xue, Jiang Qiang, Zhao Wei, Li Yongfan, Li Song. Research on Online Learning Procrastination Diagnosis and Intervention Based on Learning Analytics in the Era of Big Data (in Chinese) J. E-educ Res, 2017,38(07):51-57.
8. Hooshyar D, Pedaste M, Yang Y (2019) Mining Educational Data to Predict Students' Performance through Procrastination Behavior. Entropy 22:12
9. Yang Y, Hooshyar D, Pedaste M (2020) Predicting course achievement of university students based on their procrastination behaviour on Moodle. Soft Comput 24(24):18777–18793
10. Michinov N, Brunot S, Le Bohec O, Juhel J, Delaval M (2011) Procrastination, participation, and performance in online learning environments. Comput Educ 56(1):243–252
11. Akram A et al (2019) Predicting students' academic procrastination in blended learning course using homework submission data. IEEE Access. 7:102487–102498

Alternative Digital Credentials—UAE's First Adopters' Quality Assurance Model and Case Study

Samar A. El-Farra, Jihad M. Mohaidat, Saud H. Aldajah, and Abdullatif M. Alshamsi

Abstract As the leading applied higher education institution in the United Arab Emirates, the Higher Colleges of Technology is engaging Alternative Digital Credentials (ADCs), including micro-credentials and digital badges, as part of the nation's flagship digital transformational journey which has been propelled by expectations of governmental and business sectors, as well as learners. In addition, the diversity in alternative digital credential providers is placing higher education institutions on the verge of an "identity crisis". With the scenario of a sound technology foundation, the focus of this paper is on the educational aspect of alternative digital credentials, as the fourth generation of qualifications' alternative digital credentials are challenged by skepticism and confusion, due to the wide spectrum of providers. Quality assurance standards place higher education institutions in a privileged position for the provision of alternative digital credentials. We developed an outcome-focused quality assurance standards model and a subsequent case-study illustration, designed to pivot a nationwide application and constitute a blueprint for the future of ADCs' provisions that capture learners' "soft skill" competencies and character qualities within the profession "hard skills" context of their prospect careers. To our knowledge, no case studies, with a focus on the educational aspect of implementation, have been published.

Keywords Alternative digital credentials · Micro- credentials · Digital badges · Higher education · Academic Quality Assurance · Case study

1 Introduction

Higher Colleges of Technology (HCT) is the largest federal Higher Education Institution (HEI) in the United Arab Emirates (UAE), founded in 1988, educating 23,000 male and female students. It has awarded more than 92,000 undergraduate degrees in Applied Media, Business, Computer Information Science, Engineering Technology,

S. A. El-Farra (✉) · J. M. Mohaidat · S. H. Aldajah · A. M. Alshamsi
Higher Colleges of Technology, P.O.Box: 25026, Abu Dhabi, United Arab Emirates
e-mail: selfarra@hct.ac.ae

Health Sciences, Education, and Military and Security. In additional to the international paradigm shifts in tertiary education, the national strategies, such as the Dubai 50-year Charter, UAE Centennial 2071 and UAE Employment Strategy are placing flexibility on how HCT delivers its programs and offerings, to allow for re-skilling and up-skilling opportunities. The HCT Strategic Vision 2026 has core strategic pillars that are related to micro-credentials (MCs), including having a central education file for every citizen, evaluating and establishing MCs as a new way of flexible academic credentialling, and personalized "anytime, anywhere" life-long education [1]. Moreover, as a government funded HEI, HCT has a strong emphasis on graduate employability, a key performance indicator metric that formally ties the performance of HCT to the number of its graduates receiving job opportunities related to their credentials. As work placements have positive impact on learners and employer [2], all HCT programs are attached to work placement courses or course components.

A case for HEIs' change of identity is becoming evident, driven by the emergence of digitalization, globalization [3], automation and Artificial Intelligence [4], and life-long education imperatives [5]. While traditional transcripts continue to lag in connecting learners' capacity with job requirements, non-university, Alternative Digital Credentials (ADCs) are gaining popularity as they respond to young adults' preferences for shorter and more job outcome-based learning [3, 6]. This widespread popularity mandates regulated certification to improve the confidence of employers. Moreover, the digital component of ADCs keeps pace with the emerging reliance on digital searches for hiring practices (ICDE,2019). Furthermore, HEIs are widely accepted as the academic heartland, but the increased interest in ADCs is challenging their existing provision and business model [6]. When HEIs successfully respond to the opportunities and challenges brought about by skills gaps and unemployment, an anticipation of increased numbers of learners with relevant skills for employment, and entrepreneurship can be assumed [7]. Lately, HEIs offer ADCs for visibility and reputation; to navigate new pedagogies or technologies; to gain financial benefits of revenue or prudence; and/or to respond to learners' and employers' changing expectations [8]. Worldwide, governments are embracing ADCs and expecting HEIs to play a role in the provision and management of ADCs. For example, the Australian government allocated $4.3 million to build and run ADC opportunities [5, 9], and suggested national qualification framework adjustments to allow for flexible ADCs that are responsive to societal needs. Meanwhile, ADCs are a challenge to traditional degrees' curricula and according to International Council for Open and Distance Education (2019): "The failure to connect traditional higher education to workforce needs is an increasingly evident gap" (p8). Nonetheless, the benefits of ADCs critically depend on the readiness of providers and the quality of provision [3]. Furthermore, focus of employers on the relevant "soft-skills" or Competencies and Character Qualities (CCQs) in addition to the hard competencies is becoming more evident while some CCQ requirements and expectations are different from one job to another [10].

Since 2013, a fourth generation of qualification frameworks, relating to credentials of twenty-first century skills, implied a number of policy challenges, including new skills and ADCS as new forms of credentialing [7]. Alternative Digital Credentials is a term used to incorporate MCs, Digital Badges(DBs), and Professional Certifications and Qualifications (PCQs) [3]. Since 2017, HCT programs have had industry recognized (PCQs) embedded within undergraduate programs, offered by awarding bodies, such as Adobe ® and IBM. As new concepts, we are investigating MCs and DBs exclusively in this paper. Thus, wherever the term ADCs is used it is referring to MCs and/or DBs.

As technology and education are two quality components of ADCs' ecosystems, it is important to separate the educational from the technological concerns and follow best practices in both domains [6]. Despite the wealth of literature on the technological and the combined technological and educational concerns, to our knowledge, no case studies with a focus on the educational aspect of implementation have been published. ADCs of uncertain quality are vulnerable to exploitation by unscrupulous providers. [5]. Quality assurance (QA) standards builds confidence, recognition, portability and learner protection [5]. With emphasis on the educational aspect, we propose a model that was benchmarked against the United Nations Educational, Scientific and Cultural Organization (UNESCO) report by Chakroun and Keevy [7] where four areas of success forecast and identified stakeholders' trust and acceptance of ADC content at the academic level as the main two quality challenges to the HCT-ADC ecosystem. The proposed Value, Resources, Standards, Validation, Accessibility, Support (VRSVAS) model mitigates the content challenges, by means of building trust and acceptance via:

1. Outlining a distinct ADC value in relation to employability.
2. Outlining evidence of resources' availability.
3. Declaring transparent standards for ADCs.
4. Sound and life-long documented validation processes.
5. Incorporating individualized accessibility characteristics.
6. Offering learner support avenues.

Finally, this paper provides a step-by-step approach of ADC development while referring to real case study exemplar, within a technologically well-established HEI. The case study key construct for a hybrid ADC of an MC, integrated with a DB, was developed in response to the surge in thorax medical imaging demands due to the COVID-19 pandemic. Awarding standards of the case study are numbered and used as a reference in each of the sections of the VRSVAS model for the purpose of example illustration.

2 The Case for Change

There are three different disruption factors that are currently affecting and changing the conventional higher education as we know it. Firstly, learners' interest in full credentials has dropped by 20% and only 40% of new Bachelor enrollees graduate within the expected duration [3]. Secondly, the Organization for Economic Cooperation and Development (OECD) member countries report a growth in higher education attainment, which corresponds to no-growth in employment rates. Finally, in their 2019 report the ICDE encouraged HEIs to embrace full-scale implementation of ADCs in response to the emerging societal and employability needs [6].

2.1 Learner and Employer Centricity

Undergraduate learners report that they prefer short, focused, intense, and workplace-relevant courses. As high as 65–72.4% of millennials show moderate to strong interest in earning ADCs [11]. Now, more than before, people with existing careers, or with full credentials searching for a job, are seeking up/reskilling opportunities [3]. The short half-life of skills driven by technology [8] and maintaining/enhancing the relevancy of learners' attributes to the Fourth Industrial Revolution are also becoming motivations for earning ADCs [12]. The characteristics of ADCs, such as shorter duration and dynamic and agile provisions to specific job profiles, give ADCs the capacity to bridge the gap between HEIs' lengthy cycles of achieving and/or changing learning outcomes and the fast changing job proficiencies [3].

From the employers' standpoint, a "skill gap" is widening in the job market between what traditional credentials are offering and the market needs [8, 13]. ADCs have the potential to address market swings by providing: (1) short competency-based trainings with targeted objectives, (2) comprehensive details about the attained competencies, and 3) embedded evidence, such as ePortfolios to showcase a learners' capacity [6]. In a recent survey on adult learners, 81% of the respondents reported that their ADC was useful for keeping them marketable to employers; 82% reported those certificates were very useful in landing a job; 66% reported improved work skills and 80% reported the certification helped them maintain a current job [14]. Also, employers report a gap between resumes and transcripts [12] and the unreliability of traditional resumes as hiring tools. In a review of 5,500 resumes, over 80 percent had some inconsistencies and 12 percent had false information [15]. Moreover, job seekers' digital footprint is becoming a conclusive employment factor with the emergence of hiring processes that are Artificial Intelligence and digital search-dependent [6].

2.2 Articulating a New Social and Economic Identity for HEIs

As sole providers of educational programs that issue academic credentials, HEIs continue to supply employers with documented competencies' attainment. In the last decades, financial costs of education have encouraged learners to shift their focus to online learning and competency-based education ACDs [8]. The current ecosystem for credentialing is penetrated by many non-HEIs becoming credential issuers, thus

causing a serious challenge in the role of HEIs as the traditionally dominant credentialing entities [3, 6], while calls for HEI to incorporate ADCs in their offerings
continue to grow [12]. More than ever before, it is becoming evident to learners,
employers and policymakers, that HEIs' traditional degrees are no longer optimally
serving society, thus, posing existential questions for the HEIs [8]. Consequently,
international reports advise HEIs to implement ADCs [6], and several HEIs are
transforming their modus operandi by "unbundling" their larger construct programs
into targeted, and more "micro", episodes of learning [8].

There is wide acceptability of credit-bearing education [16]. Learners are more
inclined to prefer ADCs offered by prestigious HEIs [3] which could be attributed to
the quality standards that HEIs follow in issuing credit-bearing education outcomes
that are assessed as a whole, and represented in regulated transcripts. Furthermore,
when HEIs provide credit bearing ADCs that stack into formal qualifications, trust
and acceptance is enhanced and equivalency and recognition of previous learning
becomes possible [17]. Ultimately, ADCs resonate well with the existing HEIs' high-
profile competency-based education, prior learning assessment, and online learning
themes [8, 16] and ePortfilios [16].

With a wide spectrum of ADC providers, lack of common standards creates an
atmosphere of misperception, confusion and skepticism [3, 6, 8]. Currently, ADCs
can be issued by anyone with no QA, creating a 'jungle of badges' [7], casting a
serious threat to the credibility, notion, and acceptability of ADCs [6, 7]. Given their
existing, well-established awarding standards and QA processes, HEIs are well-
positioned as best-fit ADC providers. Similar to many other first adopters, HCT
is engaging the ADC transformational journey, propelled by the employers' and
learners' expectations. Moreover, as the leading applied HEI in the UAE, we are
driven by the need to respond to disruptions introduced by the evolving social and
economic changes, while providing a national flagship role in ADC provision with
outcome-focused quality assurance standards.

3 Scope and Aim

In the lack of a universally agreed definition and scope of ADCs [5, 12, 17], the
novelty of ADCs manifests variations in their scope, outcomes and naming conven-
tions. Thus, it is important for HEIs to establish standards and definitions that describe
the relationship, similarities, and differences between ADCs and traditional credentials [8]. The scope of this paper is limited to ADCs as part of a formal education
program at the tertiary education level, excluding PCQs because those are already
embedded within HCT curricula since 2017. To set the terminology tone for this
paper, we will refer to MCs, DBs, or the combination of both as ADCs. In literature,
DBs and MCs are sometimes used interchangeably but common characteristics that
apply to the scope of this paper are the ADC functionalities of:

1. Storing sharable/displayable information that signifies the achieved level of competency with embedded meta-information used to validate evidence of educational achievements [13, 16, 18]
2. Affording institutional online record that states expectations, standards and evidence of an achievement [7, 19].
3. Offering smaller modules of learning than conventional academic awards that can be stacked to a full credential [6].

To set the naming convention, and to have a vivid segregation between MCs and DBs, we introduced distinguishing outcome-based characteristics, which are:

1. When an HCT-offered ADC can be "independently" mapped to an in-demand job, it is considered an MC,
2. when an HCT-offered ADC represents general CCQs "soft-skill" needed for a job but does not "independently" match a job profile, it is categorized as a DB, and,
3. when work placement is part of the ADC, it is categorized as a DB that is linked to evidence of attainment.

In addition to other literature, within the context of the ICDE, [6]; Kato et al. [3], and the UNESCO report by Chakroun and Keevy [7], this paper seeks to review an ADC model suitable for the UAE context, with transparent features that enhance the future ubiquity and interoperability at the international level. The case-study example reviewed provides better understanding and constitutes a blueprint for the future of HCT and similar HEIs planning for ADCs provisions.

4 ADCs Provision Model

ADCs' provisional instructional design models should be driven by pedagogical goals coupled with rigorous assessments and corresponding embed metadata, with links to criteria and final work product of the ADC earned [20].

HCT categorises ADCs into four main types, based on content and value of employability, which are:

Category 1   MC composed of a course, or a collection of courses, in a specific job with integrated DB(s) in CCQs that are job/career specific within a work placement environment (for example, radiographer assistant, exhibiting CCQs like empathy and communication while performing medical imaging tasks on real patients) [case study 6.2].

Category 2   DB in general and widely applicable competencies not match a specific job profile requirement alone for example, communication, or basic mathematics. Such DBs can have levels of proficiency as suggested by [6], i.e., DB of communication level 1, 2, 3 etc.

Category 3   MC and/or DB in intermediate generality competencies, usually required for reskilling in response to new concepts introduced to a

Category 4   wide range of jobs/careers (e.g., basic statistics, or project management skills).

Category 4   MC and/or DB in specific competencies that match specific upskilling jobs/career demands (e.g., sets of skills required to be promoted to a managerial position).

Because CCQs are sometimes domain specific, an award in a certain CCQ, such as growth mindset, is more generalised regardless of the job. However, some CCQ requirements and expectations are different from one job to another [10]. Even within the same Faculty of Health Sciences, for example, communication skills for a social worker mandate maintaining proper eye contact with clients, while the emphasis on team members' eye contact in an emergency room, while transferring a stroke patient is almost non-existing. Furthermore, the CCQ priorities per job profile are different. For example, while empathy might be on the top of the CCQs' list for the healthcare providers, empathy for business management might not supersede leadership, and critical thinking. To differentiate CCQs within the professional context, HCT offers DBs embedded within MCs for those non-cognitive CCQs (e.g., collaboration for programmers or, empathy for nurses) within the respective ecosystem. In summary, DB(s) of CCQs are fundamental ADC components when they are job/career profile specific within a work placement environment but do stand alone for general and common CCQs. The provision, as such, presents flexible study pathways for a wide spectrum of learners up/reskilling or undergraduates and a wide range of selection ranging from highly specialised to widely generalised awards.

HCT's initial adoption of the ADC model is anchored on the three models of integration, namely, modular, embedded, and prior learning recognition models [3]. Highly dependent upon the nature of the domain and level of knowledge and competency, an ADC can be embedded within a program as one or more courses within a program. For example, the Bachelor of Science Medical Imaging (MI-BSc) program structure can be broken down to speciality "Modules", each of which is equivalent to an ADC that matches a certain job competency. In this example, the MI-BSc is broken down to ADCs of General Radiography, Mammography, Trauma Imaging, Magnetic Imaging Resonance (MRI), and Medical Imaging Leadership (see Fig. 1). The modular approach outlines the relationship between ADCS, and full HCT qualifications. As advised by Noonan 2019, and addresses inherent risk of "stack-ability" if the "stack" ADCs are greater than a full credential described by Chakroun and Keevy [7]. Finally, ICDE, report argues that modular models come with an associated risk, pertinent to the lack of stakeholders' engagement. Consequently, this avenue is approached in a careful balance between what employers need and what the professional and academic expectations and standards are, per program or domain [6].

The second model that HCT adopts is based on embedding the ADCs within the program, the best example of which is PCQs, which is beyond the scope of this paper. Instead the discussion also refers to a Mammography ADC that is embedded within the program (see Fig. 2). Previous learning, if deemed equivalent, will account for an ex-post recognition of prior learning into an ADC offered by HCT. The recognition of

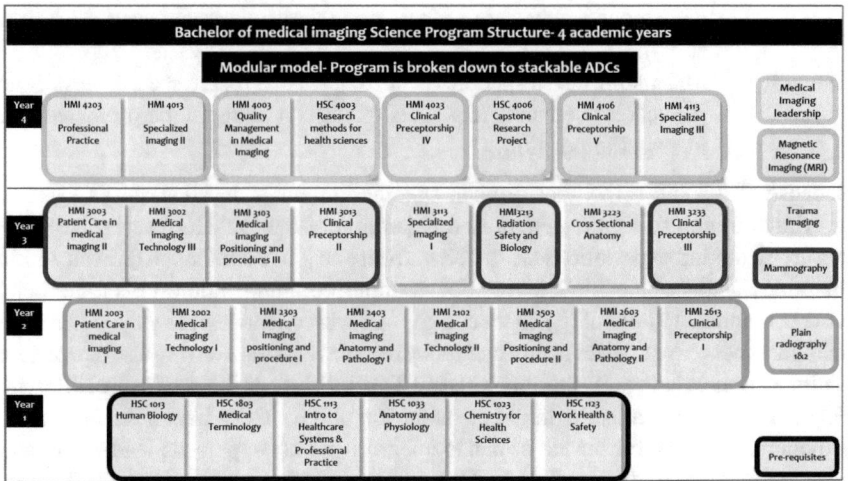

Fig. 1 Medical imaging B.Sc. program structure by course and academic year broken down to modules with five distinct ADCs (Medical imaging leadership, MRI, trauma imaging, Mammography, and two levels of plain radiography) with common pre-requisite courses in black and clinical preceptorship courses offered as DBs. Note that some courses are "double-dipped" across two different ADCs for example, the double-dipping of HMI 3233 clinical preceptorship III for ADCs of Trauma imaging and Mammography (yellow and red)

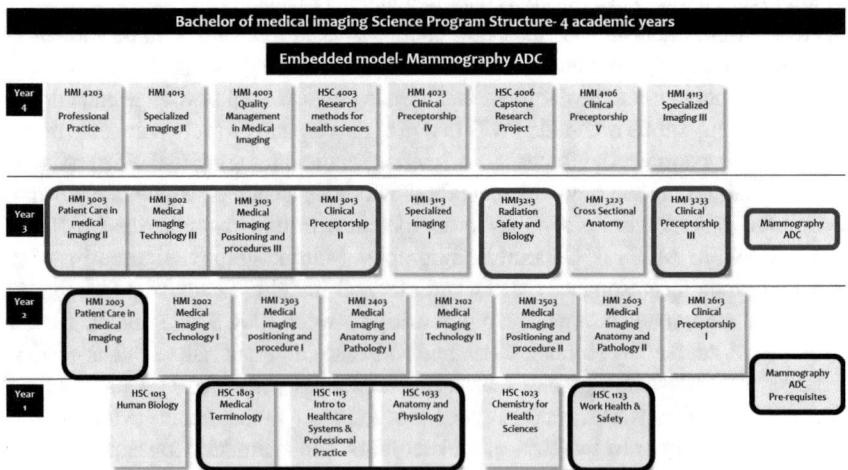

Fig. 2 Medical imaging B.Sc. program structure by course and academic year with an embedded ADC of Mammography comprised of six courses and five pre-requisite courses in black, with the corresponding two clinical preceptorship courses offered as a BD

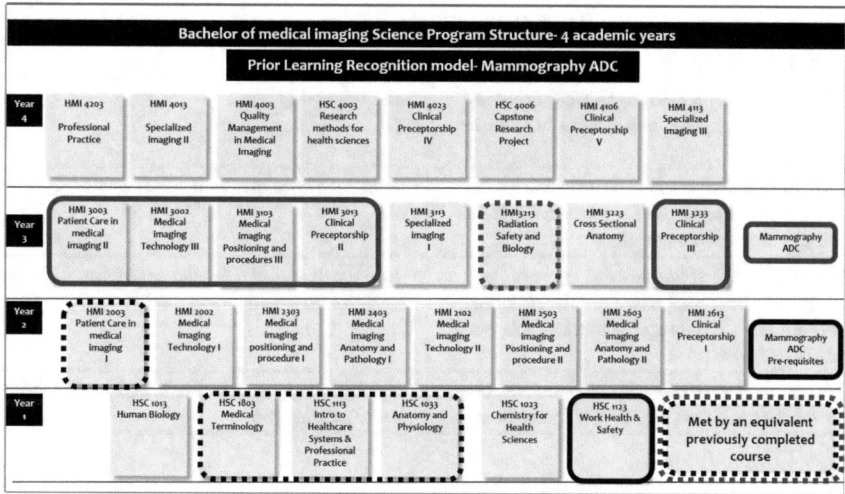

Fig. 3 Medical imaging B.Sc. program structure by course and academic year with an embedded ADC of Mammography comprised of six courses and five pre-requisite courses in black, with the corresponding two clinical preceptorship courses offered as a BD. One of which is a prior learning recognition equivalent (dotted red) and four are pre-requisite courses that are prior learning recognition equivalent (dotted black)

prior learning model adopted is based on the equivalency process of prior achieved courses, by recognized and authorized entities (see Fig. 3). In all three models, work placement clinical preceptorship courses come in DBs and are "double-dipped" courses that could potentially meet two different ADCs. This implies adjustments in training and in assessment, based upon where the course is situated; especially true for up/reskilling and less prominent for undergraduates. For example, for a learner to meet the requirements of the Mammography ADC she will be not be managing any trauma patients while attempting course HMI 3233 Clinical Preceptorship III, and vice versa for a learner who is enrolled in the same course. However, to complete an ADC in Trauma Imaging s/he will not perform any Mammograms on patients as an ADC requirement (see Fig. 1). A Summary of ADC value in HCTs' provision is provided (Table 1).

ADCs' components must be structured in a comprehensible and clear statement on how those ADCs will be addressing stakeholders' expectations [21]. Academics like to think of credentials as units of time, after all, it is a global trend to identify an award by the time spent to earn it. While having a metric, such as credit hours, might constitute a fair ground of reference, it might not serve the concept of ADCs in the best way. ADCs are built around flexibility and competency, so if a learner can perform a task with a fair amount of consistency over a fair amount of time, with an acceptable accuracy, s/he should qualify for the job. To address the opposing expectations of meeting standards while extending flexible opportunities, HCT balanced a collection of approaches. For example, HCT adopts the Unites States definition of an ADC, as a

Table 1 Summary of ADC value in HCTs' provision

ADC offering	Value
MC with integrated DB(s) in CCQs that are job/career specific within a work placement environment	• *Employability* • *Change of profession*
DB in general and widely applicable competencies but would not match a specific job profile requirement alone	• *Better performance in a job or in life skills*
MC and/or DB in intermediate generality competencies usually required for reskilling in response to new concepts introduced to a wide range of jobs/careers	• *Reskilling required for the emerging skills to perform an existing or a modified job*
MC and/or DB in specific competencies that match a specific upskilling job/career demand	• *Upskilling required to perform a higher role in an existing job*

learning experience that is more than one course and less than any of the full degrees [3, 22] offered by HCT. Also, HCT-ADCs have the potential to accumulate "stack" contributing to a full degree, an approach that is adopted by the European Higher Education Area [3, 23]. Similarly, the prescribed range of 5–40 credits adopted by the Oceania region [21] fits well with our model because HCT courses are never less than three credits, and having more than 40 credits for an ADC is counterintuitive because HCT offers diplomas that are 60 credits long, rendering ADCs more than 40 credits is not appealing to learners. Conversely, limiting ADCs' profile by credit numbers might not be practical in some instances. For that we are also adopting the more comprehensive ADCs' scope described by ICDE, [6] which envisions the focus on taught and assessed workforce competencies, within work placement settings, as the main discriminator of ADCs from the traditional HEI credentials [Case study 6.2]. In summary, the workaround of the somewhat rigid credit count is two-fold: (1) using in-demand job profiles as value indicators and, (2) offering DBs for competencies that might not stand alone to match a job profile, but are equally compatible as they stack, thus achieving a point of equilibrium between the opposing expectations.

5   Quality Assurance Attributes

The functional duality between traditional degrees and ADCs is leading to confusion amongst stakeholders calling for clear QA standards [7], and credit-bearing credentials are expected to be QA verified [5]. Lack of QA poses a significant threat to the credibility of ADCs, and contrariwise the existence of QA processes sets constraints on the flexibility of traditional degrees [5, 7]. Furthermore, calls for action on QA of ADCs, including alignment to national qualification frameworks, is communicated at government levels [21]. For decades, QA standards have been deeply rooted and practiced in the HEIs' awarding processes and modus operandi. On the contrary, the variety of other ADC providers introduced different types of standards with lack of common agreement [7].

In our approach we first identified the functionalities that must be addressed within our QA scope by mapping with the Chakroun & Keevy, proposed models of digital credentials' ecosystem. The model outlines seven interrelated functional stakeholders that makeup ADCs' environments [7], based on that HCT provisions are geared towards ensuring that ADCs are:

1. **Used** by HCT undergraduates and alumni, and UAE population seeking up/reskilling.
2. **Provide** employability outcome-based ADC components and curricula.
3. **Awarded** by HCT using the same safe and secure systems that are used for the current students' records.
4. **Evaluated** and verified based upon transparent academic assessment standards.
5. **Verified** by employers and stakeholders based upon academic standards, and evidence of attainment.
6. **QA practiced,** and enforced, by qualifications authorities, whenever adjustments to accredited curricula are in place. In addition to following QA processes used for full credentials, the application of the VRSVAS model to address the special characteristics of ADCs, and
7. **Aimed at convention** by international agencies, such as UNESCO and the OECD can only be achieved with time, and in collaboration with local HEIs, as a starting point before claiming any full success at the national level.

Additionally, we reviewed the Dowling's [7, 24] model developed to analyze the success forecast of high-level digital ecosystems for credentialing. The model identifies five architectures; central repository, exchange network, hub-and-spoke, badge framework, and public blockchain. Each architectural element is examined for benefits and challenges against four elements:

1. Scope and functionality.
2. Mobility.
3. Security, trust and privacy.
4. Participation.

As mentioned before, the focus of this paper is exclusively on the academic aspect of ADC development, so we analyzed challenges of the "credential framework" architecture against the four elements and proposed the VRSVAS quality model to mitigate the identified challenges. We found that the four success forecast areas are mainly challenged by learners' acceptance and employers' trust. Our model is a sequential top-down QA paradigm that addresses credential framework challenges by:

1. Outlining a distinct ADC **value** in relation to employability.
2. Outlining evidence of **resources** availability.
3. Declaring transparent **standards** for ADCs.
4. Ensuring sound and life-long documented **validation** processes.

5.   Incorporating individualized **accessibility** characteristics.
6.   Offering learner **support** avenues.

The Value, Resources, Standards, Validation, Accessibility, Support (VRSVAS) model mitigates the credential framework challenges by an outcome-focused dynamic provision, focused towards gaining learners' acceptance and employers' trust.

5.1   *Value*

Despite disagreement on what constitutes ADCs, consensus is observed about the linkage between the value of ADC when it is aligned to a certain job requirement by training and verifying competencies of specific workforce task(s) [3, 6, 17]. To enhance ADCs' value, the ultimate purpose and outcome must be clearly communicated to stakeholders [20]. The (NZQA) requires evidence of a need from employers to justify the offerings of ADCs [21]. For that, any HCT offered ADC will collectively or individually be compatible with at least one in-demand job/career profile.

Employees' involvement with stakeholders not only increases ADCs' ubiquity, but also promotes a sense of ownership[20]. Also, aligning with international entities, such as the International Standard Classification of Occupations (ISCO-08), provides confidence to employers that the ADC issued is not representing a job profile that was "fabricated", based on assumption [25]. To develop a sound value proposition, ADC provisions are mapped to professional endorsements, or a valid need analysis, and to an ISCO-08 classification [Case study 1.2–1.3]. Reviewing the four elements of success [7, 24], clear value establishment has the potential to improve the following success forecast areas:

1.   Scope
2.   Mobility
3.   Trust
4.   Participation is enhanced in combination with the validation and documentation discussed later.

5.2   *Resources*

While HEIs are potentially the best ADC providers, resources play an important role in the sustainability of provision [6], and capability and resources of ADCs' provision are required [21]. The current HCT human capacity, IT and logistical infrastructure puts HCT in the position of ADC hybrid provision. The full cycle of ADC creating, running, validating, documenting, authenticating and retiring is owned by HCT. HCT programs are responsible for content and learning material development, creation, delivery, learning support provision, validation, awarding and linking awards to evidence. Organizing the hybrid provision is reportedly desirable for better learning outcomes [26] and the diversification of delivery reduces workload on staff and presents flexibility for learners, who are looking for after-working-hours up/reskilling opportunities. Face-to-face delivery has a defined timetable and is required for lab sessions, some assessments, and work placement, such as clinical preceptorships, while theoretical sessions are pre-recorded and can be reviewed at the learners' pace. The final safeguarding tool used to ensure sustainability is that HCT offerings are based on demand and feasibility. For example, ADCs that are not a part of a regular academic year offering, and/or are beyond the maximum class capacity, are only initiated and offered on part-time recruitment feasibility and a cost-benefit analysis basis. Reviewing the four elements of success [7, 24], ensuring resource availability has the potential to improve the trust success forecast area.

5.3 Standards

As a new concept, standards of ADCs lack the consistency across the globe. However, maintaining transparent standards, and documented outcomes and scope, are key for stakeholders [8], while declared standards protect the learners' time, effort and money invested to obtain a certain award. Awarding standards are critically important, as full credentials, stating clear standards for awarding is used to cover useful information for all stakeholders being, the ADC's profile information, entry and completion requirements, teaching and training content, assessment and evidence [Case study 1-2-3-4-5-6]. Awarding standards are part of the published meta-data linked to the awarded ADC. Reviewing the four elements of success [7, 24], articulating clear expectations and attainment standards has the potential to improve the following success forecast areas:

1. Scope and functionality.
2. Mobility.
3. Security, trust and privacy.

5.4 Validation and Documentation

Similar to the agreement on value of relevancy is an equally strong accord on the importance of reliable validation of attainment in controlled and graded testing settings [3, 6]. Assessment processes are direct indicators of quality and recognition, and confidence by employers, [3, 5, 6, 27] Also, attendance-based certificates are less appealing, compared to criteria-based accomplishments [29]. Verification processes impact trustworthiness and ubiquity of the ADCs [7]. Moreover, NZQA requires ADCs to incorporate one or more assessment standards that certify the achievement of learners' capacity in workplace training, so competency-based assessment within authentic workplace settings are integral parts of HCT DBs [Case study 6.2] [21].

Like all other well-developed assessments, the rubrics and criteria of HCT-ADCs' assessment are designed and shared with all learners and graders to measure the competency and/or learning achievements. The characteristics of documenting metadata increase trust and value of ADCs, and benefit all stakeholders [16, 28, 29]. In HCT, undergraduates sign a pledge that contractually obliges them to abide to academic integrity policies and procedures for the entire study duration. Equally, learners enrolled for up/reskilling purposes are required to submit an academic integrity policy and procedure pledge as part of the admission process [Case study 2.1.4.]. All written assessment is proctored in HCT campuses [Case study 4.3] while assessments held in labs and clinical sites are performed under faculty and / or clinical training supervision [Case study 6.2]. HCT invests in building rapport through engagement of employers and, whenever possible, involves employers in evaluating learners within work placement settings [Case study 6.3]. Based on ICDE, 2019 recommendations, HCT declares a descriptive ADC full title that gives a clear idea about the scope of the award [Case study 1.1]. ADCs offered by HCT are unique awards that are not a duplicate, or a replacement, of HCT transcripts of any given program. If an award is part of a traditionally "transcripted" course, the award standards document will contain more granular admission requirements, and/or work placement, and/or evaluation/validation components. Moreover, the ADC issuing digital platform will be presently and securely linked (through Blockchain) to level descriptive competencies and learning achievements attained, teaching and learning methodology, validation assessments, and evidence/artifact/portfolio of successful completion, as stipulated in the award standards document [Case study 6.3]. Similar to the existing full credentials, HCT is accountable to secure issuance and maintenance of life-long immutable ADC records that ascertain and protect the validity of learners' proof of readiness for a job [2, 6]. Reviewing the four elements of success [7, 24], sound validation and documentation has the potential to improve the following success forecast areas:

1. Mobility.
2. Security, trust and privacy.

5.5 Accessibility

In order to graduate more learners with relevant skills for employment, HEIs need to ensure inclusive and equitable quality education and promote lifelong learning opportunities [7]. Obviously, HCT distinguishes offerings for up/reskilling purposes from undergraduate ADC enrollments in the sense that those learners are more vulnerable to drop out due to their life and job requirements and schedules. To enhance the accessibility of those learners, three main areas have been adjusted, namely, pace and duration, admission criteria and, attendance and assessment.

1. Pace and duration: While pace of study is limited by the time of the course/part of course offering for the undergraduates [Case study 3.3], mature learners

seeking re/upskilling awards have the option to extend their study of an ADC for a period that is up to double the time duration prescribed for undergraduates [Case study 4.4].

2. Admission: While undergraduates are expected to have a minimum admission cumulative Grade Point Average (cGPA), mature learners are exempted from this requirement if they have evidence of 5 years of experience practicing within a career field related or equivalent to job profile in question [Case study 2.1.2].

3. Attendance and assessment: Attendance rules and expectations for lab, work placement, and assessments are unified for both enrolment types. However, mature learners can opt to have a single final assessment and not the full set for the other five coursework assessments that must be completed by the undergraduates. Also, didactic attendance is not mandatory for the mature learners as they can review the recorded material at their own pace.

Furthermore, because ACDs are new to both learners and educators, if a learner fails an ADC, those results will not be available for external entities and employers to review in the learner's record, regardless of the number of attempts. This is critically important for CCQs as they tend to imply personal connotations. Finally, in the case of certain learners' physical requirements, those requirements are declared in the standards document, so learners with special needs or certain medical conditions are objectively informed about their enrolment eligibility [Case study 2.1.3]. Reviewing the four elements of success [7, 24], establishment of a clear inclusion criteria and having flexible approaches to address the diversity of learners and vulnerable learners like working mothers for example, has the potential to improve the following success forecast areas:

1. Scope and functionality.
2. Security, trust and privacy.

5.6 Learner Support and Awareness

There is evidence that coaching is becoming an integral role in learners' journeys and success in higher education [30]. We propose coaching learners as a suitable avenue in addressing the innovation and complexity of study pathways, introduced by ADC offerings, as it builds normalization, awareness and provides support to learners. Furthermore, an academic awareness program is developed to transform faculty and academics from instructors with coaching capacity training. Reviewing the four elements of success [7, 24], learner support and awareness has the potential to improve the following success forecast areas:

1. Scope and functionality.
2. Security, trust and privacy.

To this end, the VRSVAS model has to potential to enforce learners' acceptance and employers' trust through ADC offerings that have:

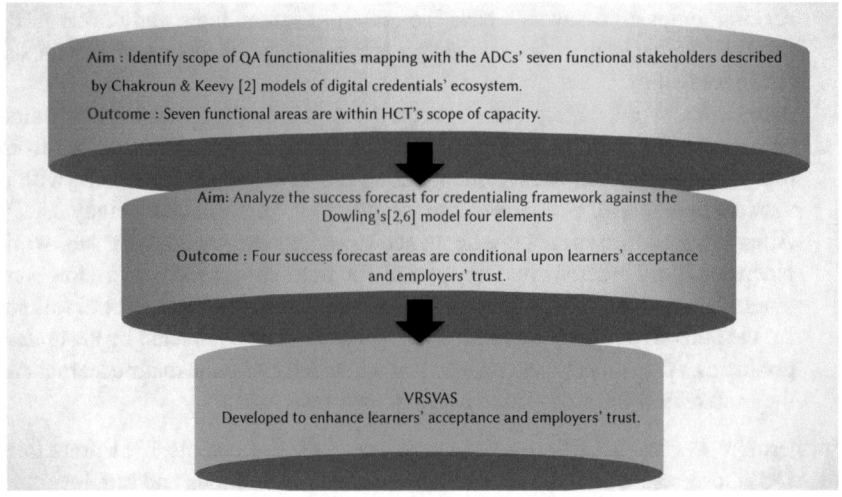

Fig. 4 Benchmarking methodology used to develop the VRSVAS model

1. Clear scope and standards, expectations, and attainment levels for better ubiquity and interoperability and potential international recognition of learning
2. Evidence of institutional capability and quality of provision.
3. Predictability of employment prospects.
4. Linkage of attainment to graded and evidence assessment incorporated in life-long immutable learner records.
5. Fixability and opportunities that address the unique individual requirements.
6. Awareness and engagement in making life-long decisions.
7. Relevancy though employers' engagement.
8. The characteristics that support the social and economic responsibility of HCT.

Figure 4 depicts the benchmarking methodology used to develop the VRSVAS model.

6   Case Study

The case study presented is presented as a summary[1] case study for an MC with an integrated DB to form an ADC, developed in response to the surge in thorax medical imaging demands, due to the COVID-19 eruption. As front-line radiographers were challenged by the extraordinary work demands the Radiographers Society of the Emirates (RASE) worked with HCT to develop a plan of training volunteers/workers for the "new normal" workloads. The ADC is aligned to the job title of "Medical Imaging Assistant" under ISCO-08 code 5329, level 1 plain radiographs of human thorax ADC. The case represents an embedded ADC of an MC and a DB within the

---

[1] For full study case information corresponding author can be contacted.

existing B.Sc. Medical Imaging program. It also features a recognition of previous learning characteristics evident under standard 2. As this ADC stands, it possesses the potential of constituting a building block within a modular model if most of the other ADC components (see Fig. 1) are required by employees. (Table 2) represents

Table 2 Brief Award Standards Document—Medical Imaging Assistant—Level 1—ADC (MC and a DB in CCQs)

1. General Profile				
1.1	Full title	Medical Imaging Assistant ADC—Level 1—Imaging plain radiographs of human thorax		
1.2	ISCO Job title	Medical Imaging Assistant—level 1	ISCO code:	5329
1.3	Endorsement	• Endorsed by Radiographers Society of the Emirates		
1.4	Full degree accumulation	• Higher diploma in medical imaging technology • Bachelor's in medical imaging sciences		
1.5	Credit	9		
1.6	Scope	"Setting up imaging and patient transfer instruments, preparing materials, and assisting radiographers during procedures" (ISCO-08) limited to human thorax		
2. Entry requirements				
2.1	Pre-requisites	Minimum course entry requirements	2.1.1 The successful completion of courses: HSC 1033 (or equivalent), and HMI 2403 (or equivalent)	
		Mandatory requirements	2.1.2 Learner has a cGPA that is not less than (3.0/4) in core courses (in the past 5 years from date of enrolment) or a minimum of 5 years of experience in a medical field that is related to or equivalent to medical imaging 2.1.3 Learner presents physical fitness certificate 2.1.4 Learner pledges, acknowledges, practices, and seeks clarification if needed about HCT academic integrity policies and procedures	

3. Completion requirements—Undergraduate enrolment

It is the HCT's responsibility to ensure that the following combination of rules are met/adhered to:

3.1 Successful completion of MC components: Concurrent learning of course codes HMI 2303 and, HMI 2003, followed by the subsequent course offering of HMI 2613. as per stated performance standards with a total of 9 credits

3.2 Completion of DB as per stated performance standards

3.3 Learner meets all the requirements within a regular semester duration of 15 weeks as stated in program sequence

4. Completion requirements: Upskilling / reskilling enrolment

(continued)

Table 2 (continued)

1. General Profile

It is the HCT's responsibility to ensure that the following combination of rules are met/adhered to:

4.1 Successful completion of concurrent learning of course codes HMI 2303 and, HMI 2003 or equivalent in the past 6 years from date of enrolment followed by the subsequent course offering of HMI 2613. If equivalency is met for HMI 2303, HMI 2003. HMI 2613 will be the first enrolled course

4.2 If courses were deemed not equivalent, learner has the option to attempt a summative written on campus proctored assessment for HMI 2303, HMI 2003 combined. If learner achieves a total score of 80% in the assessment, s/he will get a recognition of previous learning and would be able to directly enroll in HMI 2613

4.3 Learner has the option to attempt all assessment of HMI 2303, HMI 2003 or opt for a single summative written exam worth 100% of the course grade on campus

4.4 Learner meets the requirements within a maximum of 4 regular semester or 3 regular and 2 summer semesters duration within 24 months from the time of enrolment

4.5 Attendance policies are applied, while didactic sessions are not attendance mandatory, all lab and clinical training are; the absence of more than 15% will result in failing the ADC

Grading	Competent = Total score percentage 88% or more
	Not Yet Competent (unissued statement) = less than 88%

5. MC teaching and training Content		
5.1 Knowledge: Didactic theory learning	5 Credits	90 Contact hours
5.2 Skills: Laboratory	1 Credit	30 Contact hours
5.3 Application: Clinical Preceptorship in clinical setting	3 Credits	180 Contact hours
Total	9 Credits	300 contact hours

6. Digital Badge Details, assessment and evidence requirements	
6.1 Title	Competencies and character qualities "soft-skills competencies" within Medical Imaging of Human Thorax context badge
6.2 Performance Criteria	• Successfully, safely, and accurately executing human thorax medical imaging tasks within a clinical setting for: 1. 100 different patients under clinical supervision, and 2. Five different patients under assessment conditions conducted by clinical site assessor while exhibiting evidence of proficiency in CCQs
6.3 Evidence linked to learner's record-Performance Criteria	• Clinical logbook of number of hours (90 h) and patient quota (100 patients) graded and verified by clinical site trainers and assessment (5 patients) conducted, graded, and verified by clinical site trainers

a brief Award Standards Document—Medical Imaging Assistant—Level 1, ADC six standards and corresponding subsections.

7 Limitations

Our paper has limitations in the scope that it is confined to HEI with well-established technology infrastructure. Additionally, we did not investigate the effects of uncertainty of employers as to what constitutes relevant job demands post the COVID-19 pandemic.

8 Conclusion

As the fourth generation of qualifications, ADCs are becoming critically important as integral components of HEIs' business models. This is due to the gap between full credentials and skills required in the job markets. The fact that an endless number of entities can issue ADCs puts pressure on HEIs to play a role in safeguarding quality of education. In the landscape of a sound technology foundation, the AQ standards we apply to existing credentials place our institution in a privileged position for the provision of trustworthy ADCs. We developed an outcome-focused quality assurance standards model, and a subsequent case-study illustration, designed to pivot a nationwide application and constitute a blueprint for the future of ADC provisions. We first justified a trifurcated learner, employers and assuming a new social identity case for change. We then identified the scope and terminology for benchmarking, to set the tone of this paper. The model of provision we proposed was benchmarked against the UNESCO report by Chakroun & Keevy describing four areas of ADC provision success forecast. Benchmarking has revealed trust and acceptance of ADC content at the academic level as the main two quality challenges to the HCT-ADC ecosystem [7]. In response to those challenges, we proposed the VRSVAS model, designed to enhance stakeholders' trust and acceptance through declaring an associated value, availability of resources, transparent standards, validation and documentation standards, improving accessibility to learners, and ensuring learners' support and awareness. Finally, we developed an ADC exemplar case-study, applying the constituents of QA, as described, and linked the case study. Awarding standards of the case study are numbered and used as a reference in each of the sections of the VRSVAS model for the purpose of example illustration. Placing learners and employers at the heart of the proposed VRSVAS model, HCT-ADCs are engineered to be flexible and innovative, in response to the national skills' needs with key objectives to ensure that HCT provisions are aligned to job profiles and employers' needs and with accountability to focus on outcomes, pitched towards value competency, employment-based stipulation.

At least, locally, HCT is flag shipping the ADCs project, based on the size of the institution which statistically signifies the value of pilot projects. We are unaware of any published study that provides a step-by-step example of a case-study, with focus on the educational QA aspects of implementation. We are also unaware of such business model that incorporates grading CCQs "soft-skills" within work placement in the region. Therefore, we envision this paper to constitute a beneficial guideline and blueprint for HEI first adopters, with sufficient technical infrastructure, to support the digital aspect of ADCs.

References

1. Higher Colleges of Technology: HCT Strategic Vision 2026. (2020)
2. Gauthier T (2020) The value of microcredentials: The employer's perspective. J Competency-Based Educ 5:e01209. https://doi.org/10.1002/cbe2.1209
3. Kato S, Galán-Muros V, Weko T (2020) The emergence of alternative credentials. Paris. https://doi.org/10.1787/b741f39e-en
4. World Economic Forum: Realizing human potential in the Fourth Industrial Revolution: an agenda for leaders to shape the future of education, gender and work. World Economic Forum, Geneva, Switzerland (2017)
5. Department of Education: Review of the Australian Qualifications Framework Final Report 2019 - Department of Education, Skills and Employment, Australian Government., Australia (2019)
6. International council for open and distance education: The Present and Future of Alternative Digital Credentials (ADCs). (2019)
7. Chakroun B, Keevy J (2018) Digital credentialing: implications for the recognition of learning across borders. United Nations Educational, Scientific and Cultural Organization UNESCO, Paris, France
8. Gallagher S (2018) UPCEA convening on the future of credentials: post-event executive summary
9. Tehan D, Cash M (2021) Marketplace for online microcredentials | Ministers' Media Centre. https://ministers.dese.gov.au/tehan/marketplace-online-microcredentials. Last Accessed 03 March 2021
10. Devedzic V, Tomic B, Jovanovic J, Kelly M, Milikic N, Dimitrijevic S, Djuric D, Sevarac Z (2018) Metrics for students' soft skills. Appl Measur Educ 31:283–296. https://doi.org/10.1080/08957347.2018.1495212
11. Fong J (2017) Increasing millennial interest in alternative credentials young millennials middle millennials old millennials percentage of millennials with interest in alternative credentials
12. Selvaratnam RM, Sankey M (2021) An integrative literature review of the implementation of micro-credentials in higher education: Implications for practice in Australasia. J Teach Learn Graduate Employability 12: 1–17. https://doi.org/10.21153/jtlge2021vol12no1art942
13. Fong J, Janzow P, Peck K (2016) Demographic shifts in educational demand and the rise of alternative credentials. Pearson
14. Department of Education, U., Center for Education Statistics, N.: Adult Training and Education: Results from the National Household Education Surveys Program of 2016 (NCES 2017–103rev), First Look. (2016)
15. Williams H (2021) Blockchain May Offer a Résumé You Can Trust - WSJ, https://www.wsj.com/articles/blockchain-may-offer-a-resume-you-can-trust-1520820121. Last accessed 03 March 2021
16. Hickey DT, Willis JE, Quick JD (2015) Where badges work better BRIEF. Denver

17. Oliver B (2019) Making micro-credentials work for learners, employers and providers. Victoria, Australia
18. Ahn J, Pellicone A, Butler BS (2014) Open badges for education: What are the implications at the intersection of open systems and badging? Res Learn Technol 22. https://doi.org/10.3402/rlt.v22.23563
19. Lemoine PA, Richardson MD (2015) Micro-credentials, nano degrees, and digital badges: new credentials for global higher education. Int J Technol and Educ Market 5:36–49. https://doi.org/10.4018/ijtem.2015010104
20. Stefaniak J, Carey K (2019) Instilling purpose and value in the implementation of digital badges in higher education. Int J Educ Technol Higher Educ 16: 44. https://doi.org/10.1186/s41239-019-0175-9
21. New Zealand Qualifications Authority: Guidelines for applying for approval of a training scheme or a micro-credential » NZQA, https://www.nzqa.govt.nz/providers-partners/approval-accreditation-and-registration/micro-credentials/guidelines-training-scheme-micro-credential/, last accessed 2021/02/28
22. Pickard L (2021) Analysis of 450 MOOC-Based Microcredentials Reveals Many Options But Little Consistency — Class Central, https://www.classcentral.com/report/moocs-microcredentials-analysis-2018/. Last accessed 28 Feb 2021
23. MicroHE Consortium: MicroHE – Supporting Learning Excellence through Micro-Credentials in Higher Education, https://microcredentials.eu/, last accessed 2021/02/28
24. Dowling A (2018) Ten minutes "Under the Hood of Digital Qualifications, Qualifications Frameworks and Their Policy Challenges". Presentation. In: Seventh Annual Groningen Declaration Network Meeting EXECUTIVE SUMMARY. pp 14–15. Groningen Declaration, Paris
25. Organization I.L (2021) International Standard Classification of Occupations (ISCO) - ILOSTAT, https://ilostat.ilo.org/resources/concepts-and-definitions/classification-occupation/, last accessed 2021/03/03
26. Paniagua A, Istance D (2018) Teachers as Designers of Learning Environments: The Importance of Innovative Pedagogies. OECD, Paris. https://doi.org/10.1787/9789264085374-en
27. Hickey DT, Otto N, Itow R, Schenke K, Tran C, Chow C (2014) Badges design principles documentation project., Indiana
28. Dyjur P, Lindstrom G (2017) Perceptions and uses of digital badges for professional learning development in higher education. TechTrends 61:386–392. https://doi.org/10.1007/s11528-017-0168-2
29. Mah DK (2016) Learning analytics and digital badges: potential impact on student retention in higher education. Technol Knowl Learn 21:285–305. https://doi.org/10.1007/s10758-016-9286-8
30. van Nieuwerburgh C (2015) Coaching in professional contexts. SAGE Publications, London

Quality Analysis of Graduate Dissertations in Natural Science

Lin Yuan, Shuaiyi Liu, Shenling Liu, Yuheng Wang, Aihong Li, and Yanwu Li

Abstract The dissertation is an import basis for graduate students to obtain a degree and an import manifestation of scientific research and innovation. The quality of dissertations is an import yardstick for measuring the quality of graduate education. Research related on the quality of the dissertations plays an important role in establishing the monitoring system of the dissertations and improving their quality. In this paper, we use 1706 dissertations from 2012 to 2019 as the database, and we select six objective elements from the dissertations and their detecting reports as the research points. We find that the number of pages, the number of words, the number of references, the percentage of English references, and the number of author's achievements including academic papers, patents and other related awards have positive correlations with the quality of dissertations in different levels, while the percentage of coping words has a negative correlation with the quality of dissertations. Our research results can provide direction for education management departments to import the quality of dissertations.

Keywords Quality of dissertations · Influence factors · Pearson's correlation coefficient · Data analyse

1 Introduction

The dissertation is an import basis for graduate students to obtain a degree and an import manifestation of scientific research and innovation. The quality of dissertations is an import yardstick for measuring the quality of graduate education.

In recent years, the country has successively issued a series of documents [1–3], putting forward higher requirements on the quality of dissertations. Research related on the quality of the dissertations plays an important role in establishing the monitoring system of the dissertations and improving their quality.

L. Yuan (✉) · S. Liu · S. Liu · Y. Wang · A. Li · Y. Li
Graduate School, National University of Defense Technology, Changsha 410073, Hunan, China
e-mail: YL5470@nudt.edu.cn

Some researchers have evaluated the quality of dissertations from different kinds of views [4–7]. In this paper, we select some objective factors from the dissertations and their detecting reports as the research points, in order to remove the subjective assessment.

In this paper, we use 1706 dissertations from 2012 to 2019 as the database, and we select six objective elements from the dissertations and their detecting reports as the research points. We first apply "min–max" method to normalize the data and then we use Pearson's correlation coefficient to evaluate the relevance. We find that the number of pages, the number of words, the number of references, the percentage of English references, and the number achievements have positive correlations with the quality of dissertations in different levels, while the percentage of coping words has a negative correlation with the quality of dissertations. We may provide direction for education management departments to import the quality of dissertations.

2 Experimental Data

2.1 Data Acquistion

To research the influential elements of the quality of academic dissertations, we select some dissertations from 2012 to 2019 as the basement of this research, which were scored by three or five senior professors in their fields.

In this paper, the database contains two parts: the academic dissertations of graduate students, including doctors' and masters' dissertations, and their detecting reports of academic misconduct based on CNKI database. All the dissertations are open published and can be found on CNKI database. The number of empirical used dissertations is 1706. The detailed information is shown in Table 1.

These dissertations and corresponding detecting reports are in the portable document format. We use pdfminer, which is a toolkit of Python, to extract key information. We extract the page of dissertations, the total number of words, the percentage of coping words, the number of references, the percentage of English references and the number of author's achievements. All the information is only extracted from the dissertations and corresponding detecting reports.

2.2 Data Preprocessing

To remove the influence of datasets in different scales, we apply a universal approach to normalize the datasets used in our experiment. The equation of the normalization can be shown as follows:

Table 1 The constituent of experimental data

	S & E*			M & B+		
	Doctor		Master	Doctor		Master
2012	53		66	35		39
2013	53		65	30		37
2014	51		77	43		63
2015	60		73	40		61
2016	44		65	29		40
2017	41		51	23		39
2018	53		85	40		59
2019	54		98	66		73
Total	409		580	306		411

*S & E means Science and Engineering
+M & B means Medicine and Biology

$$x_{new} = \frac{x - \min(x)}{\max(x) - \min(x)} \tag{1}$$

where x means the original variable, x_{new} means the variable after normalization, $\min(x)$ means the minimal value of x and $\max(x)$ means the maximum of x. After normalization, the value scope of x_{new} is [0, 1].

3 Methods

To study the impact of objective elements on the quality of dissertations, we first plot the distribution between the elements and the dissertations' score. Then we use the linear regression to reflect the trend between the elements and the scores. After that, we apply Pearson's correlation coefficient to reflect the correlation between the elements and the scores. Pearson's correlation coefficient can be shown as follows:

$$r = \frac{\sum_{i=1}^{n}(x_i - \bar{x})(y_i - \bar{y})}{\sqrt{\sum_{i=1}^{n}(x_i - \bar{x})^2}\sqrt{\sum_{i=1}^{n}(y_i - \bar{y})^2}} \tag{2}$$

where r means the relevance between elements and scores. The sign of the r value indicates the sign of the correlation. The absolute value of r indicates the strength of the correlation.

4 Results

The relationship between the influence elements and the score of dissertations will be shown in each subsection. The blue dots indicate the distribution between the score of dissertations and the influence elements. The green line means the mean score of dissertations in each class and the red line means the fitting curve between them. 'm' indicates the mean value of the influence elements. The Pearson's correlation coefficient is also given in each subfigure.

4.1 Number of Pages

The relationship between the number of pages and the score of dissertations is shown as Fig. 1.

From the distribution in Fig. 1 (blue dots), we can find that when the number of pages is larger than 'm', the score is almost larger than the mean score and when the number of pages is less than 'm', the score is also less than the mean score.

The r are 0.89, 0.91, 0.87, 0.84 respectively and the p-values are almost 0. From the Pearson's correlation coefficient shown in Fig. 1, we can see that the number of pages and the paper score show a significant positive correlation for both doctors' and masters' dissertations.

4.2 Number of Words

The relationship between the number of words and the score of dissertations is shown as Fig. 2.

From the distribution in Fig. 2 (blue dots), we can find that when the number of words is larger than 'm', the score is almost larger than the mean score and when the number of words is less than 'm', the score is also less than the mean score.

The r values are 0.78, 0.84, 0.78, 0.74 respectively and the p-values are almost 0. From the Pearson's correlation coefficient shown in Fig. 2, we can see that the number of words and the paper score show a significant positive correlation for both doctors' and masters' dissertations.

4.3 Percentage of Coping Words

The relationship between the percentage of coping words and the score of dissertations is shown as Fig. 3.

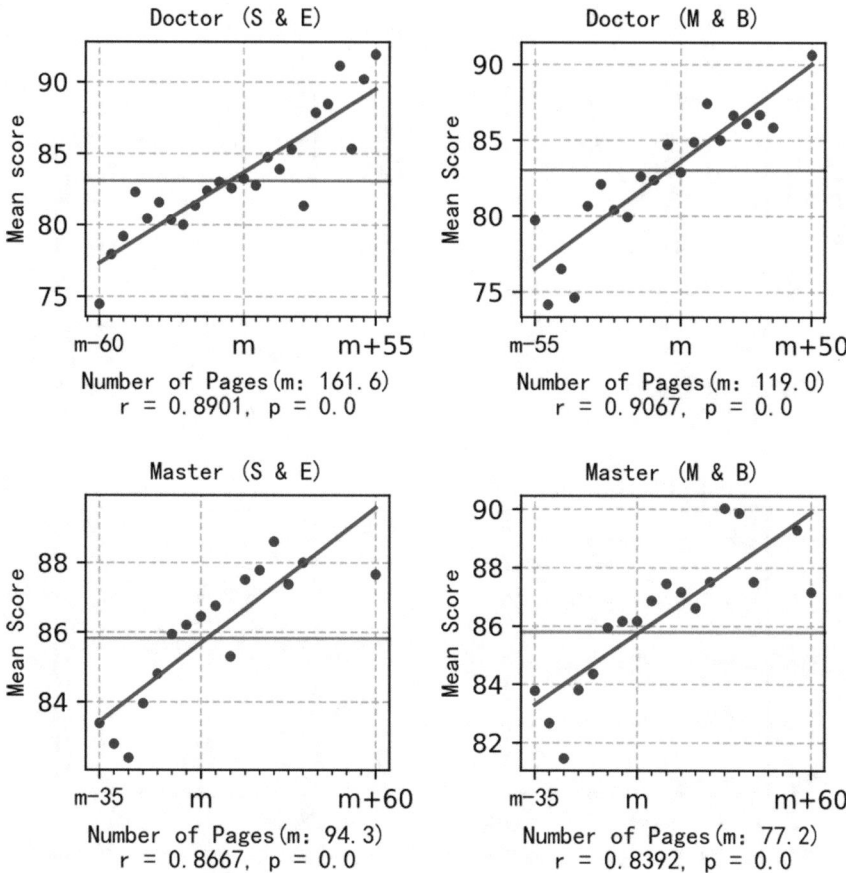

Fig. 1 The relationship between the number of pages and the score of dissertations

From the distribution in Fig. 3 (blue dots), we can find that when the percentage of coping words is larger than 'm', the score is almost less than the mean score and when the percentage of coping words is less than 'm', the score is also larger than the mean score.

The r values are −0.79, −0.24, −0.28, −0.07 respectively and the p-values are 0.0007, 0.3937, 0.1825, 0.7666. From the Pearson's correlation coefficient shown in Fig. 3, we can see that for science and engineering doctors, the percentage of coping words and the paper score show a significant negative correlation. Although the negative significance of masters' dissertations in science and engineering and doctoral and masters' dissertations in medicine and biology is not strong, it is still negatively correlated.

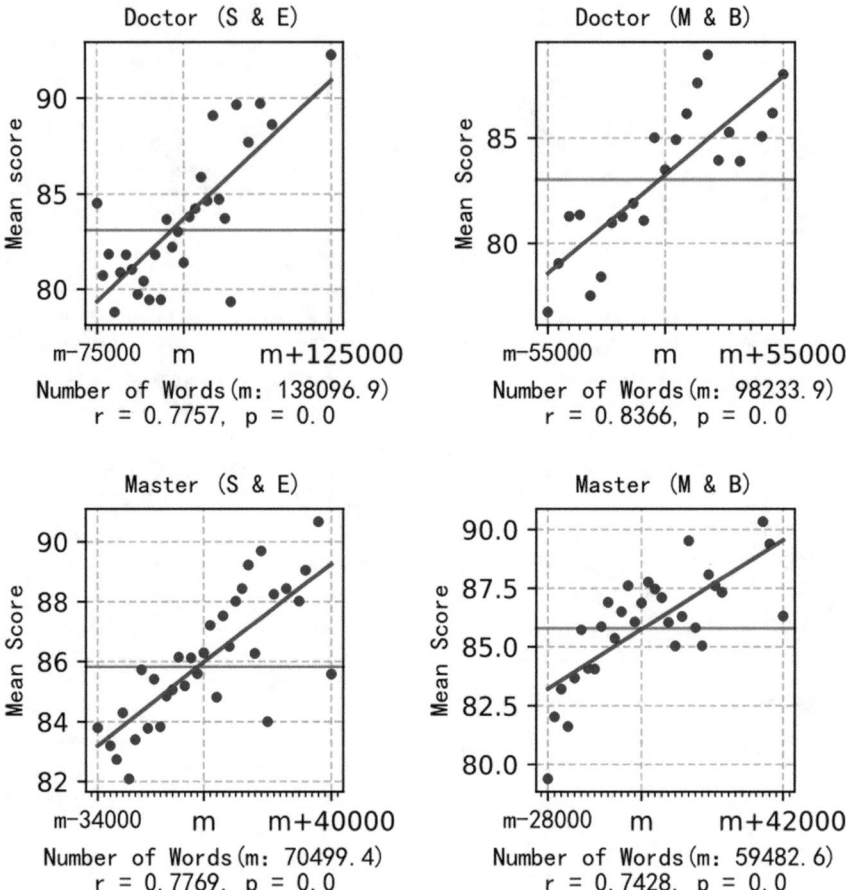

Fig. 2 The relationship between the number of words and the score of dissertations

4.4 Number of References

The relationship between the number of references and the score of dissertations is shown as Fig. 4.

From the distribution in Fig. 4 (blue dots), we can find that when the number of references is larger than 'm', the score is almost larger than the mean score and when the number of references is less than 'm', the score is also less than the mean score.

The r values are 0.81, 0.59, 0.84, 0.45 respectively and the p-values are all below 0.05. From the Pearson's correlation coefficient shown in Fig. 4, we can see that the number of references and the paper score show a significant positive correlation for both doctors' and masters' dissertations.

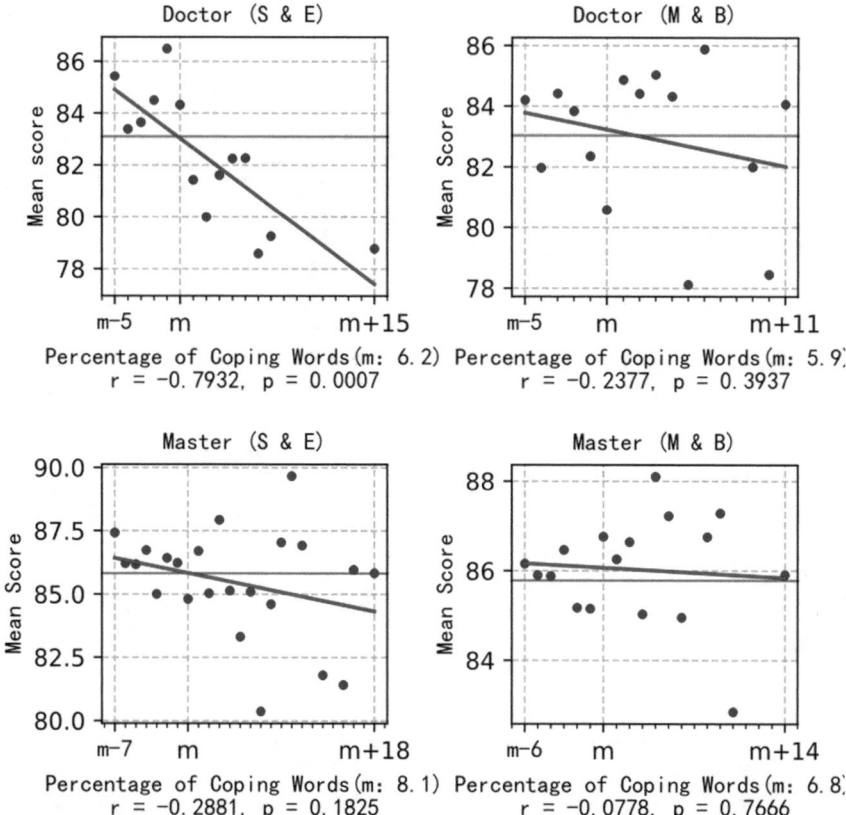

Fig. 3 The relationship between the percentage of coping words and the score of dissertations

4.5 Percentage of English References

The relationship between the percentage of English references and the score of dissertations is shown as Fig. 5.

From the distribution in Fig. 5 (blue dots), we can find that when the percentage of English references is larger than 'm', the score is almost larger than the mean score and when the percentage of English references is less than 'm', the score is also less than the mean score.

The r values are 0.81, 0.07, 0.83, 0.55 respectively and the p-values are 0.0, 0.87, 0.0, 0.05. From the Pearson's correlation coefficient shown in Fig. 5, we can see that the percentage of English references and the score of doctoral and masters' dissertations in science and engineering show a significant positive correlation. Although the positive significance between dissertations' scores in medicine and biology and their percentage of English references is not strong, it is still positively correlated.

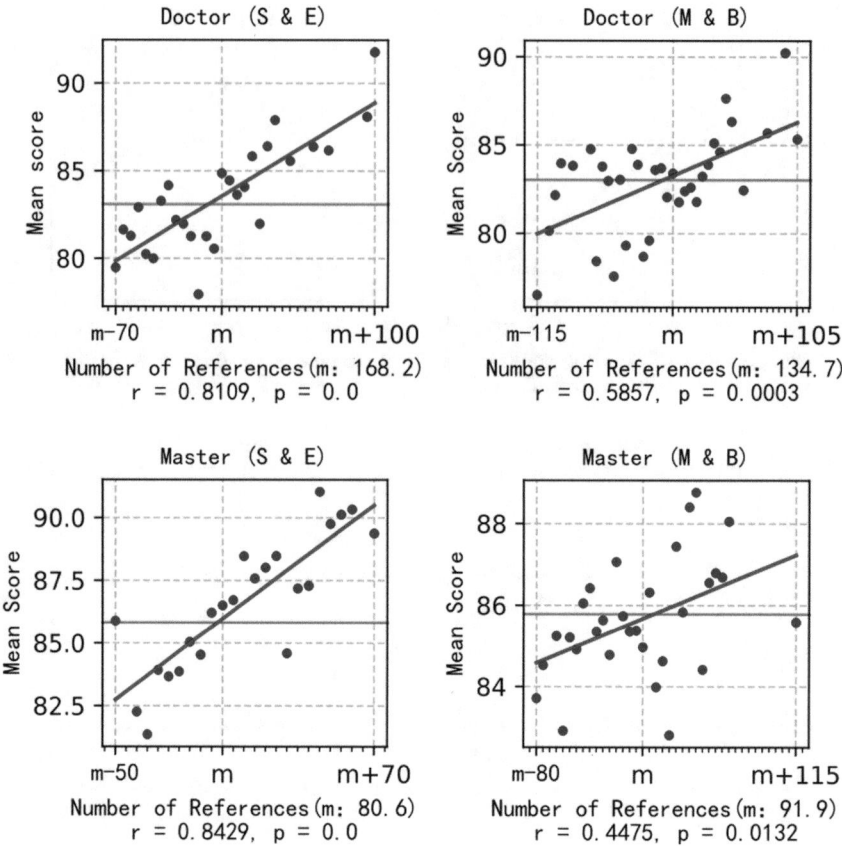

Fig. 4 The relationship between the number of references and the score of dissertations

4.6 Number of Achievements

The relationship between the number of achievements and the score of dissertations is shown as Fig. 6.

From the distribution in Fig. 6 (blue dots), we can find that when the number of achievements is larger than 'm', the score is almost larger than the mean score and when the number of achievements is less than 'm', the score is also less than the mean score.

The r values are 0.48, 0.84, 0.92, 0.62 respectively and the *p*-values are all below 0.05. From the Pearson's correlation coefficient shown in Fig. 6, we can see that the number of achievements and the paper score show a significant positive correlation for both doctors' and masters' dissertations.

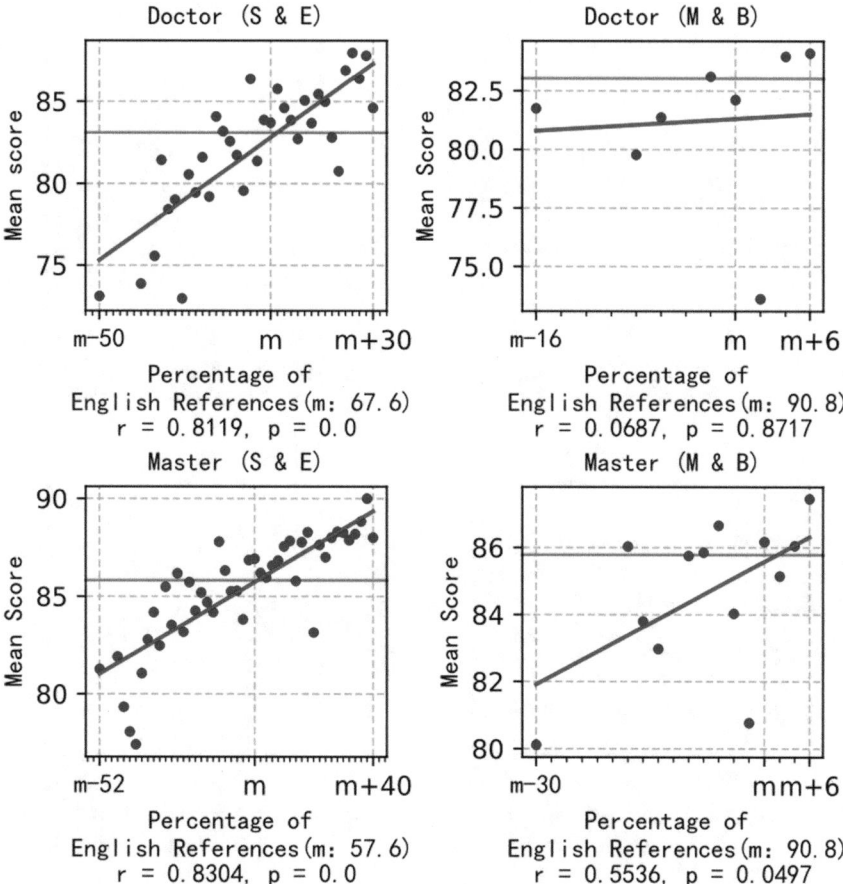

Fig. 5 The relationship between the percentage of English references and the score of dissertations

5 Discussion

In this paper, we take the objective elements from the dissertations as the starting point to research the elements affecting the quality of the thesis. We use the score of dissertations as the quality of them because the scores are the mean values that are from senior professors in the related researching field. Hence the score can be seen the quantification of the quality of dissertations. Based on this, we find that the objective elements can reflect the quality of the dissertations in different levels.

We firstly research the length of dissertations as the influence element, which contains two parts: the number of pages and the number of words. The number of pages and the number of words are the important manifestation of the workload of the dissertations. For the science and engineering and the medicine and biology, which are based on experiments, the number of pages and the number of words can

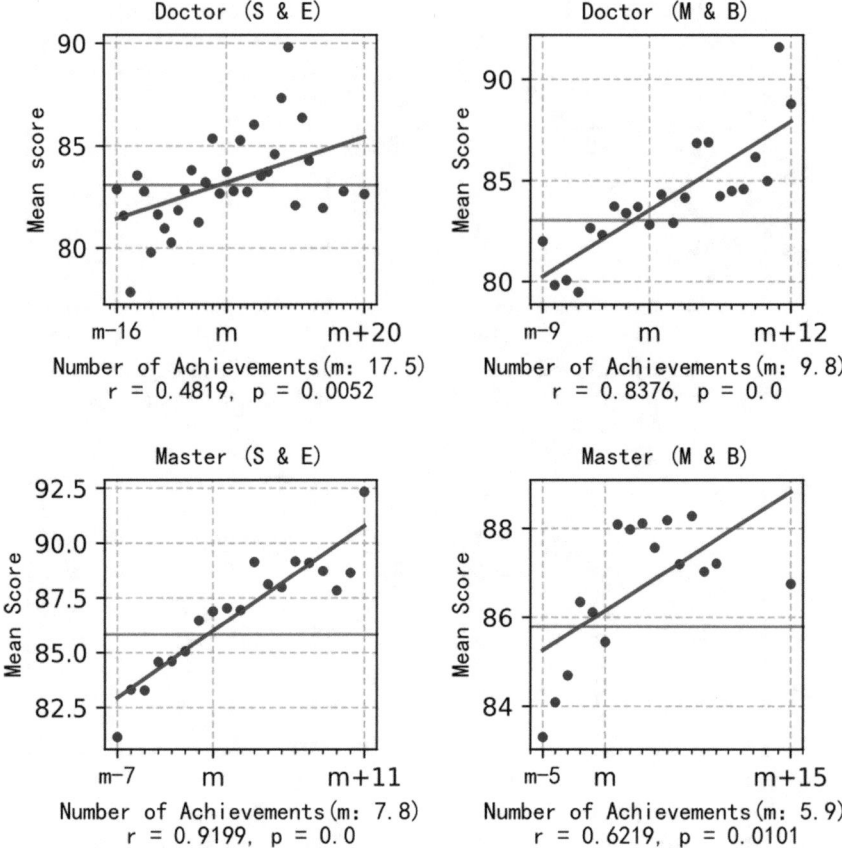

Fig. 6 The relationship between the number of achievements and the score of dissertations

directly reflect the workload of the dissertations. Our experimental results are highly consistent with our cognition, which reflects that our conclusion is believable.

Secondly, we study the originality of dissertations as the key element. The percentage of coping words reflect the degree of repetition with other articles. The higher percentage of coping words, the lower percentage of the author's workload. Hence the percentage of coping words should be negative correlation with the quality of dissertations. The experimental results is consistent with our hypothesis.

Thirdly, we discuss the references of dissertations, which contains two parts: the number of references and the percentage of English references. The number of references reflects the author's review of previous research results in related fields and the percentage of English references reflects the newest research trending in related fields. Hence the more the references including English references an author reads, the higher the quality of his/her dissertation will be. The results of our research can get the same trend as the assumption, which means that our conclusion is believable.

Finally, we plan to find the relationship between the number of achievements and the quality of dissertations. The number of achievements can reflect the author's ability of science research. The more excellent the author is, the more achievements he/she will obtain. Hence the larger the number of achievements is, the higher the quality of dissertations will be. The results of significantly positive correlation between the number of achievements and the score of dissertations can prove the our assumption and the conclusion is also believable at the same time.

6 Conclusion

In this paper, we used 1706 dissertations as the database, and we selected six objective elements from the dissertations and their detecting reports as the research points. We first applied "min-max" method to normalize the data and then we used Pearson's correlation coefficient to evaluate the relevance. We found that the number of pages, the number of words, the number of references, the percentage of English references, and the number achievements had positive correlation with the quality of dissertations in different levels, while the percentage of coping words had negative correlation with the quality of dissertations. We could provide direction for education management departments to import the quality of dissertations.

Acknowledgements The research in this paper is supported by Research Project of Degree and Postgraduate Education Reform in Hunan Province (No. 2020JGZD004), National Educational Science in 13-th Five-Year Plan Project (No. DIA180383), the Ministry of Education Science, Technology Development Center University Scientific Research Integrity Construction Fund Project(No. 2020A050110) and Equipment Technology Basic Projcet (No. 211GF45001).

References

1. Goverment Document (2014) Notice on further standardizing and strengthening the management of postgraduate training (in Chinese)
2. Goverment Document (2014) Options on strengthening the quality assurance and supervision system of degree and graduate education (in Chinese)
3. Goverment Document (2014) Sampling method for doctoral and master's thesis (in Chinese)
4. Hemlin S (1993) Scientific quality in the eyes of the scientist: a questionnaire study. Scientometrics 27(1993):3–18
5. Kyvis S, Thune T (2015) Assessing the quality of PhD dissertations: a survey of external committee members. Access Evaluc High Educ 40:768–782
6. Mullins G, Kiley M (2002) It's a PhD, not a nobel prize: how experienced examiner assess research theses. Study High Educ 27:369–386
7. Tinkler P, Jackson C (2000) Examing the doctorate: institutional policy and the PhD examination process in Britain. Study High Educ 25:167–180

Printed in the United States
by Baker & Taylor Publisher Services